ADVANCEMENTS IN CHEMICAL ENHANCED OIL RECOVERY

ADVANCEMENTS IN CHEMICAL ENHANCED OIL RECOVERY

Edited by

Tushar Sharma, PhD
Krishna Raghav Chaturvedi, PhD
Tarek Ganat, PhD
Imtiaz Ali, PhD

AAP | APPLE ACADEMIC PRESS

First edition published 2025

Apple Academic Press Inc.
1265 Goldenrod Circle, NE,
Palm Bay, FL 32905 USA

760 Laurentian Drive, Unit 19,
Burlington, ON L7N 0A4, CANADA

CRC Press
2385 NW Executive Center Drive,
Suite 320, Boca Raton FL 33431

4 Park Square, Milton Park,
Abingdon, Oxon, OX14 4RN UK

Apple Academic Press exclusively co-publishes with CRC Press, an imprint of Taylor & Francis Group, LLC

Library and Archives Canada Cataloguing in Publication

CIP data on file with Canada Library and Archives

Library of Congress Cataloging-in-Publication Data

CIP data on file with US Library of Congress

ISBN: 978-1-77491-508-0 (hbk)
ISBN: 978-1-77491-509-7 (pbk)
ISBN: 978-1-00345-372-7 (ebk)

About the Editors

Tushar Sharma, PhD
Associate Professor, Rajiv Gandhi Institute of Petroleum Technology (RGIPT), Jais, India

Tushar Sharma, PhD, is currently working as an Associate Professor at the Rajiv Gandhi Institute of Petroleum Technology (RGIPT), Jais, India. He is also the Head and Lead Instructor at the Enhanced Oil Laboratory at RGIPT. His main research areas include enhanced oil recovery, nanofluids, emulsions, and rheology. Dr. Sharma also has expertise in handling rheometers, core-flooding equipment, surface tensiometers, hydrates, well-testing, computational fluid dynamics, and rock geomechanics. He reviews articles for international peer-reviewed journals and has authored over 60 papers in leading international journals. Among other research work, Dr. Sharma has filed two patents for consideration with the Indian Patent Registry. Dr. Sharma worked as a visiting research fellow at the University of Alberta, Canada, on a project funded by the Canadian government. This project focused on solvent-heavy oil interactions and microfluidic visualization for solvent-steam-assisted heavy oil recovery. He has conducted training seminars for engineers from the Oil & Natural Gas Corporation (India). Dr. Sharma is also the faculty coordinator of the RGIPT Society of Petroleum Engineering (SPE) student chapter. Dr. Sharma is a Society of Petroleum Engineers (SPE) member. Dr. Sharma received his doctoral degree from IIT Madras for his work on pickering emulsions and their application in EOR.

Krishna Raghav Chaturvedi, PhD
Outreach Manager at the American Chemical Society

Krishna Raghav Chaturvedi, PhD, is currently an Outreach Manager at the American Chemical Society, where he works in outreach with the scientific community and leads efforts for the upskilling of the upcoming workforce. He recently obtained his PhD thesis from the Rajiv Gandhi Institute of Petroleum Technology (RGIPT), Jais (An Institute of National Importance),

India. His research focuses on synthesizing novel single-step silica nanofluids for improved CO2 flow behavior and reduced formation damage. Primarily, this work focuses on improving the efficacy of CO2-based EOR in depleted oil fields that are essential to ensure a net-zero future. Dr. Chaturvedi is a highly cited "Top 2% Elsevier List" petroleum engineer-turned carbon capture, utilization, and storage expert with a PhD and two books, more than 30 papers, and two patents to his name. He was awarded with the inaugural Best PhD Thesis Award in New Energy Studies for the year 2022 by the Federation of Indian Petroleum Industry (FIPI). He earned his bachelor's and master's degree in petroleum engineering from the University of Petroleum & Energy Studies, Dehradun and RGIPT, respectively.

Tarek Ganat, PhD
Lecturer, Petroleum and Chemical Engineering Department,
Sultan Qaboos University, Oman

Tarek Ganat, PhD, joined the Petroleum and Chemical Engineering Department at Sultan Qaboos University, Oman, as a lecturer in 2021. Dr. Tarek has BSc, MSc, and PhD degrees, all in petroleum engineering, and has more than 27 years of experience in petroleum production and reservoir engineering through his participation in integrated studies with different oil and gas companies, such as Gowoff Oil Company, Libya; Repsol Exploration Company, Spain; Petro-Canada Oil Company, Canada; and Petronas Grigali Oil Company, Malaysia. His professional experience comes from his work in the oil and gas industry in tanning centers and from his work that focused on subsurface evaluation for green fields and brownfields. He has a wide range of overseas experience gained in fractured carbonate and clastic reservoir characterization and full-field development. He also has considerable experience in the application of petroleum-industry-related packages in multinational and multicultural joint ventures. He has authored three textbooks published with Springer and has published many papers in journals and conferences (ISI [Q1 & Q2] journals and Scopus). In addition, he is one of the reviewers for the *Journal of Petroleum Exploration and Production Technology*. Dr. Tarek managed to secure two research grants pertaining to Petronas Oil Company as project leader, with a cumulative amount of about RM380,000, and two research grants as project leader, for EOR projects with a cumulative amount of about $23,670.

Imtiaz Ali, PhD
Assistant Professor, Department of Petroleum and Gas Engineering, Balochistan University of Information Technology, Engineering and Management Sciences (BUITEMS), Quetta, Pakistan

Imtiaz Ali, PhD, is currently working as an Assistant Professor in the Department of Petroleum and Gas Engineering, Balochistan University of Information Technology, Engineering and Management Sciences (BUITEMS), Quetta, Pakistan. Dr. Ali worked with the Oil & Gas Development Company Pakistan before associating with the King Saud University (KSU), Riyadh, Saudi Arabia, as a researcher, wherein he developed a rock-scratching cell to determine different rock mechanical properties. His research interests include chemical enhanced oil recovery, the development of non-damaging drilling fluids, CFDs, and formation damage. He authored five book chapters and has made 25 scholarly contributions to peer-reviewed research journals. Dr. Ali has long been associated with the Department of Petroleum and Gas Engineering, BUITEMS, and plays a crucial role in training future hydrocarbon workforce of Pakistan. He completed his PhD in Petroleum Engineering at Universiti Teknologi PETRONAS (UTP), Malaysia; his bachelor's degree in Petroleum and Gas Engineering at Balochistan University of Information Technology, Engineering, and Management Sciences (BUITEMS), Pakistan; and his master's degree in Petroleum Engineering from King Saud University, Riyadh, Saudi Arabia.

Contents

Contributors

Imtiaz Ali
Department of Petroleum and Gas Engineering, Balochistan University of Information Technology, Engineering and Management Sciences, Takatu Campus, Airport Road Quetta, Balochistan, Pakistan
Petroleum and Chemical Engineering Department, Sultan Qaboos University, Muscat, Oman

Muhammad Arif
Department of Petroleum Engineering, Khalifa University, Abu Dhabi, United Arab Emirates

Tandrima Banerjee
Department of Chemistry & Physical Sciences, Neotia University, Kolkata, India

Annapurna Baruah
Department of Petroleum and Natural Gas Engineering, University of Petroleum & Energy Studies, Energy Acres, Dehradun, Uttarakhand, India

Krishna Raghav Chaturvedi
Department of Petroleum Engineering and Geosciences, Rajiv Gandhi Institute of Petroleum Technology Jais, Amethi, Uttar Pradesh, India

Kushal Das
Enhanced Oil Recovery Laboratory, Department of Petroleum Engineering, Rajiv Gandhi Institute of Petroleum Technology, Jais, Amethi, Uttar Pradesh, India

Tarek Ganat
Petroleum and Chemical Engineering Department, Sultan Qaboos University, Muscat, Oman

Rishi Raj Goswami
Oil India Limited, Assam, India

Sankari Hazarika
Department of Petroleum and Natural Gas Engineering, University of Petroleum & Energy Studies, Energy Acres, Dehradun, Uttarakhand, India

Anshi Jain
Department of Petroleum Engineering and Geosciences, Rajiv Gandhi Institute of Petroleum Technology Jais, Amethi, Uttar Pradesh, India

Nilesh Kumar Jha
School of Petroleum Technology, Pandit Deendayal Petroleum University, Knowledge Corridor, Raisan, Gandhinagar, Gujarat, India

Ayush Joshi
Department of Petroleum Engineering and Geosciences, Rajiv Gandhi Institute of Petroleum Technology Jais, Amethi, Uttar Pradesh, India

Abhijit Kakati
School of Petroleum Technology, Pandit Deendayal Petroleum University, Knowledge Corridor, Raisan, Gandhinagar, Gujarat, India

Ravi Shankar Kumar
Department of Petroleum Engineering, Khalifa University, Abu Dhabi, United Arab Emirates
Enhanced Oil Recovery Laboratory, Department of Petroleum Engineering, Rajiv Gandhi Institute of Petroleum Technology, Jais, Amethi, Uttar Pradesh, India

Vishnu Chandrasekharan Nair
Department of Petroleum Engineering and Geological Sciences, Rajiv Gandhi Institute of Petroleum Technology, Jais, Amethi, India

Ramesh Narukulla
Enhanced Oil recovery Laboratory, Rajiv Gandhi Institute of Petroleum Technology, Jais, Amethi, India

Anjannay Pandey
Department of Petroleum Engineering and Geosciences, Rajiv Gandhi Institute of Petroleum Technology Jais, Amethi, Uttar Pradesh, India

Abhijit Samanta
Department of Chemistry & Physical Sciences, Neotia University, Kolkata, India

Tushar Sharma
Department of Petroleum Engineering and Geosciences, Rajiv Gandhi Institute of Petroleum Technology Jais, Amethi, Uttar Pradesh, India

Alpana Singh
Enhanced Oil Recovery Laboratory, Rajiv Gandhi Institute of Petroleum Technology, Jais, Amethi, India

Rishabh Tripathi
Department of Petroleum Engineering and Geological Sciences, Rajiv Gandhi Institute of Petroleum Technology, Jais, Amethi, India

Abbreviations

AA	acrylic acid
AM	acryl amide
ASP	alkali–polymer–surfactant flooding
BA	butyl acrylate
Ca	capillary number
CCS	carbon capture and storage
CCUS	carbon capture utilization and storage
CO_2	carbon dioxide
CMC	critical micellar concentration
DI water	deionized water
DLS	dynamic light scattering
EDS	energy dispersive spectroscopy
EA	ethyl acetate
EOR	enhanced oil recovery
FE-SEM	field-emission scanning electron microscopy
HMPAM	hydrophobically modified poly(acryl amide)
HPAM	partially hydrolizedpoly(acryl amide)
IFT	interfacial tension
M	mobility ratio
NPs	nanoparticles
PAM	polyacrylamide
PEO/PEG	poly(ethylene oxide)
PPO	poly(propylene oxide)
PVA	poly(vinyl alcohol)
PV	pore volume
SDS	sodium dodecyl sulfate
SP	polymer–surfactant flooding
TEM	transmission electron microscopy
TEOS	tetra-ethyl orthosilicate
TGA	thermogravimetric analysis
UV-vis	ultraviolet-visible
VA	vinyl acetate
XRD	X-ray diffraction

Foreword

The COVID-19 pandemic brought various economic activities to a halt and confined billions of people inside their homes, causing an unprecedented fall in the consumption of hydrocarbons. This slump in oil consumption and fall in revenues has forced oil and gas majors to slash exploration and production budgets and lay off experienced staff, some of whom plan to leave the industry forever. This, coupled with the rising environmental concerns about the carbon footprint of oil and gas extraction activities, has motivated several economists and policymakers to write an ode to oil and signal that the end is near for the industry that powered humanity into the digital age.

However, with technological advancement stagnating in the field of energy storage, unpredictable nature of renewable, resources and lack of any viable alternative to hydrocarbons, our reliance on oil will continue at least for the near foreseeable future. Industrialization of the developing world and economic growth will ensure that the aggregate demand for oil will rise again. Particularly, in the Indian subcontinent, the rising population, decrease in poverty, and improving standards of living will force the nations to continue to rely on fossil fuels to meet their need for cheap energy. This, along with lack of major new discoveries of oilfields, will force the oil and gas majors (in the Indian subcontinent and beyond) to coax out every oil barrel possible out of the existing brownfield reservoirs which has motivated researchers to explore new avenues in EOR, particularly the use of chemical additives to mobilize additional oil from oilfields which have already shown maximum oil production from primary and secondary oil recovery.

A cursory glance at the academic literature shows that chemical EOR is one of the most interesting and popular areas of research nowadays. Application of chemical additives like alkalis, surfactants, polymers, and their combinations have been continuously explored for EOR and published over the last decade. However, the efficacy of these chemicals has been found to decrease with the harsh subsurface conditions like salinity, temperature, heterogeneity, and presence of crude oil, and thus, alternatives must be explored to increase their efficiency in adverse conditions. Alternatives include altering the chemistry of the additive (macro) molecules (i.e., from linear to branched polymers), exploring ternary combinations, functionalization of new chemical groups (monomeric to Gemini surfactants), and alterations in the injection methodology. Recent field experience has shown that these chemical combinations have shown high potential for oil recovery in field applications due to their synergistic nature, birthing a new spectrum of techniques in chemical EOR. While various methods of chemical EOR have demonstrated high efficiency for increasing oil production in previous field implementations, the selection of the most optimum chemical EOR method is plagued by several challenges. Also, unlike the massive homogenous oilfields of the past, modern oilfields are highly heterogeneous, much smaller in size, and have complex reservoir geology. This has created additional challenges in which modern EOR practices need to be tailored for each individual producing zone, let alone scaling up the entire process to the whole field.

In this regard, the world will need more expertise in chemical EOR, particularly the need to comprehensively understand its recent advancements. In my view, this book presents a comprehensive overview of the mechanisms and chemistry behind the different chemical EOR techniques currently being practiced worldwide, the economic and ecological rationale for chemical EOR, recent advancements in this field, what opportunities these chemical EOR methods hold for implementation in modern complex reservoirs, decision-making methodology for optimizing the application, and a brief review of chemical EOR field applications. The authors have done a commendable work explaining the methods clearly and lucidly. The text is easy to follow and details the small intricacies of chemical EOR processes, which will make it easy for readers who lack sufficient understanding of petroleum geology, yet want to work in EOR. To conclude, this book presents a comprehensive work compiling information on advanced methods for oil recovery, their working, and how they have sparked a new field of research wherein chemistry, geology, and petroleum engineering play a major role. I believe, this book, with its emphasis on both theory and practice, is a must-have for both academicians and field personnel.

To conclude, this book presents a comprehensive work compiling information on advanced methods for oil recovery, their working, and how they have sparked a new field of research wherein chemistry, geology, and petroleum engineering play a major role. I believe, this book, with its emphasis on both theory and practice, is a must-have for both academicians and field personnel.

—Professor Kris Ravi
University of Leoben (Montanuniversität Leoben),
Leoben, Styria, Austria

Preface

Regardless of the rapid advancements made in the fields of other sources of energy, it is estimated that for the near foreseeable future, hydrocarbon-based fossil fuels would remain the dominant source of energy. However, given the recent decline in the discovery of new hydrocarbon reserves, it has become a challenge to satisfy the global demand of energy by economically and ecologically sustainable sources of hydrocarbon energy. In the meantime, a significant portion of oil remains underground even after the use of all conventional oil recovery practices such as pressure maintenance and artificial lift methods. This has created an opportunity for chemical enhanced oil recovery (EOR) techniques to improve the rate of oil recovery from depleted reservoirs.

Chemical EOR depends on the infusion of synthetic compounds to alter the properties of rock or crude oil to make the mobilization and recovery of oil easier. These EOR methods range from the injection of foams, surfactants, polymers, or even nanofluids into the reservoir and all of them have their unique challenges and advantages. While several books have broached the subject of EOR in the recent past, recent advancements in chemical EOR methods, especially in this low-price scenario, is missing. To the best of our knowledge, the investigation of binary mixtures of chemicals in heterogeneous reservoirs and the role of nanotechnology in improving the efficacy of existing chemical EOR methods have not been extensively reported. Also, the past studies have negated adequately explaining the role of salinity altered water injection (also known as low-salinity or smart waterflooding) in conjunction with other chemical EOR methods. Finally, a current update on the application of chemical EOR methods, especially in the oilfields of the Indian subcontinent has been missing in all recent studies.

Hence, this book discusses various methods of chemical EOR which is most often the third and final phase of economically viable oil production. Some publications have referred to EOR as the phase of tertiary production. Regardless of the nomenclature, the general guidelines and procedure of chemical EOR remains unchanged worldwide. Conventionally, chemical EOR, as the name suggests, involves the injection of any chemical additive which may be polymer, surfactant, alkali, or their combination, into the subsurface. On the field, this injection of the chemical additives is first preceded by a preflush (by water or low salinity brine) which places a reservoir compatible buffer of fluid to isolate the high-salinity reservoir fluid and the injected chemicals. This is essential as the preflush serves to minimize the adverse effect that the high salt concentration will have on the chemical additives.

Some chemical EOR methods like foam injection utilize gases like CO_2 or lighter hydrocarbons injection of gas. The injection of foams improves oil recovery by allowing for a higher oil displacement and flow diversion from highly permeable zones to zones of low permeability. It also serves to inject gas for the maintenance of pressure (gas injection)-enhanced oil displacement by miscible or solution drive (miscible EOR). However, these methods have not been detailed in this book. We acknowledge that chemical EOR is just one of the many techniques which are technically feasible and are usually used widely for field applications. However, rising complexity in modern reservoirs and the need to increase oil recovery means that the need for chemical EOR has become so gargantuan and expertise must be developed in it so that it can be used in all situations.

This book, thus, aims to improve the fundamental understanding of industry researchers, academicians, undergraduate, and postgraduate students. In this role, it may also be used as a reference book by university libraries and as a ready reference for industry professionals. To make the contents of the book easily understandable to all individuals at all levels, the text is lucid and the images easily comprehensible. The uniqueness of the book lies in the lucidly and easily comprehensive text of the sections which makes it more understandable for the undergraduate students while at the same time, retains the depth of the subject which will enable it to be used by researchers to direct their oncoming and planned research.

To improve the scope of the book, this study has not been restricted to only a structured update of the recent works of chemical EOR but also includes discussions on why such an approach may or may not work for other reservoirs. This special emphasis on understanding the various mechanisms behind the chemical EOR methods, especially the role of salinity in EOR, incite research interest, not only by petroleum engineers but also by material scientists and chemists. To this end, the role of hybrid low salinity waterflooding, that is, low salinity waterflooding in conjunction with polymer, surfactant, or nanofluid solutions has been detailed for lab and field applications.

The effect of the various reservoir parameters like temperature, pressure, and reservoir heterogeneity has been catalogued for each of the individual chemical EOR methods. Details on the synthesis of a new surfactant class (the Gemini surfactant) by modifying the basic architecture of their macromolecules has been provided in this book for use by the researchers interested in exploring synthesis of new surfactants. The book also provides details on how these chemical EOR methods may find increased use in the unconventional hydrocarbon reservoirs like shales, coal bed methane, and heterogeneous reservoirs.

The use of a new chemical additive for EOR, the revolutionary ionic liquids, is explored in this book and their properties alone and in the presence of salts are explored in detail for proposed field applications. Aspects of chemical EOR like theoretical and numerical investigations of mobility control, rheology, and various modeling considerations in porous media is also provided in this book. One prime concern which the book has tackled is the analysis of the various difficulties encountered in the use of conventional chemical EOR methods and the role of nanotechnology in improving the efficacy of these methods.

Lastly, the ongoing field investigations, pilot tests, and the industrial readiness of each chemical EOR method are reported. A special emphasis has been given to the chemical EOR methods in the Indian subcontinent due to the presence of brownfields in the region, high population, and increasing dependence on imported oil. A detailed discussion on the methods currently used for reservoir simulation of chemical EOR is provided for ensuring that laboratory-scale results can be upscaled to field scale. To further make the text more palatable for the readers, at the end of each chapter, an "In a Nutshell" style key point summaries are provided.

It is our firm belief that the book, *Advancements in Chemical Enhanced Oil Recovery,* will be of great interest to its readers, eliciting interest in both field and research personnel, to develop competence in chemical EOR.

—Tushar Sharma,
Tarak Ganat,
Imtiaz Ali,
Krishna Raghav Chaturvedi

Introduction

In the next 30 years, global energy consumption is expected to increase by nearly 60%. Multiple mature fields are nearly exhausted. Recovering difficult-to-extract residual hydrocarbons using current extraction techniques is a challenge for the oil and gas industry which seeks creative solutions. Despite an increase in renewable energy output, oil and gas continue to be the world's primary energy sources. As the world's energy consumption rises in tandem with its dwindling energy supply, enhancing oil recovery from underexploited reservoirs becomes increasingly important to satisfy rising demand. The oil recovery procedure comprises three mechanical processes: primary, secondary, and tertiary (also called enhanced oil recovery, EOR). After both primary and secondary treatment, two-thirds of the original oil in the reservoir remains (OOIP). This is the result of the oil's or fluid's resistance to capillary pressure being circumvented in another manner. Recovery is ineffective due to heterogeneity in the reservoir or an unfavorable mobility ratio between the aqueous and oleic phases. In contrast, the remaining oil is composed of discrete ganglia that form when a narrow neck develops around an oleic mass protrusion due to the combined effects of local pressure differential and interfacial tension (IFT).

Numerous EOR solutions have been conceptualized, theoretically/experimentally created, and then used in an attempt to improve the total oil displacement efficiency. During oil recovery, the total oil displacement efficiency is the sum of the macro- (volumetric sweep) and micro- (pore size) displacement efficiencies. The microscopic displacement efficiency is the capacity of displacing fluids to mobilize oil trapped inside pores. In conclusion, EOR is facilitated by any technique that increases the efficiency of oil recovery on the micro-, macro-, or both scales. The vast majority of EOR techniques employ thermal and nonthermal technologies. Thermal EOR is inefficient in reservoirs with a narrow pay zone that are extremely deep. As a result, nonthermal EOR approaches have gained a large amount of interest as a method for recovering subsurface oil.

Chemical approach , a nonthermal EOR technique, has been heralded as the most promising of all EOR methods because of its higher efficiency, economic and technological viability, and low capital cost. This EOR technique gained popularity in the 1980s as a result of increased oil prices and technological advancements that simplified its operation. Chemical EOR techniques improve the efficacy of water injected into a reservoir to displace oil, hence enhancing oil recovery. Chemicals injected with water slug modify the fluid–fluid and/or fluid–rock interactions in the reservoir, depending on the kind of chemical EOR process used. It decreases the interfacial tension (IFT) between the imbibing fluid and the oil or increases the viscosity of the injected fluid to enhance mobility and conformity control. In addition, the injected chemicals affect the wettability of the rock, improving its oil permeability.

Well-known chemical EOR techniques include polymer flooding, surfactant flooding, and alkaline flooding. There are a number of restrictions on chemical EOR. In the presence of reservoir brines and high temperatures, polymers, a recovery mechanism for mobility control, undergo a loss of viscosity. Passage of surfactants and alkalis through porous surfaces causes adsorption. Several chemical injection techniques were subsequently developed, evaluated, and implemented in EOR processes. Alkali/surfactant (AS), surfactant/polymer (SP), alkaline/polymer (AP), and alkali/surfactant/polymer (AS) are notable examples (ASP). Synergy has increased the efficacy of combination chemicals in oil wells. Recent research has centered on foam made from surfactants and polymers as an efficient EOR mobility control agent.

The performance of chemical EOR processes may be enhanced with the use of nanotechnology. In this context, nanotechnology refers to the production and use of substances with at least one dimension in the nanoscale range (1–100 nm). Using nanotechnology, several oil and gas related challenges, including drilling operations, petroleum exploration, asphaltene deposition, gas hydrate formations, hydraulic fracturing duties, and improved oil recovery, have been addressed (EOR). As a chemical EOR approach, nanofluid flooding has been developed and evaluated, and its field use in Colombia has been described. Adjustments of structural disjoining pressure, modification of wettability, IFT reduction, and enhancement of injectant viscosity have been identified as the methods for enhancing oil recovery. Recent study reveals that the inclusion of nanoparticles into typical EOR compounds produces new materials with exceptional and interesting properties. Polymeric nanofluids, such as a synergistic mixture of nanoparticles and polymers, were revealed to have increased rheological properties and stability for usage in high-temperature and high-salinity conditions. In addition, mixing nanoparticles with surfactants inhibits adsorption through a competitive adsorption process, whereas combining nanoparticles with foams produces stable foams with a longer half-life.

This book presents an overview of current chemical EOR research and is mostly a foundational study. Initial research focuses on the traditional chemical EOR approaches. Traditional EOR chemical types, as well as their uses and limitations, have been identified. The exploration and development of binary applications of typical chemical EOR technology follows. Next, the present trend of additive combinations for chemical EOR is examined. It covers the different types of polymer and surfactant solutions, their application processes, and laboratory testing. This book also addresses reservoir modeling's applicability in chemical EOR and its scalability in the field. In conclusion, the challenges associated with chemical EOR approaches are recognized, along with suggestions for further study.

CHAPTER 1

Introduction to Chemically Enhanced Oil Recovery

KRISHNA RAGHAV CHATURVEDI and TUSHAR SHARMA

Department of Petroleum Engineering and Geological Sciences, Rajiv Gandhi Institute of Petroleum Technology, Jais, Amethi, India

ABSTRACT

This chapter delves into the critical role of chemical enhanced oil recovery (CEOR) in maximizing hydrocarbon extraction from reservoirs. Examining the challenges faced by conventional methods, the chapter underscores the growing necessity for innovative approaches. This chapter discusses the origin, extraction, and stages of oil recovery in petroleum exploration. It begins by explaining the natural origin of petroleum, detailing the geological processes over millions of years that lead to the formation of oil. The focus is on the geologic carbon cycle and the conversion of organic matter into petroleum, emphasizing the conditions required for oil formation. The discussion then delves into the extraction process, providing insights into the life cycle of an oil well. It also covers the design and preparation before drilling, the trajectory planning, and the factors considered during drilling, such as casing and completion design. The chapter explains the role of drilling fluid, various drill bits, and the use of casing to maintain wellbore integrity. Subsequently, the completion process, where the well is made operational, is explored. This involves creating a flow path, introducing acids and fracturing fluids, and sealing the well above the reservoir section. The stages of oil recovery primary, secondary, and enhanced (tertiary) are outlined. Primary recovery involves the natural drive of the reservoir, while secondary recovery methods, like water and gas injection, aim to maintain

Advancements in Chemical Enhanced Oil Recovery.
Tushar Sharma, Krishna Raghav Chaturvedi, Tarek Ganat, and Imtiaz Ali (Eds.)

the reservoir pressure. Enhanced oil recovery techniques such as chemical EORs and thermal EORs are also discussed, emphasizing their role in increasing oil mobility. This chapter also advocates for the integration of chemical solutions to optimize oil recovery, ensuring sustainable and efficient utilization of global energy resources. Then the stages of a well's life cycle, from production to abandonment are discussed. It explains how wells are operated and, if necessary, artificially lifted. The economic limit, where a well becomes uneconomical, is discussed, along with the challenges of abandonment, including sealing and restoring abandoned wells. This chapter also covers historical context of surface prospecting and its integration with geologic and geophysical data. The significance of petroleum for the global economy is highlighted, emphasizing its multifaceted applications. Finally, the chapter provides a comprehensive overview of the origin, extraction, and recovery stages of petroleum, covering geological processes, drilling technologies, and the economic and environmental considerations associated with oil exploration and production.

1.1 INTRODUCTION

Without a doubt, fossil fuels (especially crude oil) have had a major influence on the course of human history during the last two centuries, sparking an explosion of economic advancements and technological achievements (Doman, 2016). Even though crude oil has been used for over four thousand years, whether for the construction of walls and towers or by the Chinese for salt manufacturing dating back to the fourth century BC, modern petroleum usage began in the 19th century A.D. when paraffin could be extracted from naturally occurring crude oil (which was often found in seepages on the ground) (Muggeridge et al., 2014). The first modern oil wells were drilled in the mid-19th century, and by the early 20th century, petroleum had firmly established itself as the future fuel of choice (Priest et al., 2012). While it was first employed mainly for the distillation of kerosene and the burning of oil lamps, its diverse properties allowed for the creation of several fractional commodities as well as a wide variety of industrial uses in the centuries that followed. Following that, crude oil's availability and accessibility became vital to a country's military success. Ships driven by crude oil were faster and could cover larger distances than ships propelled by coal. Aside from this, advancements in fuel technology allowed for the development of new engines (most notably, the internal combustion engine) for automobiles, armored vehicles, and airplanes, among other things. These technical advancements brought the world closer together, illuminated our homes, and established the framework for the modern world as we know it to come. As a result of its features as a highly concentrated, efficient, and portable source of energy, crude oil has grown to become one of the world's most precious and traded commodities. Even though crude oil combustion emits CO_2, a significant greenhouse gas, fractional distillation of petroleum yields a wide range of products, including transportation fuels (petrol and diesel), heating and cooking fuels (liquefied petroleum gas, LPG), aviation turbine fuel, asphalt, bunker fuel oil, and a variety of other feedstocks for the production of chemicals, pharmaceuticals, and other synthetic materials. Crude oil is currently utilized to power more than 92% of all autos on the road. In addition, it is used in the production of energy (in the form of power plants or as backup electrical generator sets). Currently, over 61% of all Indian families utilize LPG cylinders to cook their meals (Kumar, 2017). In contrast to prior forecasts that renewable and nuclear energy will eventually replace crude oil production, the United States' dependency on crude oil has risen throughout the first two decades of the 21st century. Nonetheless, as the third decade of the 21st century begins, the future of crude oil is jeopardized due to strong opposition from some major stakeholder groups and members of the general public concerned about the growing impact of fossil fuel consumption and its harmful impact on the environment. However, this has emboldened a lot of opponents, and the business has been constantly chastised by politicians trying an impossible balancing act between competing needs for renewable energy and the environment. In conjunction with the objective of decarburization and rising environmental worries about climate change, this has resulted in a barrage of economic and social sanctions aimed at oil and gas corporations, disrupting production cycles and delaying output. Additional advantages of the extraordinary lockdown measures put in place to combat the rare COVID-19 disease include a significant reduction in energy consumption because factories, transportation, and offices were closed and the majority of the world's population remained indoors as a result of the lockdown. As a result of the global energy revolution, oil and gas companies are finding it increasingly difficult to forecast their demand, financial, and social futures. As a result of this uncertainty, financial institutions throughout the world have been forced to reassess their support for hydrocarbon exploration and extraction projects, while present energy investors have grown increasingly concerned about the return on their investment. According to one metric, companies in the energy sector have lost 48% cumulatively since 2015, making it the worst-performing sector in the S&P 500 since 2015. However, the rapid and sustained drop in oil prices (from USD 100 to less than USD 25 before stabilizing in the USD 45–60 range) has helped hydrocarbon companies maintain market share in the face of rising competition from renewables, trade wars, overshooting production quotas, and uncertainty about demand recovery following the COVID-19 pandemic. The hydrocarbon sector's survival is also threatened by excessively ambitious politicians who are always attempting to reduce CO_2 emissions or even entirely decarbonize the business. As a result of these actions, many investors and experts have begun to contemplate the possibility of a future hydrocarbon demand ceiling (Ferguson et al., 2009; Muggeridge et al., 2014).

This does not mean, however, that the oil and gas industry is condemned to a lengthy demise. Rather than that energy consumption is expected to increase, raising doubts about the potential of alternative energy sources (such as solar, wind, or nuclear) to supply this demand affordably and equally. This has created an enormous potential for oil and gas-producing companies to fulfill the projected energy demand by exceeding the deployment of alternative, non-fossil energy sources and devising strategies to mitigate the industry's carbon footprint. By doing so, hydrocarbon companies will be able to meet multiple political, financial, and societal demands, enabling a transition to a low-carbon future while continuing to meet global energy demand for the foreseeable future, at least until other sources of sustainable fuel are developed (Andersson and Grönkvist, 2019).

Confidence in the hydrocarbon business's future is not only hyperbole. Even the most ardent industry skeptics acknowledge that oil and gas will continue to play a critical part in the world's energy balance. This is especially true for emerging and developing economies, which will continue to rely on cheap fossil fuels to power their economic growth endeavors. Even the International Energy Agency's (IEA) Sustainable Development Scenario (SDS) and Shell's Sky Scenario, which are both considered aggressive in their decarbonization projections, acknowledge oil and gas's long-term importance. This need for hydrocarbons will remain strong even if consumption levels decline from present highs (some industry leaders believe that we are already past the peak demand and the oil demand is not expected to rebound to pre-pandemic levels). Due to the diversity and ease of their energy generation, fossil fuels (especially natural gas) will continue to dominate India's and China's energy mix (the two most populous nations). Even in the United States and the European Union, where anti-hydrocarbon sentiment and environmental concerns are strongest, energy demands are met by cheaper shale deposits (in the US) or Russian natural gas supplies (for the European Union). While there are plans to phase out crude-powered vehicles on our roads shortly and significant funding is being directed toward newer technologies, such as hydrogen fuel cells, photovoltaic, and nuclear fusion reactors, several industries such as aviation and shipping lack viable alternatives and will continue to rely on crude oil for the foreseeable future. Even if crude oil's use in transportation and power generation declines, it will continue to be necessary for the chemical industry, as it provides various vital by-products. Needless to say, crude oil's use will not be completely lost in the next decades, and there will continue to be a need to locate and extract crude oil from its reserves deep inside the Earth's belly.

1.2 ORIGIN OF PETROLEUM

1.2.1 NATURAL ORIGIN OF PETROLEUM

The origin of petroleum has been described in detail in previous work (Barker, 1985). However, a brief discussion is provided as follows. The fossil fuel from which we extract, petroleum (or crude oil), is the end product of a series of basic Earth processes that transpired over millions of years (Zhixue et al., 2010). Petroleum, like all other organic matter, is produced and accumulated as part of the geologic carbon cycle (Fig. 1.1), and like all other organic matter on Earth, it is derived from the photosynthetic reaction system (the basic driving force behind the biological carbon cycle) (Kutcherov and Krayushkin, 2010). Petroleum contains the complete range of ancient organic molecules, from the smallest molecule (methane) to complicated high-molecular weight asphaltenes. As a result, oil exists in the vaporous, fluid (condensate and unrefined petroleum), and semi-solid states (strong bitumen or black-top). Oil is a mixture of saturated hydrocarbons (n-alkanes and spread non-cyclic (paraffinic) as well as alicyclic (naphthenic) hydrocarbons), aromatics (benzene derivatives, naphthene aromatics, and polycyclic aromatics including sulfur, primarily thiophene-type aromatics), and a hetero-(fundamentally O, S, N-) compound division (for example, natural acids, ketones, phenols, heterocyclics). Carboxyl, hydroxyl, carbonyl, and other functional groups are dispersed randomly throughout the particle as the molecular weight of the crude increases (particularly the resins and asphaltenes). Thus, petroleum production is the process by which highly volatile natural substances found in source rocks are converted to liquid or vaporous hydrocarbons (Fig. 1.2). Because the earth is covered by layers of solid (or, at significant depths, liquid) stone, the oil contained inside it cannot survive in a self-contained "lake," but must instead choose to reside inside the small spaces (or pores) found in these stones. As with the wipe in your kitchen sink (albeit less pliable and slightly heavier), certain types of rock (most notably sandstone and limestone) possess pores large enough and connected enough to act as storage and relocation locations for oil, water, or any other liquid desiring to call them home. Due to the fact that the majority of hydrocarbons are lighter than water and rock, those found inside the earth will generally migrate upward until they reach the surface or are trapped by an impermeable layer of rock (Fig. 1.1).

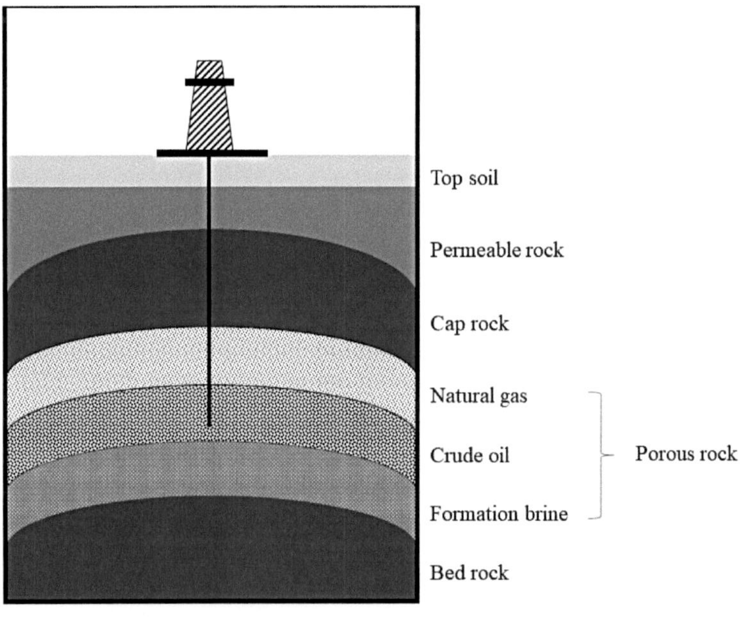

FIGURE 1.1 Location and extraction of a crude oil from its reservoir.

The zooplankton must locate a precise temperature window to frame oil. If the temperature is too low, the oil will remain trapped as kerogen; nevertheless, if the temperature is too high, the oil will be converted to petroleum gas (through "warm splitting"). As a result, establishing an oil supply requires the far-fetched assembly of three specific conditions: first, a source rock rich in natural material (framed during diagenesis) must be covered to the appropriate depth to locate an attractive window; second, a

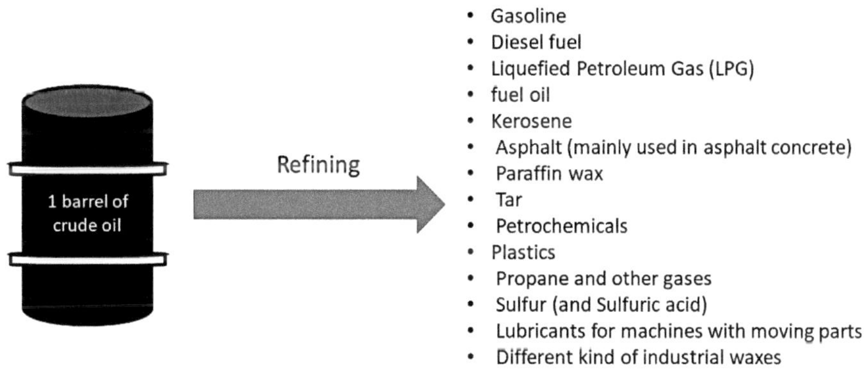

FIGURE 1.2 Various end products obtained from refining crude oil.

permeable and penetrable (associated pores) repository rock must be available for it to collect in; and third, a top stone (seal) or another instrument must be available to prevent it from vanishing to the surficial layer. Certain locations on the planet have a geologic history that makes them substantially more likely to possess the essential combination of circumstances. The oil age begins under typical conditions at temperatures of roughly 70–80°C, which frequently corresponds to an entombment depth of 2.5–3 km, depending on the geothermal angle. The depth at which hydrocarbons begin to form denotes the beginning of the oil window. Between the surface and the head of the oil window, natural issue undergoes an unexpected evolution as a result of the combined effects of increased temperatures, bacteria, and so forth; these changes referred to as diagenesis, result in the age of natural acids, CO_2, and H_2O. They induce the residual kerogen (the insoluble natural substance in similar natural solvents) to be dynamically augmented in hydrogen and carbon, at the expense of non-hydrocarbon mixtures. The kerogen is considered immature at this stage of development, and so no hydrocarbon or associated chemical may be formed.

Oil is extracted from lucrative source beds that contain a large amount of natural carbon. This high natural carbon content resulted from the encapsulation of plants and animals. Because not all plants and critters have the same carbon content, the C range of the oil fluctuates as the carbon source changes. The potential of a stone to produce hydrocarbons indicates its growth. Thus, the higher the grade of the stone, the more mature it is. Development occurs over an extended period, involving diagenesis and catagenesis. Between 10% and 20% of oil is framed during diagenesis. The majority of oil is generated during the catagenesis and metagenesis of biogenic natural matter that remains. Development is the process of converting biomolecules into the oil. "Diagenesis is defined as the synthetic reactions that occurred in the first few thousand years after confinement at temperatures below 50°C." During diagenesis, oxygen, nitrogen, and sulfur are removed from the natural issue, increasing the sedimentary natural issue's hydrogen content. Additionally, the presence of iron during diagenesis increases the possibility of the sulfur ejection, since the iron reacts with sulfur over a long period of entrapment, which increases the monetary value of oil. Diagenesis is a process of compaction that occurs at low temperatures and weights. When natural amphibian residues (proteins, lipids, and carbohydrates) are retained, they are submerged in water and a mineral-rich environment. During internment, water is limited out and proteins and carbohydrates separate to form new structures such as a waxy material termed "kerogen" and a black tar-like substance called "bitumen." This occurs entirely within the first several hundred meters of entombment. Although bitumen includes the heaviest portions of oil, kerogen will undergo additional transformation to become hydrocarbons. The reactions that occur between 60°C and 200°C are considered catagenetic in nature. During catagenesis, natural mixtures are subjected to a variety of heated corruption responses, which include two-fold security reduction via the addition of sulfur or hydrogen particles, splitting reactions, and buildup reactions. With increasing temperatures and weights (deeper encapsulation), the cycle of catagenesis begins, which is the thermal corruption of kerogen to shape hydrocarbon chains. Critically, the catagenesis cycle is accelerated by the minerals that are preserved and persist in marine diagenesis. Catagenesis states determine the item, with greater temperatures and weights resulting in more complete "splitting" of the kerogen and hence lighter and smaller hydrocarbons. Oil development, then, requires a specific window of circumstances; too hot and the item will sustain combustible gas (small hydrocarbons), but too cold and the little fish will remain trapped as kerogen. This behavior is in direct conflict with the coal arrangement. Due to earthy entrapment, natural silt is mostly replaced by cellulose and lignin, and mineral division is significantly reduced. In this case, the natural issue's degradation is slowed differently. The natural problem is thick enough to form peat, and if sufficient warmth (geothermal energy) and weight are applied, it consolidates and undergoes catagenesis to form coal. Increased temperatures and weights generally result in increased coal locations." Metagenesis occurs at temperatures more than 200°C and is regarded as a sort of degraded transformation. Following cooking or the formation of petroleum, its major migration occurs inside the source rock as a result of weight increment caused by the age of oil and gas. Because gases and fluids are less dense than solids, they absorb more volume, resulting in pressure escalation. Similarly, a lucrative source rock must be efficiently shattered to allow liquids to flow through it. During critical movement, gas and oil travel in unison during the fluid stage due to the large weights in the source rock, which typically exceed the air pocket point pressure at which gases begin to separate from fluid. Ejection from the source rock may occur in beats because of the closure of pores and fractures following the first removal. With time, the oil creates sufficient strain to reopen fractures and pores, necessitating additional removal, and the beats continue as long as the oil can recreate sufficient strain to restart its route out of the source rock. Following the movement, the weight of the body decreases and the fissures and pores shut. Finally, as oil is extracted from the source rock, pressures decrease, especially if the oil is vertically relocated. Secondary migration is the process by which oil is developed away from the source rock and into a repository bed. The gas and oil separate during auxiliary movement, with the gas moving ahead of the oil. The largest oil reserves are the result of sidelong relocation, which results in a greater volume of source rock seeping than vertical migration. Contrasts in porousness between adjacent stratigraphic strata restrict migration in the majority of situations. This results in the flow of oil within geological units. To collect oil in a trap, it must come into contact with a top stone or seal, such as gas hydrates, which form the ideal seals due to their poor penetrability.

Surface prospecting for subsurface collections of oil and gas is an idea that has been around for a long time in some structures. For instance, the old Egyptians were utilizing a type of surface prospecting when they recuperated squares of black-top from the Dead Sea two or three thousand years prior. The beginning of contemporary surface geochemical prospecting was fairly spoiled with businesspeople overselling to some degree questionable methods. A great part of the overselling brought about customer encounters giving surface prospecting a terrible name in the business. All the more as of late, with the overall improvement of strategies, the coordination of surface prospecting with geologic and geophysical data and the utilization of petroleum (due to its superior properties, see Fig. 1.3), the importance of petroleum has increased manifold for the global economy.

1.2.2 EXTRACTION OF PETROLEUM

This is the life cycle of a well, which begins with careful design and preparation. Before drilling a well, a geologist or geophysicist chooses a geology

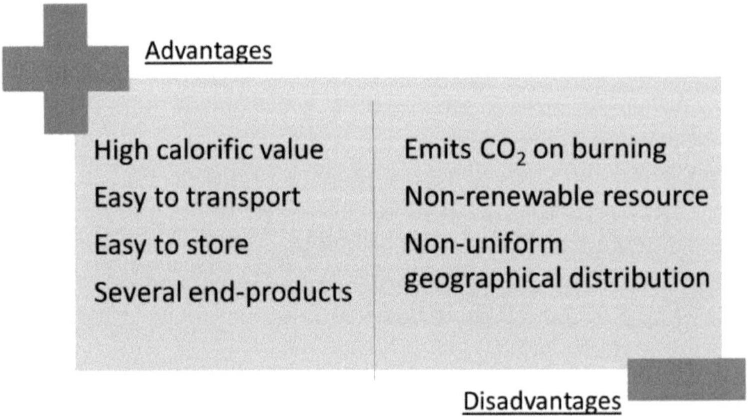

FIGURE 1.3 Primary advantages and disadvantages of crude oil.

target that will meet the objectives of the well and is suitable for drilling. When drilling a production well, the goal is to maximize well output while also controlling the reservoir outflow. To determine the existence or extent of a possible hydrocarbon resource, the target for an exploration or appraisal well is selected. If a target is selected for an injection well, it is done so that the injection point can be located in a permeable zone capable of displacing water or gas and/or driving hydrocarbons into nearby producing wells. Once the objective (the well's endpoint) and the starting point (the well's beginning point) have been identified, a trajectory between them will be generated. When designing the trajectory, a variety of factors must be taken into consideration, including clearance to any nearby wells (anti-collision) or whether this well will obstruct future wells, avoiding faults whenever possible, and the fact that certain formations may be easier or more difficult to drill at specific inclinations or azimuths. An investigational well path will be determined by a team of geoscientists and engineers, who will then develop a set of putative subsurface characteristics that will be drilled through to reach the intended target. One or more of these characteristics include porous pressure, fracture gradient, wellbore stability, porosity, and permeability. Other characteristics include lithology, faults, and the presence or absence of clay. When a good engineering team performs casing and completion design, they also perform detailed planning, which includes the selection of drill bits, design of a back-hole assembly (BHA), selection of drilling fluid, and writing of step-by-step procedures for safely and efficiently executing the well. Trajectories and designs may undergo multiple changes before a plan is authorized due to the interplay of several components in a good design and the fact that altering one factor has a knock-on effect on a large number of other parameters.

A well is then dug as a result of this procedure. Tradition has it that a well is made by drilling a hole into the earth with a diameter ranging from 12 cm to 1 m (5 × 40 in) and utilizing a drilling rig that circles a drill string with a bit attached to bore the hole. Following the drilling of the hole, lengths of steel pipe (casing) with a diameter that is somewhat smaller than the borehole are placed into the hole. Cement can be put between the outside of the casing and the annulus of the borehole. Cement is used to seal the borehole. It maintains the structural integrity of a newly drilled wellbore while also separating potentially dangerous high-pressure zones from one another and from the surrounding environment at the same time. Due to the safe separation of these zones, and protection provided by the casing, the well may be bored farther (into potentially more unstable and violent formations) and cased with a smaller bit and casing. Typically, modern wells feature between two and five sets of gradually smaller hole sizes, each of which is sealed with casing. The weight of the drill string above the drill bit assists the drill bit in cutting through the rocky terrain. It is possible to employ numerous different types of drill bits; some cause rock to disintegrate by compressive failure, while others shear the rock away as the bit rotates. Once the drill pipe has been filled with

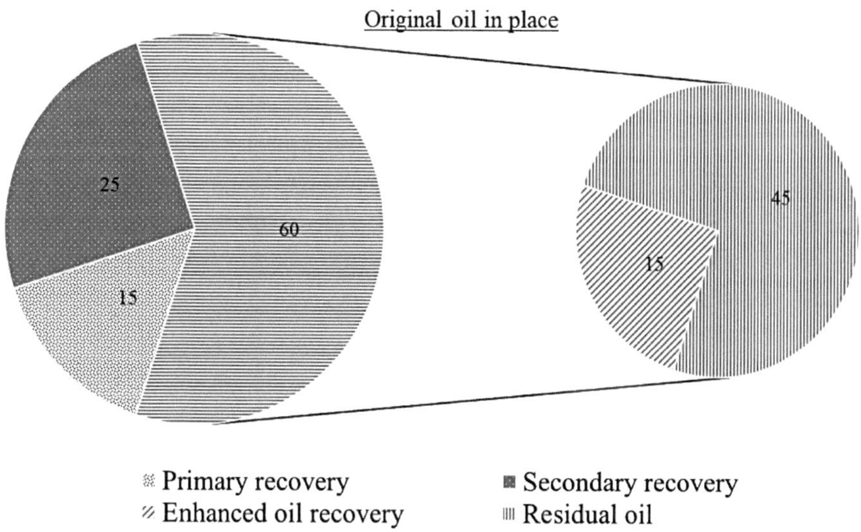

FIGURE 1.4 Various mechanisms of oil recovery with recovered oil (at the end of process) as a function of original oil in place.

fluid and the drill bit has been reached, the fluid is pumped down the inside of the pipe until it reaches the drill bit. Though largely consisting of water and clay, drilling fluid also contains a complex mixture of fluids, particles, and chemicals that must be properly calibrated to provide the physical and chemical properties essential for effectively digging the hole. In addition to freezing the bit and elevating rock cuttings to the surface, drilling mud is also useful in preventing rock instability in the wellbore walls and resisting the pressure of fluids trapped within the rock to prevent them from entering the wellbore. Certain oil wells are excavated using nothing more than air or foam as the drilling fluid, and the results are spectacular. As the drilling fluid goes back to the surface outside of the drill pipe, it picks up the rock "cuttings" that were generated during the drilling process. Following that, the fluid is strained to separate the cuttings from the good fluid, which is then returned to the pit. Monitoring for anomalies in the cuttings (which are returning to the surface) as well as the pit volume or rate of returning fluid is crucial for recognizing "kicks" early in their progression. "Kick" occurs when the pressure in the formation at a depth greater than that of the hydrostatic head of the mud above, and it can occur if the blowout preventers are not closed or the drilling fluid's density is not increased sufficiently to prevent the formation fluids and mud from rising uncontrollably through the annulus. To drill deeper into the ground, additional 9 m (30 ft) sections or "joints" of pipe are screwed into the pipe or drill string connecting to the bit at the surface. This process is repeated until the well is dug to the desired depth and then stopped. "Tripping" is the term used to describe this. Stands with many joints may be made by combining several different joints, allowing for more efficient tripping while pulling out of the hole. Using the example of a standard triple, the pipe would be extracted from the hole three joints at a time and stacked in the derrick to save time. One by one, several sophisticated rigs, collectively referred to as "super singles," detonate the pipe and rack it as they go. A drilling rig, which is designed to circulate drilling fluid, lift and spin the pipe, regulate downhole operations, remove cuttings from the drilling fluid, and generate on-site energy for these activities, makes the entire process much easier to complete.

Following the drilling and casing of the well, it must be "completed." In the oil and gas industry, completion is the process through which a well is made operational and capable of producing oil or gas. During a cased-hole completion, small holes known as perforations are drilled into the section of the casing that passed through the production zone to allow oil to flow from the surrounding rock into the production tubing and the production tubing. "Sand screens" or "gravel packs" are typically used to fill the remaining drilled, uncased reservoir part of an open hole throughout an open hole completion project. These help to maintain the structural integrity of the wellbore in the absence of casing yet allowing reservoir flow into the wellbore to occur. Screens also keep formation sands from getting into production tubulars and surface equipment, which can cause washouts and other problems, especially in offshore fields with unconsolidated sand formations. The introduction of acids and fracturing fluids into the well after the creation of a flow path is intended to clean, fracture, or otherwise prepare and stimulate the reservoir rock to maximize hydrocarbon output into the wellbore. Once this has been accomplished, the part of the well above the reservoir section is sealed inside the casing and connected to the surface by a smaller diameter pipe known as tubing. In addition to acting as a redundant barrier against hydrocarbon leaks, this system allows for the repair of damaged components. Additionally, the tubing's smaller cross-sectional area accelerates reservoir fluids, reducing liquid fallback that would otherwise result in increased back pressure, and protecting the casing from corrosive well fluids, according to the manufacturer. In many wells, the inherent pressure of the subterranean reservoir is sufficient to allow the oil or gas to flow to the surface of the ground. In other cases, this is not the case, particularly in depleted regions where pressures have been decreased by other producing wells or in oil reservoirs with poor permeability, as previously stated. The installation of tube with a smaller diameter may be sufficient to help in production, while artificial lift methods may also be necessary in some circumstances. Downhole pumps, gas lift, and surface pump jacks are all typical techniques of extracting water from the ground. Throughout the preceding decade, a slew of innovative well finishing techniques were introduced. Multiple packer systems used in conjunction with frac ports or port collars have decreased completion costs while simultaneously increasing the output, notably for horizontal well completions. For optimal hydrocarbon recovery, casings should be run into the lateral zone with the packer/frac port located in the proper place.

Producing oil and gas is the most essential stage of a well's life cycle since it is at this stage that the oil and gas are produced. They control pressures, regulate flow, and allow access to the wellbore in case more completion work is necessary. Using the output valve of the production tree, the flow may be connected to a distribution network of pipes and tanks, which can then transport the product to refineries. If the reservoir pressure remains sufficient, the production tree will be sufficient to produce the well for as long as it is in operation. If the pressure falls below a given level and it is determined that it is economically feasible, the artificial lift method mentioned in the conclusion section may be employed to restore pressure. The use of enhanced recovery techniques, such as reservoir pressure enhancement through water flooding, steam flooding, or CO_2 flooding, can result in the ejection of hydrocarbons from the reservoir by increasing reservoir pressure and causing a "sweep" effect, which causes hydrocarbons to be ejected from the reservoir. When a reservoir's pressure is depleted, the oil viscosity is too high, or even early in a field's existence, such processes entail the use of injection wells (which are typically picked from existing production wells in a carefully set sequence), and they are employed. Depending on the geomechanics of the reservoir, reservoir engineers may discover that utilizing a water flooding approach early in the field's development, rather than later, can increase the amount of oil ultimately recoverable. Secondary recovery is a term used to describe the expanded healing procedures that are employed. When a well's most efficient production rate is insufficient to cover its operating expenses, which include taxes, it is said to have reached an "economic limit." By increasing the economic limit, the well's life is reduced, and proven oil reserves are lost as a result of the increased economic limit. When the economic constraint is lifted, on the other hand, the well's useful life is lengthened significantly.

It is abandoned when an oil well reaches the point when it is no longer economically feasible to continue operating the well. Some abandoned wells are ultimately sealed and their locations restored, but such procedures are extremely expensive, often costing millions of dollars to carry out. In this technique, tubing from the well is removed and portions of the hole are filled with concrete to create a flow channel between the gas and water zones as well as the surface and to divide the flow channel from the surface. It is necessary to remove the wellhead and casing following the excavation of the region surrounding the wellhead. A cap is then welded to the wellhead, and the wellhead is subsequently buried. An enormous amount of unrecoverable oil is usually left in the reservoir when the economic limit has been reached. In North America, thousands of "abandoned" wells are waiting for the right market circumstances before being completely abandoned. Frequently, lease agreements and government regulations necessitate rapid abandonment; liability and tax considerations may also be in the favor of surrendering. Reentering and restarting an abandoned well (or converting it to injection service for further recovery or downhole hydrocarbon storage) is theoretically conceivable, but it is often difficult and expensive due to the complexity of the mechanics and other factors involved. Historically, elastomer and cement plugs have been used effectively and consistently in a variety of applications to varying degrees of success. Depending on the materials used in their construction, they may deteriorate with usage, particularly in corrosive environments.

1.2.3 STAGES OF OIL RECOVERY

Three phases are involved in the extraction of oil from a well (Fig. 1.5). This category includes primary, secondary, and enhanced oil recovery (i.e., tertiary) (Shafiai and Gohari, 2020). When it comes to reservoir drive during the primary recovery phase, a lot of natural variables are in action. The pressure in a well decreases with time (Trivedi and Babadagli, 2009). At some point, the pressure beneath the earth becomes insufficient, and the oil is forced to the surface. After the natural drive of the reservoir has been exhausted, further recovery methods are used to replenish it (Fig. 1.5). External energy is given to the reservoir by fluid injection, increasing or replacing the natural reservoir drive. To boost reservoir pressure, secondary recovery procedures like water injection, gas reinjection, and gas lift are used. Gas reinjection and lift are both methods for boosting the mobility of an oil-gas mixture by diluting it with related gas, carbon dioxide, or another inert gas and so increasing its mobility. The typical recovery factor from

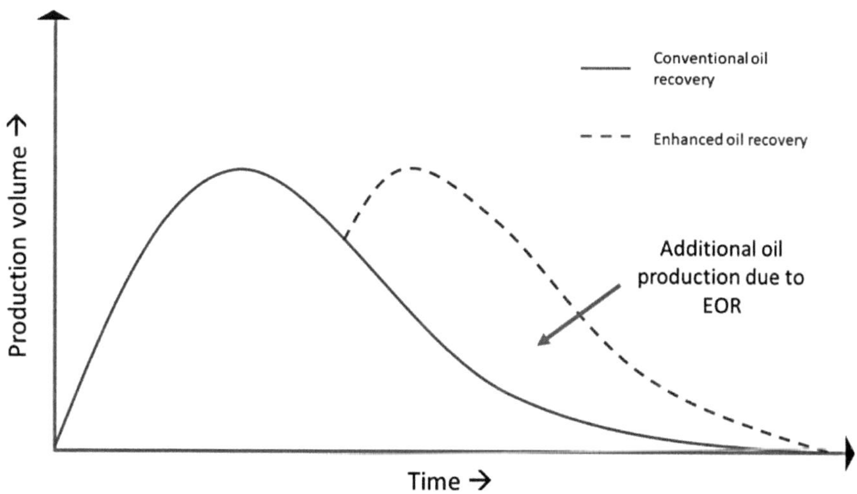

FIGURE 1.5 Oil recovery as a function of time. The implementation of EOR increases the productivity and the life of the well.

water-flood operations is around 30%, depending on the quality of the oil and the reservoir rock properties. According to industry standards, the recovery factor following primary and secondary oil recovery operations is between 35% and 45%.

Enhanced oil recovery techniques, sometimes referred to as tertiary oil recovery techniques are used to increase the mobility of oil and therefore its extraction. Temperature-assisted thermal enhanced oil recovery (TEOR) procedures are tertiary recovery techniques that include heating the oil to reduce its viscosity and facilitate extraction (Fig. 1.6). The most often used technique for TEOR is steam injection, which is typically combined with a cogeneration facility. This type of cogeneration facility generates electricity using a gas turbine while utilizing waste heat to generate steam that is then supplied to a reservoir. Oil extraction in the San Joaquin Valley, which produces extraordinarily heavy oil but accounts for just 10% of the total US oil production, has exploded in recent years as a result of this type of extraction. Fire flooding (a.k.a. in situ burning) is a kind of TEOR in which, rather than steam, a part of the oil is burnt to heat the surrounding oil similarly to how steam is used (Sheng, 2013). Surfactants (detergents) are periodically injected into reservoirs to alter the surface tension between the water and the oil in the reservoir, allowing for the mobilization of oil that would otherwise remain trapped as residual oil. Carbon dioxide flooding is another means of reducing viscosity. Tertiary recovery enables the extraction of an additional

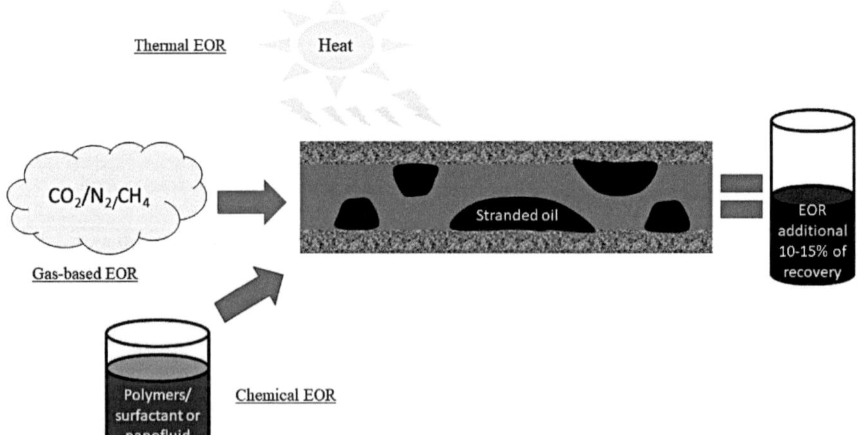

FIGURE 1.6 Methods of enhanced oil recovery for increasing stranded oil production from a depleted oil reservoir.

5–15% of the oil contained inside the reservoir. In certain California heavy oil fields, steam injection increased oil reserves and eventual oil recovery by a factor of two or three, depending on the field. Consider California's Midway-Sunset Oil Field, the state's largest oilfield. Tertiary oil recovery occurs when secondary oil recovery is no longer adequate to allow for further extraction, but only if the oil can still be produced economically at that stage. This is the scenario based on the current price of crude oil and the cost of the extraction technology. When oil prices are high, formerly unprofitable wells are brought back into operation, whereas extraction is curtailed when oil prices are low. Another tertiary healing strategy is microbiological treatment. Special microbe combinations are employed to treat and degrade the hydrocarbon chain in oil, making the extraction process easier and more efficient. Additionally, it is less costly than other traditional approaches. Employing these bacteria in what is known as secondary tertiary recovery is tax favored in some areas, such as Texas, where tax benefits are offered. These incredibly sophisticated microorganisms, which are critical for oil field applications, are only available from a few businesses. Several factors, however, affect oil recovery. Among these factors are the rock's permeability, the strength of natural drives (such as the presence of associated gas, pressure from adjacent water, or gravity), the porosity of the reservoir rock, which is the rock's capacity to store oil, and the viscosity of the oil. Oil cannot flow through "tight" reservoir rocks, such as shale; but, oil may flow through permeable reservoir rocks, such as sandstone. Even though the recovery of a well cannot be determined precisely until it reaches the point of no return, petroleum engineers often calculate an estimated ultimate recovery (EUR) based on years-out decline rate forecasts (Fig. 1.5). Numerous models, mathematical approaches, and approximations are employed.

1.3 ADVANCEMENTS IN CHEMICAL ENHANCED OIL RECOVERY

In previous investigations, scientists have used polymers, alkalis, and surfactants in EOR in multiple oil fields (Fig. 1.7) (Sheng, 2011). Adding water to polymers makes the aqueous phase more viscous, which makes the polymers more mobile as they move from the injection well to the producer (Kumar and Mandal, 2017). The polymer solution also helps with oil recovery by reducing the water permeability of the well. Surfactant solutions cut down on the IFT between water and crude oil by reacting with certain components of crude oil. This makes interfacial coatings solubilized and emulsified (Sagir et al., 2020). Having less IFT means that the capillary

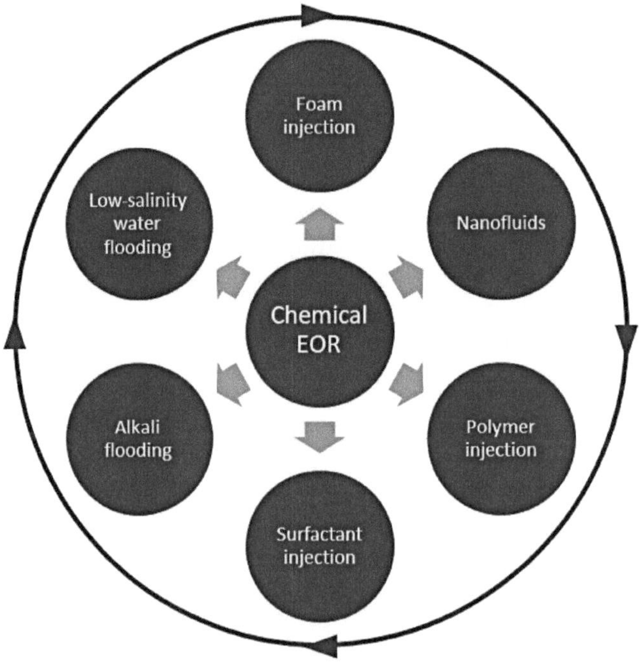

FIGURE 1.7 Various methods of chemical enhanced oil recovery.

forces of trapped and residual oil are less strong. Surfactants also stick to rocks in the ground, which makes them less wet and increases the amount of oil that can be retrieved from them (Raffa et al., 2016; Webb et al., 2005). An alkali flood is like a surfactant solution, except that the substance that is injected is different (Samanta et al., 2011). It also must be made sure the injected fluid does not get into thief zones and goes into areas with low permeability. Foam flooding is used to do this (Telmadarreie et al., 2016). Recently, a binary chemical combination was made and tested on a pilot and field scale to help more oil be found. This method of EOR was born out of a desire to maximize the synergy of the different chemicals and to compensate for their flaws during flooding. As a consequence of the binary combination, alkali–surfactant (AS), alkali–polymer (AP), surfactant–polymer (SP), and alkali–surfactant–polymer (ASP) flooding processes were established (Samanta et al., 2012). These floods were also made to add alkali, surfactant, and polymer solutions to the OOIP to make the pores bigger and make it easier for them to be swept away by the water. In the field of chemical EOR, new things are happening that use nanotechnology, test new chemical compounds like ionic liquids, and make chemical EOR more efficient (Khan and Khan, 2018).

Nanotechnology has helped solve several problems in the oil and gas industry, including problems with petroleum exploration, drilling, and completion, flow assurance, hydraulic fracturing, and better oil recovery (Rafati et al., 2018; Yousefvand and Jafari, 2015). In recent years, nanoparticles have been used in chemical EOR because they can react with the amphiphilic and macromolecular parts of EOR chemicals. With polymers, surfactants, and foams, they make polymeric nanofluids, nano-surfactants, and nanostabilized foams. This is how they work. The newly formed suspension (or composite material, as the case may be) has unique and interesting properties that are not found in any single component (Chaturvedi and Sharma, 2020). Nanoparticles can be added to polymer molecules to improve their rheological properties and protect them from degradation. Nanoparticle-supported foams have a longer half-life. In addition, the lamellae bubbles are bigger in diameter, which helps keep them stable in the thermodynamic sense. Nanosurfactants have a very low IFT because nanoparticles permanently stick to the surface (Chaturvedi et al., 2021). Nanochemicals, on the whole, have a hard time sticking to porous materials when they are tested. As a result, they have been called the next big thing in chemical EOR use. Polymer nanofluids are made up of a mix of polymer nanoparticles and macromolecules, and so they have unique and interesting properties for EOR applications. Electrostatic and van der Waals interactions, hydrogen bonding, hydrophobic interaction, steric repulsion, and electrosteric repulsion are some of the forces that work together when they are mixed together. In addition, a lot of research has been done on the combination of nanoparticles and surfactants for EOR applications. Each combination changes the reservoir's properties by changing its wettability and adsorbing at the oil–water interface because of the surface-active groups that are in each part. So, surfactant nanofluids, which are also called nanosurfactants, can achieve low IFT and stable emulsions that precursor nanoparticles or surfactants alone cannot do on their own (Lu et al., 2016). Because the stability of a foam's lamellae is so important in porous media, this is how it will hold up. Nanoparticles have recently been shown to be a good way to make a foam that is stable in a very wet environment, either by putting them in the foamy surfactant solution or by changing the surface wettability of the nanoparticles (Yang et al., 2017). Nanoparticles had the same effect on nanoparticle-stabilized foams and nanoparticle–surfactant mixtures. Most of the problems with surfactant-stabilized foams can be solved if nanoparticles are used to keep the foam in place. Because nanoparticles are solid, the foams they support should be very resistant to things like high salt levels and high temperatures in the reservoir. Nanoparticle-stabilized foams do not need to change the chain of the polymer or the chemical structure of the surfactant like chemically stabilized foams do. It is called "smart nano-waterflooding" when nanoparticles are spread out in water, brine, and alcohol and injected into the ground (nanodispersion) (Peng et al., 2017). This method makes the most of the nanoparticle's large surface area and other features to get more oil out of the ground. SiO_2 nanoparticles, for example, are very good at lowering IFT at the oil–water interface. Because of the high cost and environmental problems with traditional chemical EOR, ionic liquids, a type of molten salt, have become a popular choice for EOR applications because they are environmentally safe. Ionic liquids are usually made up of organic cations and organic or inorganic anions. They are then classified based on their chemical make-up. They have a lot of good qualities, like high solubility, high ionic conductivity, reusability, and good thermal stability. Ionic liquids have different features and abilities based on their anions, cations, and alkyl groups (Ren et al., 2017). For example, when ionic liquids are used as injectants, the amount of viscosity they add depends on the length of the alkyl group that is present (Khan et al., 2016). In addition, ionic liquids can be used to change wettability and reduce IFT because of their surface-active nature and ability to form micelles. Laboratory tests say that ionic liquids are better than surfactants because they can reduce IFT at high salt and temperature levels. Ionic liquids also have a better ability to self-organize during the process of putting together an interface. Also, ionic liquids have a better ability to keep emulsions stable. However, the downside of this chemical's new use in EOR is that when it has halogen-anions in it, it goes over the green limit. When these ionic liquids break down, they make dangerous by-products like hydrogen fluoride (HF) and hydrogen chloride (HCl). It was also found that ionic liquids adsorb on kaolinite clay when the salt level was high enough. This means that more research on ionic liquids is needed to solve these problems in the field.

However, because well-researched and previously used chemical EOR techniques do not work as well as they should, researchers have been looking for new and practical chemical EOR strategies that can increase oil recovery without having negative effects (Sheng, 2013).

KEYWORDS

- **oil exploration**
- **EOR**
- **chemical EOR**
- **thermal EOR**
- **reservoir life cycle**

REFERENCES

Andersson, J.; Grönkvist, S. Large-Scale Storage of Hydrogen. *Int. J. Hydrogen Energy* **2019,** *44*, 11901–11919. https://doi.org/10.1016/j.ijhydene.2019.03.063

Barker, C. Origin, Composition and Properties of Petroleum. In *Developments in Petroleum Science*; Elsevier, 1985; pp. 11–45. https://doi.org/10.1016/S0376-7361(08)70564-8

Chaturvedi, K. R.; Narukulla, R.; Amani, M.; Sharma, T. Experimental Investigations to Evaluate Surfactant Role on Absorption Capacity of Nanofluid for CO2 Utilization in Sustainable Crude Mobilization. *Energy* **2021,** *225*, 120321. https://doi.org/10.1016/j.energy.2021.120321

Chaturvedi, K. R.; Sharma, T. Carbonated Polymeric Nanofluids for Enhanced Oil Recovery from Sandstone Reservoir. *J. Pet. Sci. Eng.* **2020,** *194*, 107499. https://doi.org/10.1016/j.petrol.2020.107499

Doman, E. L. International Energy Outlook 2016, U.S. Energy Information Administration, 2016.

Ferguson, R. C.; Nichols, C.; Leeuwen, T. Van; Kuuskraa, V. A. Storing CO2 with Enhanced Oil Recovery, In *Energy Procedia*, 2009; pp 1989–1996. https://doi.org/10.1016/j.egypro.2009.01.259

Khan, M. B.; Khan, M. S. Experimental Investigation on Effect of Alkali, Salts on the Rheological Properties of Polyacrylamide and Mixed Assembly of Polymer-Surfactant. *Chem. Data Collect.* **2018,** *13–14*, 60–72. https://doi.org/10.1016/J.CDC.2018.01.004

Khan, M. S.; Lal, B.; Partoon, B.; Keong, L. K.; Bustam, A. B.; Mellon, N. B. Experimental Evaluation of a Novel Thermodynamic Inhibitor for CH4 and CO2 Hydrates, In *Procedia Engineering*; Elsevier Ltd, 2016; pp 932–940. https://doi.org/10.1016/j.proeng.2016.06.433

Kumar, G. S. Anatomy of Indian Energy Policy: A Critical Review. *Energy Sources B Econ. Plan. Policy* **2017,** *12*, 976–985. https://doi.org/10.1080/15567249.2017.1336814

Kumar, S.; Mandal, A. A Comprehensive Review on Chemically Enhanced Water Alternating Gas/CO2 (CEWAG) Injection for Enhanced Oil Recovery. *J. Pet. Sci. Eng.* **2017,** *157*, 696–715. https://doi.org/10.1016/j.petrol.2017.07.066

Kutcherov, V. G.; Krayushkin, V. A. Deep-Seated Abiogenic Origin of Petroleum: From Geological Assessment to Physical Theory. *Rev. Geophys.* **2010,** *48*. https://doi.org/10.1029/2008RG000270

Lu, T.; Li, Z.; Fan, W.; Zhang, X.; Lv, Q. Nanoparticles for Inhibition of Asphaltenes Deposition During CO2 Flooding. *Ind. Eng. Chem. Res.* **2016,** *55*, 6723–6733. https://doi.org/10.1021/acs.iecr.5b04893

Muggeridge, A.; Cockin, A.; Webb, K.; Frampton, H.; Collins, I.; Moulds, T.; Salino, P. Recovery Rates, Enhanced Oil Recovery and Technological Limits. *Philos. Trans. R. Soc. A Math. Phys. Eng. Sci.* **2014,** 372. https://doi.org/10.1098/rsta.2012.0320

Peng, B.; Zhang, L.; Luo, J.; Wang, P.; Ding, B.; Zeng, M.; Cheng, Z. A Review of Nanomaterials for Nanofluid Enhanced Oil Recovery. *RSC Adv.* **2017,** *7*, 32246–32254. https://doi.org/10.1039/c7ra05592g

Priest, T.; Black, B.; Carlos Boué, J.; Hakes, J.; Linenthal, E.; Mccormick, T.; Merrill, K.; Pratt, J.; Storrs, L. The Dilemmas of Oil Empire. *J. Am. Hist.* **2012,** *99*, 236–251. https://doi.org/10.1093/JAHIST/JAS065

Rafati, R.; Smith, S. R.; Sharifi Haddad, A.; Novara, R.; Hamidi, H. Effect of Nanoparticles on the Modifications of Drilling Fluids Properties: A Review of Recent Advances. *J. Pet. Sci. Eng.* **2018,** *161*, 61–76. https://doi.org/10.1016/j.petrol.2017.11.067

Raffa, P.; Broekhuis, A. A.; Picchioni, F. Polymeric Surfactants for Enhanced Oil Recovery: A Review. *J. Pet. Sci. Eng.* **2016,** *145*, 723–733. https://doi.org/10.1016/j.petrol.2016.07.007

Ren, J.; Musyoka, N. M.; Langmi, H. W.; Mathe, M.; Liao, S. Current Research Trends and Perspectives on Materials-Based Hydrogen Storage Solutions: A Critical Review. *Int. J. Hydrogen Energy* **2017,** *42*, 289–311. https://doi.org/10.1016/j.ijhydene.2016.11.195

Sagir, M.; Mushtaq, M.; Tahir, M. S.; Tahir, M. B.; Shaik, A. R. Surfactants for Enhanced Oil Recovery Applications, In *Surfactants for Enhanced Oil Recovery Applications*, 2020; pp 1–129. https://doi.org/10.1007/978-3-030-18785-9

Samanta, A.; Bera, A.; Ojha, K.; Mandal, A. Comparative Studies on Enhanced Oil Recovery by Alkali-Surfactant and Polymer Flooding. *J. Pet. Explor. Prod. Technol.* **2012,** *2*, 67–74. https://doi.org/10.1007/s13202-012-0021-2

Samanta, A.; Ojha, K.; Mandal, A. The Characterization of Natural Surfactant and Polymer and their Use in Enhanced Recovery of Oil. *Pet. Sci. Technol.* **2011,** *29*, 765–777. https://doi.org/10.1080/10916460903485819

Shafiai, S. H.; Gohari, A. Conventional and Electrical EOR Review: The Development Trend of Ultrasonic Application in EOR. *J. Pet. Explor. Prod. Technol.* **2020,** *10*, 2923–2945. https://doi.org/10.1007/s13202-020-00929-x

Sheng, J. J. In *Modern Chemical Enhanced Oil Recovery*; Elsevier Inc., 2011. https://doi.org/10.1016/C2009-0-20241-8

Sheng, J. In *Enhanced Oil Recovery Field Case Studies*; Elsevier Inc., 2013. https://doi.org/10.1016/C2010-0-67974-0

Telmadarreie, A.; Doda, A.; Trivedi, J. J.; Kuru, E.; Choi, P. CO2microbubbles - A Potential Fluid for Enhanced Oil Recovery: Bulk and Porous Media Studies. *J. Pet. Sci. Eng.* **2016,** *138*, 160–173. https://doi.org/10.1016/j.petrol.2015.10.035

Trivedi, J. J.; Babadagli, T. Oil Recovery and Sequestration Potential of Naturally Fractured Reservoirs During CO2 Injection. *Energy Fuels* **2009,** *23*, 4025–4036. https://doi.org/10.1021/ef900361n

Webb, K. J.; Black, C. J. J.; Tjetland, G. In *A Laboratory Study Investigating Methods for Improving Oil Recovery in Carbonates*, 2005 International Petroleum Technology Conference Proceedings; Society of Petroleum Engineers (SPE), 2005; pp 785–791. https://doi.org/10.2523/iptc-10506-ms

Yang, W.; Wang, T.; Fan, Z.; Miao, Q.; Deng, Z.; Zhu, Y. Foams Stabilized by in Situ-Modified Nanoparticles and Anionic Surfactants for Enhanced Oil Recovery. *Energy Fuels* , **2017,** *31*, 4721–4730. https://doi.org/10.1021/acs.energyfuels.6b03217

Yousefvand, H.; Jafari, A. Enhanced Oil Recovery Using Polymer/nanosilica. *Procedia Mater. Sci.* **2015,** *11*, 565–570. https://doi.org/10.1016/j.mspro.2015.11.068

Zhixue, S.; Zhilei, S.; Hongjiang, L.; Xijie, Y. Characteristics of Carbonate Cements in Sandstone Reservoirs: A Case from Yanchang Formation, Middle and Southern Ordos Basin, China. *Pet. Explor. Dev.* **2010,** *37*, 543–551. https://doi.org/10.1016/S1876-3804(10)60054-7

CHAPTER 2

Viability of EOR When Oil Prices Are Low

KRISHNA RAGHAV CHATURVEDI, ANJANAY PANDEY, and TUSHAR SHARMA

Department of Petroleum Engineering and Geological Sciences, Rajiv Gandhi Institute of Petroleum Technology, Jais, Amethi, India

ABSTRACT

This chapter scrutinizes the feasibility and strategic implications of employing chemical enhanced oil recovery (CEOR) methodologies during periods of low oil prices. Addressing the economic challenges faced by the petroleum industry, especially in face of the COVID-19 pandemic, the chapter explores how CEOR technologies can remain viable and even advantageous in a cost-sensitive climate. This chapter also explores the reasons behind the drastic fall in oil prices, focusing on two key events: the 2014 oil price crash and the impact of the COVID-19 pandemic on the oil market. The 2014 crash is attributed to a surge in oil supply, particularly from non-OPEC production, leading to a market excess. The subsequent sections delve into the economic ramifications of this supply-demand imbalance, affecting both oil-exporting and oil-importing nations. Whereas, the COVID-19 pandemic resulted in a lockdown situation, causing a sharp decline in oil demand due to restrictions on travel and business closures. The rejection of production cuts by OPEC and Russia, coupled with the pandemic-induced economic slowdown, results in a significant drop in oil prices. Moreover, this chapter also emphasizes the negative consequences for oil-dependent economies. Subsequently, this study explores the future of enhanced oil recovery (EOR), particularly carbon dioxide (CO_2) EOR technology. It also discusses how the price of

---

Advancements in Chemical Enhanced Oil Recovery.
Tushar Sharma, Krishna Raghav Chaturvedi, Tarek Ganat, and Imtiaz Ali (Eds.)

oil plays a crucial role in the viability of CO_2 EOR, impacting factors such as production costs and carbon footprints. The financial constraints faced by the oil and gas sector, especially the U.S. shale sector heavily reliant on debt, are highlighted, with implications for future investments and output. The chapter describes the evaluation of oil production, EOR scenarios, and their future potential, presenting valuable insights for petroleum engineering researchers and serving as a reference for understanding the current state and prospects of oil production. The chapter concludes by highlighting the significance of EOR techniques in shaping the future of oil production, with the focus on their potential impact on the developing world. Finally, through a nuanced analysis of cost-effectiveness, risk management, and industry trends, this chapter aims to provide insights into the resilience and adaptability of CEOR strategies as a valuable tool for sustaining production efficiency during downturns in oil markets, especially in the future.

2.1 WHY HAVE THE OIL PRICES CRASHED?

2.1.1 OIL PRICE CRASH OF 2014

Numerous folks were caught aback by the late 2014 decrease in crude oil prices. Brent crude oil prices have halved in less than a year, with little sign of a bottom in sight (Fantazzini, 2016; Lammerding et al., 2013). Until June 2014, oil prices had remained consistently above the $100 per barrel barrier. Prices recently fell to this level during the financial crisis, peaking at $150 per barrel in July 2008 but plummeting to less than $50 per barrel by year's end, a record low (Baffes et al., 2015). Commodity prices plummeted significantly in several nations as a result of a severe recession. By early 2011, oil prices had recovered to levels above $100 per barrel, indicating that the dip was only transitory. The most recent price decline does not begin as abruptly as the previous one. In recent memory, there has not been a significant global economic shock (Baffes et al., 2015). In actuality, the global economy is growing, albeit at a slower rate and with fewer participants than many predicted. As a result, how is it possible that oil prices have nearly halved in recent years? This begs the issue of how they will affect the financial health of businesses. After a period of relative stability that fostered trust and investment in the oil and gas business, the financial statements of oil and gas corporations today indicate significant uncertainty (Diaz and Gracia, 2017; Vătavu et al., 2018).

What triggered this? Quite simply – an increase in supply (Salygin et al., 2019). Even though global demand growth has slowed in recent years, oil supply has increased significantly, owing mostly to non-OPEC production, increasing market excess. Years of historically high and steady crude oil prices aided in the development of commercially viable shale oil resources, mostly in North Dakota and Texas, but also in other states. According to the International Energy Agency (IEA), prices decreased by $10/bbl in the 3 weeks following the Vienna summit, resulting in a 10% drop in oil inventories in a single day. Onshore oil production has helped some OPEC member nations in the Middle East, notably those in the Gulf area, due to its low cost. The average cost of extracting onshore Middle Eastern resources is $27 a barrel that is barely a third or half of the US shale extraction costs (which costs $65). While OPEC's decision to retain current output levels may help the organization preserve market share, it will raise pressure on high-cost producers, notably those in North America, which have quickly grown output over the last decade. OPEC currently controls nearly one-third of global oil output (Gupta, 2016).

While the global economy has recovered, the recovery has been slower and more sluggish than many analysts predicted. Poor Chinese statistics, for example, continue to cause concerns, while the World Bank reports that Japan's industrial and economic Growth along with the European Union languish and subdued at best. European crude oil consumption fell from 15.3 million barrels per day in 2009 to 14.3 million barrels per day in 2013, with 2014 statistics projected to be considerably lower. The European Union has warned that Europe's economy is on the approach of entering its third recession in 6 years as a result of slow growth and low inflation. In actuality, December inflation was –0.2%, according to the European Central Bank figures. Daily consumption in Japan has declined from 5 million barrels in 2007 to four million barrels in 2014. Individuals would be encouraged to save in expectation of future price decreases, while businesses would be discouraged from doing so (Diaz and Gracia, 2017).

2.1.2 THE COVID-19 PANDEMIC

As a result of a combination of supply and demand worries, as well as future uncertainty, global oil prices have fallen drastically recently (Fig. 2.1), culminating in a financial market meltdown (Stocker et al., 2018). What caused this decline, though, and what influence will it have on economies that export oil (such as Saudi Arabia and Russia) and those that consume oil (such as India) (Nguyen et al., 2017)?

As the impacts of COVID-19 spread around the world, the demand for energy sources, such as oil, natural gas, and coal is decreasing (Wheeler et al., 2020). Because oil is mostly used in transportation, which has been impeded by limitations on domestic and international travel, business closures as a consequence of country-wide lockdowns, and quarantines in several countries, the fall in oil use has been particularly severe (Garg et al., 2021; Ghosh, 2020). Several factors have an impact on the oil market's demand side. The world's oil consumption has declined for the first time since the global financial crisis of 2009. Demand is expected to decrease by a record 9.3 million barrels per day in 2020, according to forecasts (mbd). The IEA estimates that demand will be 23.1 million barrels per day lower in the second quarter of 2020 than it was in the same period the previous year, according to the agency. Even though the real impact of the COVID-19 pandemic on oil consumption will be determined by how quickly the virus can be managed by government protective measures and economic stimulus operations, the IEA's prediction for 2020 is hopeful (Jefferson, 2020).

FIGURE 2.1 Oil price fluctuations during the COVID months.

Source: Bloomberg and Vox EU.

2.1.2.1 THE FLEETING NATURE OF THINGS

On March 5, 2020, the Organization of the Petroleum Exporting Nations (OPEC) requested a 1.5 million barrels per day (mbd) reduction in production for the second quarter of 2020, with 1 million barrels coming from OPEC members and 0.5 million barrels coming from non-OPEC but allied countries (most prominently, the Russian Federation). The next day, the Russian

Federation rejected the proposal, while Saudi Arabia offered special discounts of up to 20% in key markets in response to the rejection. As a result, oil prices plunged by more than 30%, reaching as low as $31.1 a barrel (West Texas Intermediate crude (WTI)) on March 9, further aggravating the situation of the oil sector. As a consequence of a worsening recession coupled with concerns over the COVID-19 agreement, global oil prices fell to $11.57 per barrel on April 21, according to the IEA. As a result of the price reduction, global oil production is likely to decline by 12 million barrels per day in May, with OPEC+ agreeing to cut 9.7 million barrels per day from an agreed baseline level and further cutbacks from other countries, including the United States and Canada (Norouzi, 2021). However, given the market's present low demand, it appears that these supply reductions are minimal (if not insignificant), and the market will continue to face an excess of supply. Increased production, decreased demand, and uncertainty about the future all had a role in the decline in the price of oil. Low oil prices are a source of concern for both Russia and Saudi Arabia. Although both countries have extremely low oil production costs, their governments' finances and exports are heavily reliant on oil money. At a barrel price of $78.30, Saudi Arabia's government budget would have to be balanced by 2020 for the country to be self-sufficient. There is a breakeven threshold in the Russian Federation's budget that is somewhere in the neighborhood of $40 (Sabitova and Shavaleyeva, 2015). As a result, the low oil price will have a significant negative impact on the government budgets of the affected nations. Oil price declines will also have a detrimental impact on the fiscal health of governments in oil-producing countries as on the fiscal health of the oil producers, as well. This will jeopardize these nations' economic ambitions, and more importantly, their capacity to give appropriate government assistance to the healthcare sector to battle COVID-19 and other viruses. The additional economic loss would be caused to economies that are reliant on energy commodities such as natural gas or coal, as well as oil and petroleum goods. Petroleum commodities account for more than 30% of Indonesian exports and 20% of Malaysian exports, resulting in a reduction in government tax revenue in the current low oil price situation (Adedokun, 2018; Chen et al., 2016).

2.1.2.2 THE EONOMIC RAMIFICATIONS OF A DROP IN OIL PRICES, BOTH DIRECT AND INDIRECT

Because of the loss of oil income, it is predicted that a net oil exporter's GDP will have a direct negative impact when global oil (energy) prices fall rapidly (i.e., when a negative price shock happens), as has occurred recently (Chen et al., 2016; Shaikh, 2021). A net oil importer, on the other hand, sees the big decrease in oil prices as a good supply shock, but a net oil exporter does not. It is anticipated that this will lead to an increase in the demand for energy from the energy importer, which will be beneficial to the oil-exporting country. This is true in a regular situation; nevertheless, the current situation is out of the ordinary. Because non-essential businesses and activities are being shut down to halt the spread of the virus in the afflicted nations throughout the world, demand from oil importers is dropping. Government officials have also advised inhabitants to stay at home to avoid spreading the infection further. With these moves, virtually all of America's economic activity came to an end, cutting oil demand, particularly in the transportation sector. Therefore, the "indirect effect" on net oil exporters will very certainly be negative as well, and the aggregate effect, which is the total of both the direct and indirect effects, will be far greater. The cumulative negative effect will cause the income of energy exporters to drop at a faster rate than that was previously projected.

Against this, the current COVID-19 epidemic is significantly decreasing worldwide demand, particularly in the transportation sector, which consumes the greatest amount of oil in the United States, the European Union, China, and Japan. The current epidemic is expected to continue for the foreseeable future. It is also impossible to predict how much-reduced oil prices would benefit consumers in the aftermath of the coronavirus pandemic, given the absence of consumer demand. Finally, a one-sided study of the COVID-19 and the impact of recent oil price drops on economies is not enough. Increased pauses in production induced by forced business closures, interruptions in trade and commerce caused by border closures, travel bans, and other restrictions on the flow of goods, people, and money all have an impact on the global supply chain and impede growth in the manufacturing sector.

2.1.3 WHERE ARE WE HEADING?

The oil and gas sector is confronted with major financial constraints, which may have an immediate impact on output. Debt is the principal source of funding for the US shale sector. The great majority of businesses invest more than they produce and make up the difference by issuing bonds. The overall debt in the US shale business increased in December 2021 to $260 billion, more than double the amount in 2009. Since January 2011, the debt-to-gross cash flow ratio has more than quadrupled. Shale enterprises are growing more leveraged as debt growth outpaces improvements in cash flow from operations. More investment is projected to be rejected by investors and financial markets, perhaps leading to a significant fall in the number of fracking projects in the United States. Because shale oil wells deteriorate quicker than conventional wells – output may decline by as much as 60–70% in the first year – any halt in investment can result in an almost immediate drop in output. Valuation models take into account the various sensitivity and scenario elements. The $60 per barrel oil price, which was previously viewed as the low end of the range in many organizations' models, is now a reality. Lowering operating cash flows will eventually result in lower capital expenditures and the impairment of exploratory assets. Reduced operational cash flows can also lead to a corporation's demise; therefore, corporate boards will need to spend more time determining whether the going concern basis of accounting (i.e., whether the organization can continue trading shortly) is appropriate for their decision-making. Furthermore, debt renegotiating may be a challenging undertaking for businesses. When oil prices are high, refinancing current debt, whether through bank borrowings or the issue of new bonds, is an easy operation. Lower asset values and increased default risks, on the other hand, indicate that borrowers will face increasing challenges, such as the need to pay higher interest rates or increase the security packages they give, if they can borrow at all. When cash flow issues arise, corporate transactions may be undertaken at a discount as a result of investors being pushed to liquidate their assets as quickly as possible and acquire cash at any cost to avoid additional losses. However, if they are not offered what they regard to be a fair price, the majority of vendors choose to hold onto their investment until they receive a higher offer, which is exceptional and occurs only in extreme cases. When a buyer pays a discount for a business, the buyer realizes a day one profit on the resulting "negative goodwill," which is immediately recorded as profit (as opposed to positive "normal goodwill," which occurs when a buyer pays a premium for the business rather than a discount and is recorded on the balance sheet).

Hedge accounting, which gained popularity in the aftermath of the financial crisis, may be used more in the future. Because of consistently high oil prices (above $100 per barrel), the bulk of exploration and production companies have profited without hedging their assets (Diaz and Gracia, 2017). Despite the economic slump, many firms, particularly major corporations, continued to pay for price hedging programs as "insurance" against price drops. Reduced prices may encourage more producers, including small ones, to "lock in" their output through the use of hedging instruments to safeguard their profits. As a result, enterprises who have hedged their pricing (say, by setting a price floor of $80/bbl) may find themselves "in the money." Following a long period of tolerating inefficient volatility via the profit and loss account, firms that have adopted a hedge accounting approach may be forced to take a portion of their earnings through hedging reserves for the first time in years. Fuel purchasers, such as airlines and bus firms, who have hedges in place to regulate price, would have the opposite effect, rendering their hedges useless. Finally, lower pricing increases counterparty credit risk in circumstances when counterparties' cash flow is influenced by oil prices. When a company becomes a joint venture partner, the danger of joint venture partners failing to satisfy their share of duties, including decommissioning expenses, rises. As a result, other partners may be required to assume their respective responsibilities, putting further strain on their cash reserves. The reduction in oil prices has created a lot of uncertainty, which is reflected in company financial statements (Antonakakis et al., 2018; Diaz and Gracia, 2017). Despite several years of relatively steady oil and gas prices, investor confidence in the oil and gas industry has plummeted, and there is no reason to believe it will ever return. The most often asked question in the industry is whether this is the start of a long-term downward structural adjustment in oil prices or a short-term corrective measure. According to recent expert forecasts, oil prices will soar beyond $80 per barrel by the end of 2017. Even if they are accurate, the persistent ambiguity suggests that, regardless of their findings, 2021 will almost surely be a challenging year for oil companies, oil-producing states, and global policymakers (Antonakakis et al., 2018; Apergis and Apergis, 2020). The question is now not about how much oil can be produced? But is now about how much can be produced inexpensively with minimal extra cost to the shareholders of the oil and gas company.

2.2 WHAT THE FUTURE HOLDS FOR EOR?

According to the EIA's 2014 Annual Energy Outlook (AEO, 2014), the price of oil will have a significant impact on the deployment of carbon dioxide (CO_2) enhanced oil recovery (EOR) technology to boost crude oil extraction from existing producing locations (EIA, 2014). CO_2 gas is injected under high pressure into oil reservoirs, allowing it to mix with the oil. This decreases the viscosity of the oil and causes it to expand in volume (swell) (Emera and Sarma, 2007). As a result, both the total amount of oil produced cumulatively and the proportion of oil recovered in situ grow. A producer's choice to employ this method is influenced by a variety of factors, including the geophysical properties of the reservoir, the amount of oil contained inside, the cost of CO_2 EOR, and the money generated by enhanced production as well the considerations of the carbon footprint of the project (Schleiniger, 2016). Injecting miscible (mixable) CO_2 into mature oil fields to extract more oil in situ is a costly endeavor. CO_2 emissions alone may increase the cost of producing a barrel of oil by \$20–\$30 but have the potential to decarbonize various industrial segments (Núñez-López and Moskal, 2019). Additionally, the producer is responsible for the cost of surface infrastructure used to remove and compress CO_2 from the production stream before its reintroduction into the oil reservoir. Additionally, the producer bears the financial burden of the time required to repressurize decommissioned reservoirs. Thus, oil prices have a significant effect on whether the increased output from CO_2 EOR is adequate to make the process economically and commercially viable (Muggeridge et al., 2014).

By 2040, the WTI crude oil price will have increased to \$202 per barrel (2012 USD), a 45% increase over the Reference scenario. Due to the higher price of oil, there are more mature locations where CO_2 EOR technology may be profitable. By 2040, this scenario's oil output would reach 960,000 barrels per day, a 30% increase over the Reference case (740,000 barrels per day). By 2040, WTI prices will have fallen to only \$73 per barrel, reducing producers' incentive to build CO_2 EOR on existing assets. In the mid-1908s, EOR gas injection programs were terminated, and thermal and chemical EOR procedures were phased off in the 1980s. The number of GEOR projects, on the other hand, has more than doubled since 2000, owing to the increased number of CO_2 EOR projects executed. There has been a little increase in thermal projects in recent years as a result of the growing use of high-pressure air injection (HPAI) projects. Between 2004 and 2014, the most widely used EOR techniques globally were gas EOR (miscible flooding) and thermal EOR. Except for a brief time between 1979 and 1983, global oil demand has been consistently increasing over the last three decades. Simultaneously, conventional oil reserves have been steadily depleted. In 2014, global oil consumption climbed by 1.2 million barrels per day, averaging 92.4 million barrels per day. By the end of 2015, oil output had climbed to about 97 million barrels per day, a rate comparable to demand growth of 1.4 million barrels per day. Global oil consumption is expected to grow by 1.4 million barrels per day in 2016, and by 1.5 million barrels per day in 2017. Additionally, from 2014 until date, the oil supply has exceeded demand, as seen by the drop in oil prices. The present oil supply surplus, on the other hand, is the outcome of unconventional oil resource extraction (i.e., tight reservoirs). The exponential growth in global oil consumption over time has fueled the search for breakthrough technologies to boost oil output. In 1998, global EOR initiatives inundated Indonesia's Duri field with steam at a rate of 2.3 million barrels per day, which was the world's largest steam flooding project at the time. Canada and China also reported EOR output of 480,000 barrels per day, while Canada and China reported 280,000 and 200,000 barrels per day, respectively. The global EOR market has maintained an unbroken growth trend in recent years. The worldwide EOR market was worth 2095 million barrels in 2012. By 2013, it had risen to 2681.6 million barrels. EOR accounts for around 3.5% of worldwide daily oil output or approximately 3 million barrels per day. Between 2014 and 2020, a compound annual growth rate of 29.9% has been forecast; hence, a much greater growth rate of more than 16,000 million barrels by 2020, totaling about US$ 283 billion, has been anticipated. In 1988, Russia, the world's second-largest producer of oil, produced an estimated 4.6 million barrels per day. Russian oil recovery ratios have been declining over the previous 30 years due to the depletion of key reserves. The breakup of the Soviet Union in 1991 also resulted in a 2.2 billion barrels per day decline in oil production. The oil recovery factor was 50% in the 1960s, but had plunged to b/d by 1993. The oil recovery factor was 50% in the 1960s; by 1993, it had plummeted to 40%. Between 2014 and 2020, it is predicted that Russian oil output will continue to fall (560 kb/d). Russia's total liquid hydrocarbon production is predicted to fall from 10.9 million barrels per day in 2014 to 10.4 million barrels per day in 2020. However, as a result of the increasing gas production rate, natural gas liquids (NGL) and condensate production are predicted to increase by 2020, reaching 975 kb/d on average. Numerous EOR techniques (TEOR, GEOR, and MEOR) have been used in Russian oil fields, even though MEOR applications are limited. GEOR, specifically miscible gas flooding, is regarded as the most promising EOR technology. Even though various EOR procedures have been implemented in Russian oil fields, reliable estimates of finished EOR projects are difficult to obtain. Between 1996 and 2000, Russian oil output grew 2.5–2.8 times, according to oil industry figures. Russian EOR oil output has grown to around 17–18 million tons, with chemical EOR operations accounting for approximately 81–82% of the incremental oil production and gas flooding and WAG applications contributing to approximately 0.35–0.4 million tons. According to projections, gas EOR, WAG, and combination technologies will all be deployed in Russia shortly. The Permian Basin's incremental oil production from CO_2 EOR activities was estimated to reach 186,000 and 241,000 barrels per day in 2012 and 2015, respectively, with additional forecasts of 301,000 barrels per day by 2020. While EOR techniques can considerably boost oil recovery, the amount of recoverable oil is highly dependent on the type of EOR technology used. Miscible gas injection and thermal EOR are now the most widely used EOR procedures in oil fields. Shortly, EOR processes should be expanded through a synergistic strategy using a variety of EOR technologies (Afolabi, 2019).

As a result, EOR techniques will have a significant impact on the developing world's oil output in the coming years. This book discusses the different EOR approaches in great detail. This article discusses a range of EOR technologies, as well as their ability to recover oil from depleted oil reservoirs. The impact of EOR technologies on oil output after they have been fully deployed in depleted reservoirs is also being investigated more thoroughly. A final part addresses the benefits and drawbacks of EOR approaches in the Indian Subcontinent, as well as their future potential, to offer a favorable appraisal of the existing and future status and usability of EOR techniques in the Indian Subcontinent. This work is unique in that it is the first comprehensive evaluation of oil production and EOR scenarios, and it will be useful to future petroleum engineering researchers. It is critical to appropriately reflect India's current and future oil production status, as well as the deployment of EOR technology to improve oil recovery and supply to energy sector firms, to stimulate the economy. Finally, this evaluation will serve as a quick reference for researchers interested in the present state of Indian oil production and increased oil recovery (EOR).

KEYWORDS

- **crude oil price**
- **Brent**
- **West Texas Intermediate (WTI)**
- **hydrocarbon industry**
- **COVID-19**

REFERENCES

Adedokun, A. The Effects of Oil Shocks on Government Expenditures and Government Revenues Nexus in Nigeria (with exogeneity restrictions). *Futur. Bus. J.* **2018,** *4*, 219–232. https://doi.org/10.1016/J.FBJ.2018.06.003

Afolabi, R. O. Enhanced Oil Recovery for Emergent Energy Demand: Challenges and Prospects for a Nanotechnology Paradigm Shift. *Int. Nano Lett.* **2019,** *9*, 1–15. https://doi.org/10.1007/s40089-018-0248-0

Antonakakis, N.; Cunado, J.; Filis, G.; Gabauer, D.; Perez de Gracia, F. Oil Volatility, Oil and Gas Firms and Portfolio Diversification. *Energy Econ.* **2018,** *70*, 499–515. https://doi.org/10.1016/J.ENECO.2018.01.023

Apergis, E.; Apergis, N. Can the COVID-19 Pandemic and Oil Prices Drive the US Partisan Conflict Index? *Energy Res. Lett.* **2020**. https://doi.org/10.46557/001C.13144

Baffes, J.; Kose, M. A.; Ohnsorge, F. L.; Stocker, M. Down the Slide: The Collapse in Oil Prices Since 2014 is the Most Recent of Several in the Past Three Decades and may Portend a Long Period of Low Prices. *Finance Dev.* **2015,** 0052. https://doi.org/10.5089/9781475537161.022.A007

Chen, H.; Liu, L.; Wang, Y.; Zhu, Y. Oil Price Shocks and U.S. Dollar Exchange Rates. *Energy* **2016,** *112*, 1036–1048. https://doi.org/10.1016/j.energy.2016.07.012

Diaz, E.; de Gracia, F. P. Oil Price Shocks and Stock Returns of Oil and Gas Corporations. *Financ. Res. Lett.* **2017,** 20.

EIA. Annual Energy Outlook, 2014.

Emera, M. K.; Sarma, H. K. Prediction of CO_2 Solubility in Oil and the Effects on the Oil Physical Properties. *Energy Sources A Recover. Util. Environ. Eff.* **2007,** *29*, 1233–1242. https://doi.org/10.1080/00908310500434481

Fantazzini, D. The Oil Price Crash in 2014/15: Was There a (Negative) Financial Bubble? *Energy Policy* **2016,** *96*, 383–396. https://doi.org/10.1016/J.ENPOL.2016.06.020

Garg, K. D.; Gupta, M.; Kumar, M. The Impact of COVID-19 Epidemic on Indian Economy Unleashed By Machine Learning. *IOP Conf. Ser. Mater. Sci. Eng.* **2021,** *1022*, 012085. https://doi.org/10.1088/1757-899X/1022/1/012085

Ghosh, J. A Critique of the Indian Government's Response to the COVID-19 Pandemic. *J. Ind. Bus. Econ.* **2020,** *47*, 519–530. https://doi.org/10.1007/s40812-020-00170-x

Gupta, K. Oil Price Shocks, Competition, and Oil and Gas Stock Returns — Global Evidence. *Energy Econ.* **2016,** *57*, 140–153. https://doi.org/10.1016/J.ENECO.2016.04.019

Jefferson, M. A Crude Future? COVID-19s Challenges for Oil Demand, Supply and Prices. *Energy Res. Soc. Sci.* **2020,** *68*, 101669. https://doi.org/10.1016/J.ERSS.2020.101669

Lammerding, M.; Stephan, P.; Trede, M.; Wilfling, B. Speculative Bubbles in Recent Oil Price Dynamics: Evidence from a Bayesian Markov-Switching State-Space Approach. *Energy Econ.* **2013,** *36*, 491–502. https://doi.org/10.1016/j.eneco.2012.10.006

Muggeridge, A.; Cockin, A.; Webb, K.; Frampton, H.; Collins, I.; Moulds, T.; Salino, P. Recovery Rates, Enhanced Oil Recovery and Technological Limits. *Philos. Trans. R. Soc. A Math. Phys. Eng. Sci.* **2014,** *372*. https://doi.org/10.1098/rsta.2012.0320

Nguyen, H.; Nguyen, H.; Pham, A. Impact of Oil Price Fluctuations on Financial Markets since 2014, 2017. https://doi.org/10.1596/1813-9450-7957

Norouzi, N. Post-COVID-19 and Globalization of Oil and Natural Gas Trade: Challenges, Opportunities, Lessons, Regulations, and Strategies. *Int. J. Energy Res.* **2021,** *45*, 14338–14356. https://doi.org/10.1002/ER.6762

Núñez-López, V.; Moskal, E. Potential of CO_2-EOR for Near-Term Decarbonization. *Front. Clim.* **2019,** *1*, 5. https://doi.org/10.3389/fclim.2019.00005

Sabitova, N.; Shavaleyeva, C. Oil and Gas Revenues of the Russian Federation: Trends and Prospects. *Procedia Econ. Financ.* **2015,** *27*, 423–428. https://doi.org/10.1016/S2212-5671(15)01016-3

Salygin, V.; Guliev, I.; Chernysheva, N.; Sokolova, E.; Toropova, N.; Egorova, L. Global Shale Revolution: Successes, Challenges, and Prospects. *Sustainability* **2019,** *11*, 1–18. https://doi.org/10.3390/su11061627

Schleiniger, R. Implicit CO_2 Prices of Fossil Fuel use in Switzerland. *Energy Policy* **2016,** *96*, 411–420. https://doi.org/10.1016/j.enpol.2016.06.022

Shaikh, I. On the Relation Between the Crude Oil Market and Pandemic Covid-19. *Eur. J. Manag. Bus. Econ.* **2021,** *30*, 331–356. https://doi.org/10.1108/EJMBE-08-2020-0223/FULL/PDF

Stocker, M.; Baffes, J.; Some, Y. M.; Vorisek, D.; Wheeler, C. M. The 2014–16 Oil Price Collapse in Retrospect: Sources and Implications, 2018. https://doi.org/10.1596/1813-9450-8419

Vătavu, S.; Lobont, O. R.; Para, I.; Pelin, A. Addressing Oil Price Changes Through Business Profitability in Oil and Gas Industry in the United Kingdom. *PLoS One* **2018,** *13*, e0199100. https://doi.org/10.1371/JOURNAL.PONE.0199100

Wheeler, C. M.; Baffes, J.; Kabundi, A.; Kindberg-Hanlon, G.; Nagle, P. S.; Ohnsorge, F. Adding Fuel to the Fire: Cheap Oil during the COVID-19 Pandemic, Policy Research Working Papers, 2020. https://doi.org/10.1596/1813-9450-9320

CHAPTER 3

Low-Salinity Water Flooding: Mechanisms and Status

RISHABH TRIPATHI[1], ABHIJIT KAKATI[2], NILESH KUMAR JHA[2], and VISHNU CHANDRASEKHARAN NAIR[1]

[1]Department of Petroleum Engineering and Geological Sciences, Rajiv Gandhi Institute of Petroleum Technology, Jais, U.P., India

[2]School of Petroleum Technology, Pandit Deendayal Petroleum University, Knowledge Corridor, Raisan, Gandhinagar, Gujarat, India

ABSTRACT

Low salinity water flooding is a chemical EOR techniques in which reduced concentration of brine solutions are injected into oil and gas reservoir for increased oil production. Since its inception, low salinity water flooding has been a very attractive technique due to its high cost effectiveness. There are numerous mechanisms that are associated with low salinity water flooding. These mechanisms mainly depend on the rock, brine solution and crude oil interactions. There is a slight difference between smart water flooding and low salinity water flooding. Smart water flooding is a chemical EOR technique in which the composition of the brines may be altered so that the reservoir becomes more water wet and extra oil can be produced. The mineralogical composition of the oil and gas reservoir rock, temperature, composition, concentration and temperature of the injection brines are some of the most important factors behind the success of low salinity water flooding. The crude oil composition also plays a pivotal role in the success of the low salinity water flooding. The crude oil composition affects the overall wettability alteration of the reservoir rock in the presence of low salinity brines. There

Advancements in Chemical Enhanced Oil Recovery.
Tushar Sharma, Krishna Raghav Chaturvedi, Tarek Ganat, and Imtiaz Ali (Eds.)

are numerous pilot tests that have been conducted on the field in order to see the viability of the technique, most of which have been successful in producing increased oil recovery. The ion composition analysis of the low salinity water flooding provides some important insights into the in-situ adsorption of a particular salt during the flooding process. These results help in correlating the dominant mechanism responsible for producing increased oil recovery during low salinity water flooding. In this chapter, the underlying mechanisms associated with low salinity water flooding, difference in the low salinity water flooding and smart water flooding, and the importance of ion composition analysis in study related to the low salinity water flooding has been discussed in detail.

3.1 INTRODUCTION

Oil production is basically classified into three main stages: primary recovery, secondary recovery, and enhanced/tertiary oil recovery (Green and Willhite, 1998). During the primary stage, oil production happens due to the natural energy existing in the reservoir, such as water drive, gas drive, gravity drainage, etc. Secondary recovery techniques are used when primary production declines. In this stage, reservoir pressure maintenance techniques such as gas injection and waterflooding are performed. The cumulative oil recovery from both primary and secondary recovery is approximately 33% Original Oil in Place (OOIP) (Godec, 2011). The tertiary recovery stage commences after the depletion of the primary and secondary recovery stages. The recovery methods during this stage basically fall under enhanced oil recovery (EOR) methods. However, EOR methods can be implemented at any stage of oil recovery (primary, secondary, or tertiary). EOR methods alter the original properties of the formation and formation fluid. The recovery from EOR methods can be up to 65% OOIP (Tunio et al., 2011). Among the available EOR methods (thermal, chemical, miscible, etc.), chemical EOR depends on adding one or more chemicals (polymer, surfactant, caustic, micellar/emulsion flood, or combination thereof) to the injected fluid in order to alter the wettability of the rock, reduce the interfacial tension (IFT), or improve the sweep efficiency of the injected fluid. Low-salinity waterflooding is a comparatively new EOR method that falls under the category of chemical methods. This process requires the injected brine to have a low salinity in order to increase oil recovery.

3.2 THE ROLE OF SALINITY

In oil and gas reservoirs, equilibrium has been established between brine, rock, and oil over many years. Low-salinity waterflooding is an EOR technique in which low-saline brine is flooded in a reservoir for oil recovery by virtue of its interaction between rock, oil, and formation water (Tang and Morrow, 1999). The injection water properties, however, are improved by changing the composition of the ions in it to increase the oil recovery. It all started when Bernard and his coworkers introduced this technique when they found increased oil recovery using low-saline NaCl brine with a range of 0–1% salt concentration compared to deionized water (Bernard, 1967). This EOR technology sparked the industry's interest since it was simple and less costly. Later in 1997, Morrow and his coworkers recognized the potential of the low-saline waterflooding (Yildiz and Morrow, 1996; Tang and Morrow, 1997).

The low- salinity effect (LSE) is actually better for the oil-wet reservoir conditions because low-salinity waterflooding alters the wettability to a more water-wet condition, and therefore the oil recovery is increased. Under different circumstances, the increased oil recovery may or may not be observed by low-salinity waterflooding. One or more mechanisms are responsible for the improved oil recovery using low-salinity waterflooding (Cissokho et al., 2010; Morrow and Buckley, 2011). There is no consensus regarding the governing mechanism that is applicable for low-salinity waterflooding in EOR. It is obvious that particular conditions are required for the working of low-salinity waterflooding, which are stated in the literature in some form or another. However, there is no experimental evidence that proves one dominant mechanism that is universally applicable in different reservoir conditions. The complications arise due to many factors, like interactions between minerals, crude oils, and aqueous phase compositions, which contribute to the difficulties faced by the researchers in finding a mechanism suitable for all low-salinity applications in different reservoir conditions. In the last couple of decades, a large number of mechanisms have been proposed to explain the workings of low-salinity water injection for better oil recovery. Boussour et al. (2009) studied the mechanism of low-salinity waterflooding and gave experimental examples that questioned the universal applicability of almost all of the mechanisms.

3.3 PROPOSED MECHANISMS OF LOW-SALINITY WATERFLOODING

Apart from investigating low-salinity waterflooding efficiency, the researchers have also made an effort to understand the underlying working mechanism. The proposed mechanisms behind low-salinity waterflooding fall under any of the following categories:

1. Formation fine migration
2. In situ soap generation
3. Multicomponent ion exchange
4. Mineral dissolution
5. Salting-in effect
6. Electrical double-layer expansion
7. Osmosis
8. Increase in pH
9. Viscoelastic liquid interface
10. Formation of microdispersion

3.3.1 FORMATION FINE MIGRATION

This mechanism was explained by Tang and Morrow (1999) based on their observation of core flooding experiments performed on sandstone cores with and without formation fines. It hypothesized that improved oil recovery from low-salinity waterflooding is the result of detachment and migration of formation fines. Formation fines are very small, loose solid particles (< 40 micron) that are present in the pores of rock formations. The crude oil containing polar molecules and brine in the reservoir makes the formation fines exhibit mixed wet characteristics. During low-saline water injection, fines detach from the pore lining due to the expansion of the electrical double layer. The fines strip away the oil adsorbed on them, leading to increased recovery. The presence of mixed wet particles at the interface was also suspected to be a cause for recovery improvement. These solid particles make the interface semirigid and inhibit the trapping of oil in pores by snapping off.

During the low-salinity waterflooding, fines migration and their movement along with the low-saline water are shown in Figure 3.1. The fines particles that are moved along with the oil will increase the microscopic efficiency by forcing the crude oil and water to flow through narrow openings, which will increase the overall oil recovery.

However, the fines migration mechanism is questioned by many studies. Clay or fines were either absent or were very few in those experiments, and no reduction in permeability was observed (Morrow and Buckley, 2011; Rezaeidoust et al., 2009; Zhang and Morrow, 2006; Larger et al. 2008; Zhang et al., 2007). There was no necessary relationship found between fines

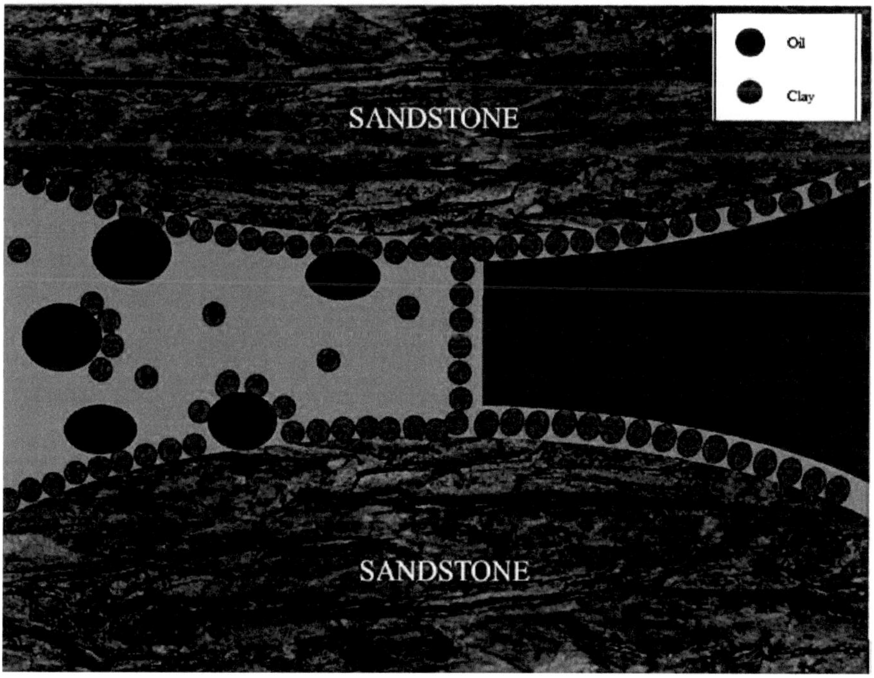

FIGURE 3.1 Stripped fines are present at the oil-brine boundary during low-salinity waterflooding.

migration and increased oil recovery by low-salinity waterflooding. Therefore, fines migration is not considered a major low-salinity waterflooding mechanism.

3.3.2 IN-SITU SAPONIFICATION

McGuire et al. (2005) proposed a hypothetical mechanism that describes how low-salinity water increases recovery in a similar fashion to alkaline flooding. According to this hypothesis, low-saline water increases the reservoir pH, and at elevated pH, the alkaline reactive components of the crude generate in-situ surfactant. The results of effluent water analysis of low-salinity core flood tests by Tang and Morrow (1999) were referred to supplement this mechanism. It was observed that the effluent pH increased when the injected brine (15,150 ppm) was changed to low-saline brine with a 1515 ppm concentration. The generated surfactant can act in three different ways: reduce IFT, alter reservoir rock wettability, and cause emulsification.

3.3.3 MULTICOMPONENT ION EXCHANGE

This mechanism was described by Lager et al. (2008) based on experiments performed using sandstone cores. The effluent analysis was performed on low- and high-salinity coreflooding experiments and studied the concentration of different ions in the effluent water. The study showed a drop in $CaCl_2$ and $MgCl_2$ concentrations in comparison to their concentrations in injecting brines. This observation indicated the adhesion of $CaCl_2$ and $MgCl_2$ to the rock surface. This mechanism describes how, on the clay surface, the multivalent cations attach to the polar oil molecules. The formation of organometallic complexes resulted in the oily wetness of the rock surface. This phenomenon is depicted in Figure 3.2. These organometallic complexes are replaced by inorganic cations during low-salinity waterflooding.

3.3.4 MINERAL DISSOLUTION

This mechanism was introduced by Pu et al. (2010), which describes the mechanism of low-salinity waterflooding in carbonate formation. The experimental observations of injecting low-saline CBM water (1316 mg/L) into clay-free carbonate cores in Wyoming were the basis for this mechanism. Inferred from the experimental results, it has been concluded that low-salinity water can cause the dissolution of carbonate minerals like anhydrite, dolomite, or calcite and the release of particles consisting of such minerals. This working hypothesis states that the recovery improvement

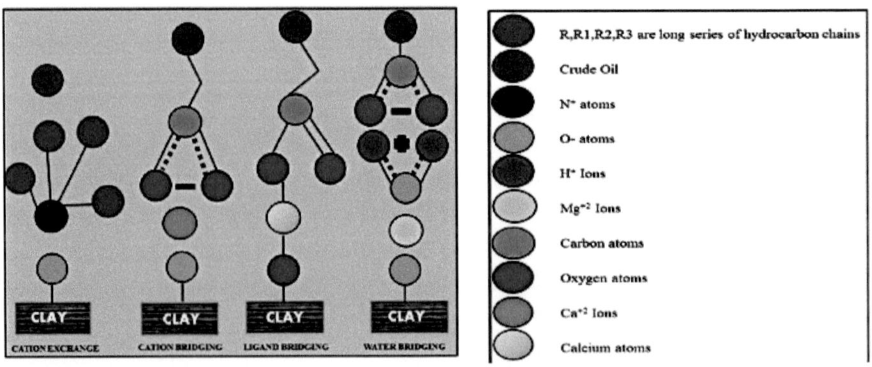

FIGURE 3.2 Different associations of polar compounds onto the clay surface.

is attributed to two effects: (1) the release of fines like dolomite particles causes an increase in pressure that promotes recovery and (2) wettability alteration occurs through detaching the outer molecular layer of the pore lining and linked organic matter.

3.3.5 SALTING-IN EFFECT

The role of injected water's salinity and composition can be explained by salting-in and salting-out effects on the solubility of polar organic material in it. RezaeiDoust et al. (2009) (Austad et al., 2010) proposed a mechanism to explain LSE in sandstone reservoirs with the help of the salting-in effect. Polar organic compounds dissolved in water always interact with water structures. Inorganic cations disturb this water structure and decrease the organic compounds solubility in water. Therefore, according to this theory, the solubility of organic material increases drastically as the salinity decreases. Increasing the solubility of organic polar compounds by reducing the salinity will cause the desorption of these compounds from the surface of rocks and their dissolution in water. This results in increasing the water wetness of the reservoir.

3.3.6 ELECTRICAL DOUBLE-LAYER EXPANSION

Ligthelm et al. (2009) suggested a mechanism based on electrical double-layer expansion in order to explain the LSE. The surface of sandstone has a negative charge due to the presence of silanol (–OH) groups on its surface, and crude oil also contains negatively-charged polar compounds. Lager et al. (2008) explained that the multivalent cations present in the brine connect the sandstone and polar compounds present in the crude oil. The intervening layer containing high-salinity brine has a sufficient amount of cations, which screen out the negative charges on both the sandstone surface and crude oil. Screening of charges reduces the electrostatic repulsion between the two layers, providing an ease for crude oil to react with the silicate surface. When low-salinity brine is introduced, the charge screening decreases with a reduction in the amount of cations. It results in increased repulsion due to the presence of similar electrostatic charges between the two layers and also due to the electrical double-layer expansion. The complete mechanism is shown in Figure 3.3.

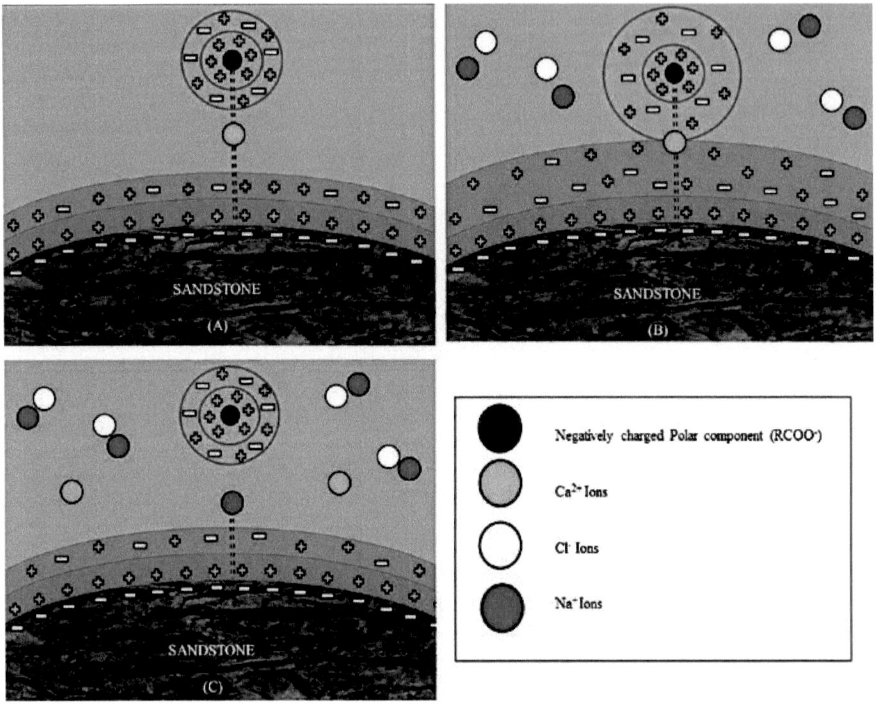

FIGURE 3.3 Schematic shows the electrical double layer expansion that helps in the increase in oil recovery. (A) Shows that before injecting low-salinity water, the double layers of polar components and negatively-charged surfaces do not interact. (B) Shows that in the presence of low-salinity water, the double layer interacts, causing the monovalent ions to bond with the surface and altering the wettability. (C) Shows how the electrical double layer expansion and its interaction desorb the adsorbed oil molecules in the presence of low-salinity waterflooding.

3.3.7 INCREASE IN PH

Rezaeidoust et al. (2009) (Austad et al., 2010) described an increase in pH as the mechanism behind low-salinity waterflooding. Large surface areas along with cation exchange capacity (CEC) make clays very important for low-salinity waterflooding. Polar compounds are adsorbed on the clays via bridging through inorganic cations. Low-salinity water disturbs the chemical equilibrium, resulting in the desorption of cations as well as the subsequent attachment of hydrogen (H^+) ions from the thin water film present near the clay surface.

$$\text{Clay surface-}Ca^{2+} + H_2O = \text{Clay surface-}H^+ + OH^- + Ca^{2+}$$

A decrease in H^+ ions increases the pH of the thin water film around the surface. This increase in pH causes a reaction between the adsorbed polar components on the clay surface and the hydroxyl ions present in the thin water film that is present close to the surface. The desorption of cations that are attached to the clay surface is very important for increasing the pH of the water layer close to the clay surface.

$$\text{Clay surface-}NHR_3^+ + OH^- = R_3N + \text{Clay surface} + H_2O$$

$$\text{Clay surface-}RCOOH + OH^- = RCOO^- + \text{Clay surface} + H_2O$$

In Figure 3.4, it has been schematically shown that the pH near the surface of the rock is increased when both types of polar components are present near the surface. This mechanism is independent of the low-saline brine

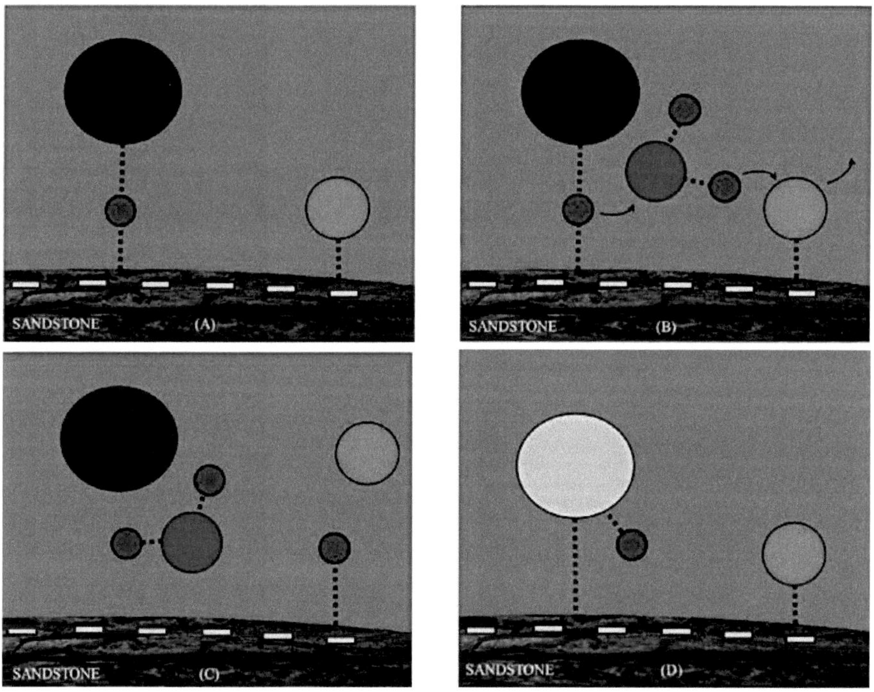

(Continued)

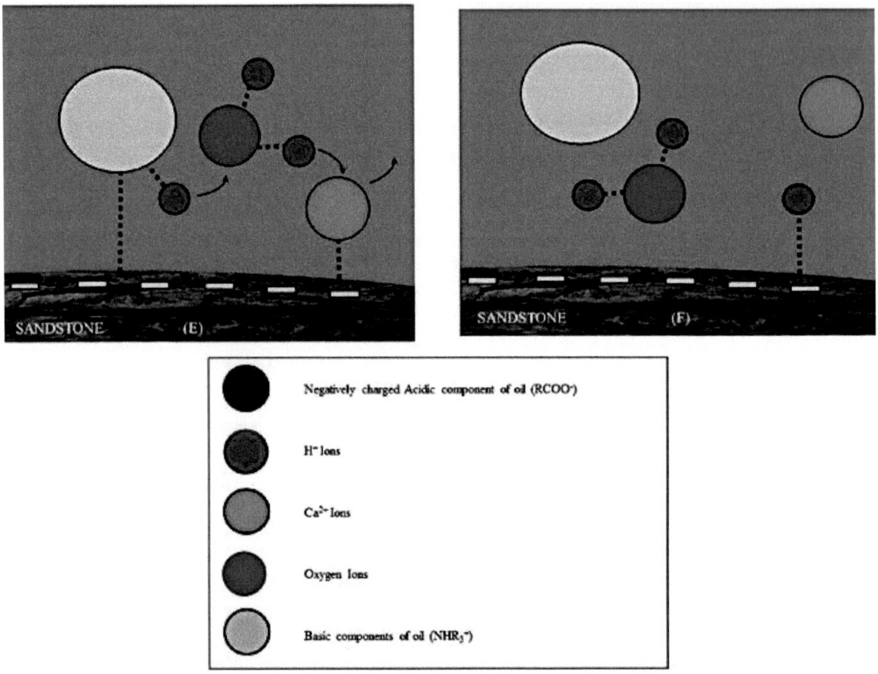

FIGURE 3.4 (A–C) Shows the proposed mechanism of increase in pH for acidic components. (D–F) Shows the proposed mechanism of increase in pH for basic components.

composition until the active inorganic ion concentrations are low enough to promote desorption of the cations taking part in the bridging phenomenon. The necessary condition for this mechanism to work requires the formation water to contain active calcium ions. Optimal low-salinity concentration depends on the following factors: (1) adsorption of active inorganic cations and polar components onto the clay surface; (2) the pH and compositions of the formation water, that is, the amount of divalent cations; (3) the rock should be oil wet or mixed wet to get increased oil recovery, since if a rock is already water wet, then the oil wetted rock part will be very less and then the oil recovered due to wettability alteration will be very less.

3.3.8 OSMOSIS

Fredriksen et al. (2016) have explained the role of osmosis and how more oil is produced due to this process. The rock should be in an oil-wet condition so that the connate water is not continuous and the water is separated by thin oil films. The salinity affects the IFT between the oil-water phases, and the IFT decreases as the salinity decreases until the IFT reaches a minimum value; the salt concentration at this value is called the critical salt concentration. When the oil phase separates two different salinity solutions, it leads to convective water transport. Osmosis is not dependent on oil/rock/brine interactions, and therefore this phenomenon is applicable to both types of rocks (oil-wet and water-wet). The pore-scale osmotic effect, in which oil production depends on capillary pressure and not on viscous fluid flow, becomes important for spontaneous imbibition. Osmosis works by expanding the high-saline brine present inside the water in the oil emulsion and moving the oil out of the core. In a strongly water-wet condition, the expansion of water in oil emulsion during low-salinity waterflooding is shown in Figure 3.5. The water in oil emulsion stability depends on the oil phase composition, that is, crude oil containing asphaltenes stabilizes emulsions (Kokal, 2005). Oils containing nonpolar alkanes do not form stable oil films around water. Water diffusion is considered the main transport mechanism for water-wet systems, thereby connecting fracture bulk flow and matrix brine (Fredriksen et al., 2016). Dilution of high-saline brine takes place in the matrix in order to make the chemical potential difference between low-saline water and high-saline brine zero. It has been found to cause the expansion of the oil film (Fredriksen et al., 2016).

Even though this mechanism is applicable to both water-wet and oil-wet rocks, oil-wet reservoirs are the best candidates for osmosis because large amounts of water in oil emulsion films can be formed that will expand during low-salinity waterflooding and push the oil out of confined volumes. This expansion causes the oil to move to the nearby flow channel.

3.3.9 VISCOELASTIC LIQUID INTERFACE

The viscoelasticity of the interface present between oil and water is also a mechanism behind low-salinity waterflooding. The low-saline water affects the oil–water interface, and in turn, increases its viscoelasticity, irrespective of the low-salinity water composition. An increase in the viscoelasticity of the oil–water interface increases the overall recovery (Alves et al., 2014). For cores that are water-wet, snap-off is common, and the presence of snap-off reduces the amount of overall recovery during flooding. There is no constant relationship between the viscoelasticity of the oil–water interface and the salinity of the injected brine (Chávez-Miyauchi et al., 2016). Since other mechanisms will also take place during low-salinity waterflooding

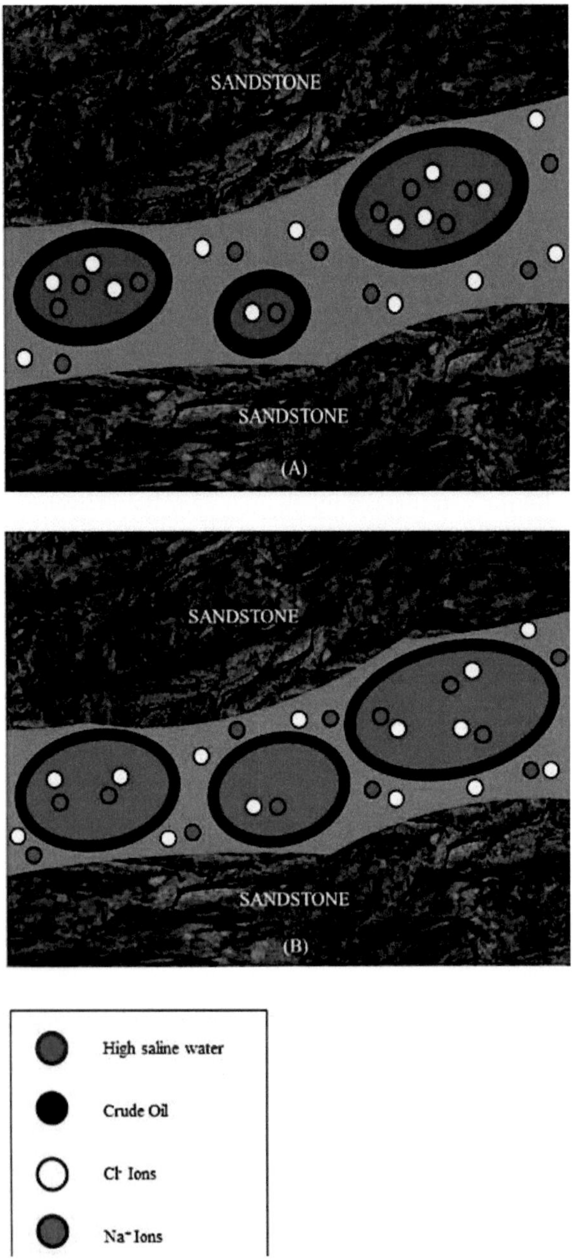

FIGURE 3.5 It shows the osmosis mechanism that comes into play when low-salinity waterflooding is done. (A) Shows the presence of water in oil emulsions containing highly saline formation water. (B) Shows osmosis that takes place through oil layer causing expansion of water in oil emulsions.

to improve the overall oil recovery, more proof and experimental work are required before clearly stating interface viscoelasticity as a mechanism for better oil recovery. Alvarado et al. (2015) (Bidhendi et al., 2018) performed the experiments and stated that the increased oil recovery is because the viscoelasticity of the oil–water interface increases at 40°C temperature at which the effect of other mechanisms will be negligible (Alvarado et al., 2015). This conclusion is not valid since wettability alteration occurs even at ambient temperatures (Agbalaka et al., 2009; Derkani et al., 2018).

3.3.10 FORMATION OF MICRODISPERSION

When low-salinity waterflooding takes place, the microbrine dispersion also gets formed near the oil/brine interphase, and researchers thought it to be a contributing factor for changing the wettability (Emadi and Sohrabi, 2013). The polar compounds will get attached to these microbrine dispersions, and this movement will break the bonds formed between polar compounds and the rock surface, which will ultimately push toward a more water-wet system (Emadi and Sohrabi, 2013; Mahzari and Sohrabi, 2014; Al-Saedi, 2019).

Sohrabi et al. (Al-Saedi, 2019) performed the experiments in which the microbrine dispersions were removed and the resulting recovery of the original crude oil was tested. It was found that oil recovery decreased, and therefore it was proven that the microbrine dispersion formation forced the rock surface toward a more water-wet condition. Thus, the microbrine dispersion formation is also considered a mechanism for increased oil recovery using low-salinity waterflooding. The microbrine dispersion formation could also impact the multiphase flow (Al-Saedi, 2019).

3.4 ION COMPOSITION ANALYSIS OF PRODUCED WATER

Produced water during low-salinity waterflooding helps in the determination of the geochemical rock–brine interaction. The compositional analysis not only tells us the undergoing mechanism but also tells us the properties that have changed in the produced water itself and how much it will be difficult to separate impurities from it. Experimental studies performed on produced water by Mousa and Arafat (2015) clearly explained in detail the factors that are critical to the produced water's further treatment and proper disposal. The solubility of oil decreases in water as the salinity in it increases. Similarly, an increase in temperature decreases the solubility of oil since the viscosity decreases and the velocity of separation increases, although after a certain value, the oil again starts forming emulsion due to very high collision rates. The pH also affects the oil solubility present in the produced water. Oil removal percent increases as the pH increases. It reaches its maximum at 7–8 pH.

The effluent analysis can also determine the best method to execute an EOR technique. Zeinijahromi and Bedrikovetsky (2016) described how the huff and puff method was used instead of a simple injection of polymer gel to reduce the high permeability streaks that are present in the reservoir. The water–oil ratio in the produced water proved that the huff and puff method for low-salinity waterflooding should be used since the 90% water–oil ratio was reached in more time than using polymer gel injection.

The effluent analysis was also used in the investigation of brine and rock interaction during low-salinity waterflooding in carbonate rock by Sharma and Mohanty (2018). In this study, experiments were performed in which various modified brines were used in limestone cores at a temperature of 120°C to gain more in-depth knowledge of the interactions between the PDI (potential determining ions) present in the modified brines and the rock. Single-phase static experiments involved mixing crushed limestone in seawater, seawater/50 and four times SO_4^{2-} seawater solutions. For the single-phase transport experiment, a flow rate of 0.5 ft/d was used for single-phase brine core flooding. For these experiments, a limestone core and brine were preheated at 120°C. In the B1 experiment, seawater was used to flood the core filled with formation brine, which was further flooded with seawater containing half of the composition used for seawater. In the B2 experiment, 0.1M NaCl brine was used to saturate the core, which was flooded with seawater. After seawater flooding, the core was further flooded with brine containing half of the composition used for seawater flooding. This was done to prevent anhydrite precipitation that must have been caused by the intermixing of formation water and seawater in the B1 experiment. In the B3 experiment, seawater was used to flood the core filled with formation water, which was further flooded with seawater containing four times SO_4^{2-} ions.

The result of the single-phase static experiment showed that there were large changes in the calcium and magnesium ion concentrations; calcium ions increased and magnesium ions decreased. The analysis of the effluent showed that calcium ion concentration increased and magnesium ion concentration decreased, along with an initial decrease in sulfate concentration. It indicated that surface dolomitization is dominant than calcite dissolution. The result of single-phase transport experiments for B1 showed that there was an initial decrease in sulfate ion concentration in the effluent in comparison to other PDIs, which meant that sulfate ions were interacting with the rock surface. The B2 results showed that probably due to the very slow geochemical interactions, the PDIs on the rock surface delayed their concentration to reach the injected brine concentration. The results of B3 were similar to those of B1; these results also showed an initial decrease in sulfate ion concentration as compared to chloride ion concentration. It can be inferred that it was due to the adsorption of sulfate ions on the limestone rock surface. In experiment B4, the result also showed a delay, like the result of all other experiments, but the delay was less due to the high sulfate ion concentrations present in the brine. The three important results from the experimental work on altering the wettability of carbonate rock are calcite and anhydrite dissolution, surface dissolution, and sulfate dissolution.

3.5 RECENT TRENDS IN LOW-SALINITY WATERFLOODING

Low-salinity waterflooding is an EOR technique that was first reported by Bernard et al. (1967). In 2004, Webb presented his work, in which he provided the field application of low-salinity waterflooding to enhance the oil recovery for a particular well (Webb et al., 2003). In the mid-90s, Dr. Morrow, along with his peers at the University of Wyoming, published many papers on laboratory core floods using low-saline water. British Petroleum (BP) started an active research on low-salinity waterflooding. Scientists at BP performed many core floods at reservoir conditions for secondary and tertiary low-salinity waterflooding using SWCTT (single well chemical tracer test) and log-inject-log tests. This series of events finally led to LoSal™ by BP.

Oil companies as well as universities started working on this topic. Until 2005, the interest in low-salinity waterflooding was low, but from 2005 to 2010, an increased number of papers were published in this area of research (Morrow and Buckley, 2011). According to Morrow and Buckley (2011), a proper explanation of the underlying mechanism for low-salinity waterflooding is still not confirmed.

Major factors that affect low-salinity waterflooding are:

- The quantity and type of clay
- Brine concentration
- Formation temperature
- Oil properties like viscosity and API gravity

- Rock properties like porosity, pore size, and permeability

Decreased polymer consumption and increased oil recovery have been the results of many experiments when low-salinity water is used for mixing polymers. Rheology data showed that 50% less polymer is required to achieve the same mobility ratio during polymer flooding. Low-salinity water, when used for polymer flooding, possesses an attractive and economical method for extra-heavy oil production (Hernández et al., 2020). When experiments were done to inspect the prospects of low-salinity polymer flooding Kuwait with a permeability of <10 mD, high temperature (113°C), high salinity, and carbonate mineralogy, it was found that the thermal stability and initial viscosity of polymers were retained at 113°C when low-salinity brine was used for mixing polymers (Al-Murayri et al., 2019). In carbonate rocks, specifically in a low-temperature (35°C) dolomite reservoir, 37 surfactants were tested along with low-salinity waterflooding. Two cationic surfactants produced 41–53% OOIP recovery, and one AOS (alpha olefin sulfonate)

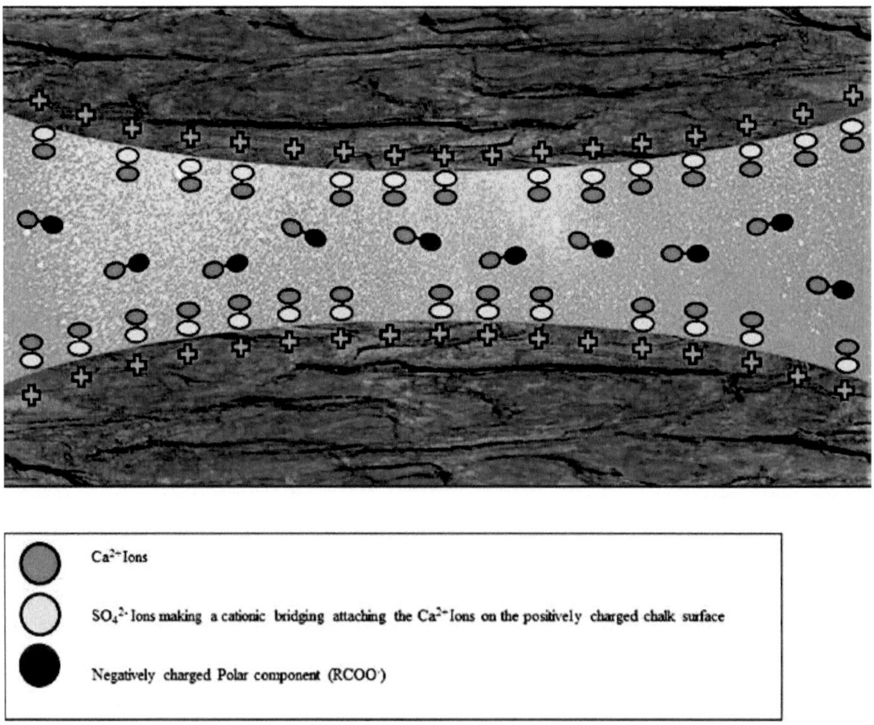

FIGURE 3.6 In this diagram, it is shown that how smart water containing only sulfate and calcium ions on chalk surface alters the wettability and increases the oil recovery.

anionic surfactant produced 39% of OOIP in a month (Shi et al., 2020). Fractures and oil-wet conditions are the main concerns in carbonate reservoirs. Gel treatment is done to reduce fracture permeability, and low-salinity waterflooding shifts the reservoir wettability toward a more water-wetting condition for improved oil recovery. Both processes do not work well when used for the carbonate reservoir, but when low-salinity water is used after gel placement, its propagation becomes more efficient into the fractures and increases oil recovery since more gel propagates into the fractures. This limits the water production as well as efficiently reduces heterogeneity since the gel strength gets reduced when the brine concentration decreases (Alhuraishawy et al., 2019).

Surface-modified pyroxene nanoparticles were studied to find the brine/rock reactions for different low-salinity solutions. The results have shown that using a 0.005 wt% nanoparticle with low-saline brine improves the nanofluid stability and EOR, and the electrical double layer thickness also gets increased, thereby increasing the zeta potential magnitude. Oil recovery increases up to 15% of the OOIP (Sagala et al., 2020).

Low-salinity waterflooding along with surfactant flooding is also a hybrid EOR method. The addition of surfactant not only reduces the IFT to mobilize the capillary-trapped residual oil but also prevents trapping of oil initially mobilized due to low-salinity waterflooding (Tavassoli et al., 2016). A variation of low-salinity waterflooding was studied, named low-salinity hot waterflooding, for application in heavy oil reservoirs. The addition of nanoparticles along with LSHW provides a combined effect of the thermal energy of hot water. The addition of nanoparticles (SiO_2 and Al_2O_3) for enhancing oil recovery from heavy oil and low-saline water changes the wettability and decreases the IFT. It has been experimentally found that in heavy oil reservoirs, the addition of nanosized particles in low-salinity hot waterflooding increases the recovery of oil (by nearly 40%) (Ding et al., 2018). Successively injecting alkaline, low-saline water, surfactant, and CO_2 proved to be better than WAG, CO_2, and foam flooding. It has the advantages of every process, but on top of that, it also inhibits precipitation, which leads to an increase in the performance of the alkaline foam EOR process by injecting a low-saline water slug between foam and alkaline flooding. The oil recovery was up to 20.1% OOIP (Al-Saedi et al., 2019). A novel EOR technique named low-salinity water-assisted foam flooding comprises injecting low-salinity water with alternating injections of CO_2 gas and an aqueous solution of surfactant. The surfactant solution reduces the IFT and also provides stability to the foam in the reservoir conditions. It produced

87.45% of OOIP when KCl composition was used in low-salinity water-assisted foam flooding (Shabib-Asl et al., 2019).

Smaller-scale experiments that are smaller than the core scale are best for understanding the mechanism responsible for LSE. These experiments have opened new applications of sciences like colloid science chemistry, surface science, and molecular dynamics in the petroleum industry. This not only gives in-depth knowledge for a better understanding of the mechanism but also produces possibilities for confusion and diversion from the topic. Increased oil recovery is only said to be valid when the oil recovery at that scale can be measured, and therefore it is not useful when the scale is at the molecular level.

KEYWORDS

- **low salinity water flooding**
- **smart water flooding**
- **wettability alteration**
- **zeta potential**

REFERENCES

Agbalaka, C. C.; Dandekar, A. Y.; Patil, S. L.; Khataniar, S.; Hemsath, J. R. Coreflooding Studies to Evaluate the Impact of Salinity and Wettability on Oil Recovery Efficiency. *Transp. Porous Media* **2009,** *76*, 77–94. https://doi.org/10.1007/s11242-008-9235-7

Alhuraishawy, A. K.; Bai, B.; Wei, M.; Almansour, A. Preformed Partial Gel Injection Chased by Low-Salinity Waterflooding in Fractured Carbonate Cores. *SPE Reserv. Eval. Eng.* **2019,** *22*, 110–121. https://doi.org/10.2118/191364-PA

Al-Murayri, M.T.; Kamal, D. S.; Al-Sabah, H. M.; AbdulSalam, T.; Al-Shamali, A.; Quttainah, R.; Glushko, D.; Britton, C.; Delshad, M.; Liyanage, J.; Dean, R. M. In *Low-Salinity Polymer Flooding in a High-Temperature Low-Permeability Carbonate Reservoir in West Kuwait*, Society of Petroleum Engineers - SPE Kuwait Oil and Gas Show and Conference 2019, KOGS 2019; Society of Petroleum Engineers, 2019. https://doi.org/10.2118/198000-ms

Al-Saedi, H. N. Scholars' Mine Scholars' Mine Doctoral Dissertations Student Theses and Dissertations Novel Insights into Low Salinity Water Flooding Enhanced Oil Novel Insights into Low Salinity Water Flooding Enhanced Oil Recovery in Sandstone Reservoirs Recovery in Sandstone Reservoirs, 2019.

Al-Saedi, H. N.; Al-Jaberi, S. K.; Flori, R. E.; Al-Bazzaz, W.; Long, Y. In *A New Design of Low Salinity-CO2-Different Chemical Matters*, Society of Petroleum Engineers - Abu

Dhabi International Petroleum Exhibition and Conference 2019, ADIP 2019; Society of Petroleum Engineers, 2019. https://doi.org/10.2118/197118-ms

Alvarado, V.; Garcia-Olvera, G.; Manrique, E. J. In *Considerations of Adjusted Brine Chemistry for Waterflooding in Offshore Environments*, OTC Brasil 2015: The Atlantic: From East to West - An Ocean of Innovation; Offshore Technology Conference, 2015; pp. 1961–1978. https://doi.org/10.4043/26293-ms

Alves, D. R.; Carneiro, J. S. A.; Oliveira, I. F.; Façanha, F.; Santos, A. F.; Dariva, C.; Franceschi, E.; Fortuny, M. Influence of the Salinity on the Interfacial Properties of a Brazilian Crude Oil Brine Systems. *Fuel* **2014,** *118*, 21–26. https://doi.org/10.1016/j.fuel.2013.10.057

Austad, T, Rezaeidoust, A.; Puntervold, T. In *Chemical Mechanism of Low Salinity Water Flooding in Sandstone Reservoirs*, SPE Improved Oil Recovery Symposium; Society of Petroleum Engineers: Tulsa, OK, 2010. https://doi.org/10.2118/129767-MS

Bernard, G. G. Effect of Floodwater Salinity on Recovery of Oil from Cores Containing Clays. SPE California Regional Meeting, CRM, 1967. https://doi.org/10.2523/1725-ms

Bidhendi, M. M.; Garcia-Olvera, G.; Morin, B.; Oakey, J. S.; Alvarado, V. Interfacial Viscoelasticity of Crude Oil/Brine: An Alternative Enhanced-Oil-Recovery Mechanism in Smart Waterflooding. *SPE J.* **2018,** *23*, 803–818. https://doi.org/10.2118/169127-pa

Boussour, S.; Cissokho, M.; Cordier, P.; Bertin, H.; Hamon, G. In *Oil Recovery by Low Salinity Brine Injection: Laboratory Results Outcrop and Reservoir Cores*, Proceedings of the SPE Annual Technical Conference and Exhibition; OnePetro, 2009; pp. 1595–1606. https://doi.org/10.2118/124277-ms

Chávez-Miyauchi, T. E.; Firoozabadi, A.; Fuller, G. G. Nonmonotonic Elasticity of the Crude Oil-Brine Interface in Relation to Improved Oil Recovery. *Langmuir* **2016,** *32*, 2192–2198. https://doi.org/10.1021/acs.langmuir.5b04354

Cissokho, M.; Boussour, S.; Cordier, P.; Bertin, H.; Hamon, G. Low Salinity Oil Recovery on Clayey Sandstone: Experimental Study. *Petrophysics* **2010,** *51*, 305–313.

Derkani, M.; Fletcher, A.; Abdallah, W.; Sauerer, B.; Anderson, J.; Zhang, Z. Low Salinity Waterflooding in Carbonate Reservoirs: Review of Interfacial Mechanisms. *Coll. Interf.* **2018,** *2*, 20. https://doi.org/10.3390/colloids2020020

Ding, Y.; Zheng, S.; Meng, X.; Yang, D. SPE-190132-MS Low Salinity Hot Water Injection with Addition of Nanoparticles for Enhancing Heavy Oil Recovery under Reservoir Conditions, 2018. https://doi.org/https://doi.org/10.2118/190132-MS

Emadi, A.; Sohrabi, M. Visual In *Investigation of Oil Recovery by Low Salinity Water Injection: Formation of Water Micro-dispersions and Wettability Alteration*, Proceedings of the SPE Annual Technical Conference and Exhibition; Society of Petroleum Engineers (SPE), 2013; pp. 4168–4182. https://doi.org/10.2118/166435-ms

Fredriksen, S.B.; Rognmo, A. U.; Fernø, M. A. In *Pore-Scale Mechanisms During Low Salinity Waterflooding: Water Diffusion and Osmosis for Oil Mobilization*, SPE Bergen One Day Seminar; Society of Petroleum Engineers, 2016. https://doi.org/10.2118/180060-MS

Godec, M. L. Global Technology Roadmap for CCS in Industry - Sectoral Assessment - CO2 Enhanced Oil Recovery, United Nations Industrial Development Organization, 2011.

Green, D. W.; Willhite, G. P. In *Enhanced Oil Recovery (Willhite)*, Second Ed.; Richardson, T. X., Henry, L., Ed.; Doherty Memorial Fund of AIME, Society of Petroleum Engineers, 1998.

Hernández, E.; Valero, E.; Rodríguez, I.; Guerra, E.; Espinoza, J.; Briceño, R.; Veliz, A. In *Enhancing the Recovery of an Extra Heavy Oil Reservoir by Using Low Salinity Polymer*

Flooding, SPE Latin American and Caribbean Petroleum Engineering Conference, 2020. https://doi.org/10.2118/199014-ms

Kokal, S. Crude-Oil Emulsions: A State-of-the-Art Review. *SPE Prod. Facil.* **2005,** *20*, 5–12. https://doi.org/10.2118/77497-pa

Larger, A.; Webb, K. J.; Black, C. J. J.; Singleton, M.; Sorbie, K. S. Low Salinity Oil Recovery - An Experimental Investigation1 - OnePetro 49. *Petrophysics* **2008,** *49* (01). https://doi.org/SPWLA-2008-v49n1a2

Lager, A.; Webb, K. J.; Collins, I. R.; Richmond, D. M. In *LoSal™ Enhanced Oil Recovery: Evidence of Enhanced Oil Recovery at the Reservoir Scale*, Proc–eedings of the SPE Symposium Improved Oil Recoverery, 2008; vol *3*, pp 1313–1324. https://doi.org/10.2118/113976-ms

Ligthelm, D.; Gronsveld, J.; Hofman, J.; Brussee, N.; Marcelis, F.; Van Der Linde, H. In *Novel Waterflooding Strategy by Manipulation of Injection Brine Composition (SPE-119835),* 71st European Association of Geosciensists and Engineers Conference and Exhibition 2009, Balancing Globoal Resources Incorposrating SPE Europe 2009; vol 2, pp 1313–1335.

Mahzari, P.; Sohrabi, M. In *Crude Oil/Brine Interactions and Spontaneous Formation of Microdispersions in Low Salinity Water Injection*, Proceedings of the SPE Symposium on Improved Oil Recovery; Society of Petroleum Engineers (SPE), 2014; pp. 731–745. https://doi.org/10.2118/169081-ms

McGuire, P. L.; Chatham, J. R.; Paskvan, F. K.; Sommer, D. M.; Carini, F. H. In *Low Salinity Oil Recovery: An Exciting New EOR Opportunity for Alaska's North Slope*, SPE Western Regional Meeting; Society of Petroleum Engineers, 2005. https://doi.org/10.2118/93903-MS

Morrow, N.; Buckley, J. Improved Oil Recovery by Low-Salinity Waterflooding. *J. Pet. Technol.* **2011,** *63*, 106–112. https://doi.org/10.2118/129421-JPT

Mousa, K. M.; Arafat A. S. Experimental Study on Treatment of Produced Water. *J. Chem. Eng. Process Technol.* **2015,** 06. https://doi.org/10.4172/2157-7048.1000261

Pu, H.; Xie, X.; Yin, P.; Morrow, N. R. In *Low Salinity Waterflooding and Mineral Dissolution*, Proceedings of the SPE Annual Technical Conference and Exhibition; Society of Petroleum Engineers (SPE), 2010; pp. 1574–1590. https://doi.org/10.2118/134042-ms

Rezaeidoust, A.; Puntervold, T.; Strand, S.; Austad, T. Smart Water as Wettability Modifier in Carbonate and Sandstone: A Discussion of Similarities/Differences in the Chemical Mechanisms. *Energy Fuels* **2009,** *23*, 4479–4485. https://doi.org/10.1021/ef900185q

Sagala, F.; Hethnawi, A.; Nassar, N. N. Integrating Silicate-Based Nanoparticles with Low-Salinity Water Flooding for Enhanced Oil Recovery in Sandstone Reservoirs. *Ind. Eng. Chem. Res.* **2020,** *59*, 16225–16239. https://doi.org/10.1021/acs.iecr.0c02326

Shabib-Asl, A.; Abdalla Ayoub, M.; Abdalla Elraies, K. In *Combined Low Salinity Water Injection and Foam Flooding in Sandstone Reservoir Rock: A New Hybrid EOR*, SPE Middle East Oil Gas Show and Conference, 2019. https://doi.org/10.2118/194975-ms

Sharma, H.; Mohanty, K. K. An Experimental and Modeling Study to Investigate Brine-Rock Interactions During Low Salinity Water Flooding in Carbonates. *J. Pet. Sci. Eng.* **2018,** *165*, 1021–1039. https://doi.org/10.1016/j.petrol.2017.11.052

Shi, Y.; Miller, C.; Mohanty, K. In *Surfactant-Aided Low-Salinity Waterflooding for Low Temperature Carbonates*, Proc–eedings of the SPE Annual Technical Conference and Exhibition, 2020. https://doi.org/10.2118/201754-ms

Tang, G. Q.; Morrow, N. R. Salinity, Temperature, Oil Composition, and Oil Recovery by Waterflooding. *SPE Res. Eng.* **1997,** *12*, 269–276. https://doi.org/10.2118/36680-PA

Tang, G. Q.; Morrow, N. R. Influence of Brine Composition and Fines Migration on Crude Oil/Brine/Rock Interactions and Oil Recovery. *J. Pet. Sci. Eng.* **1999,** *24*, 99–111. https://doi.org/10.1016/S0920-4105(99)00034-0

Tavassoli, S.; Korrani, A. K. N.; Pope, G. A.; Sepehrnoori, K. Low-Salinity Surfactant Flooding - A Multimechanistic Enhanced-Oil-Recovery Method. *SPE J.* **2016,** *21*, 744–760. https://doi.org/10.2118/173801-pa

Tunio, S. Q.; Tunio, A. H.; Ghirano, N. A.; Mohamed, Z.; Adawy, E. Comparison of Different Enhanced Oil Recovery Techniques for Better Oil Productivity. *Int. J. Appl. Sci. Technol.* **2011**.

Webb, K. J.; Black, C. J. J.; Al-Ajeel, H. In *Low Salinity Oil Recovery - Log-Inject-Log*, Proceedings of the Middle East Oil Show; Society of Petroleum Engineers (SPE), 2003; pp. 367–374. https://doi.org/10.2118/81460-ms

Yildiz, H. O.; Morrow, N. R. Effect of Brine Composition on Recovery of Moutray Crude Oil by Waterflooding. *J. Pet. Sci. Eng.* **1996,** *14*, 159–168.

Zeinijahromi, A.; Bedrikovetsky, P. SPE-182386-MS Water Production Control Using Low-Salinity Water Injection Formation Damage as a Result of Low-Salinity Water Injection, 2016, pp 25–27.

Zhang, Y.; Morrow, N. R. In *Comparison of Secondary and Tertiary Recovery with Change in Injection Brine Composition for Crude Oil/Sandstone Combinations*, Proceedings - SPE Symposium on Improved Oil Recovery; Society of Petroleum Engineers (SPE), 2006; pp. 745–758. https://doi.org/10.2118/99757-ms

Zhang, Y.; Xie, X.; Morrow, N. R. Waterflood Performance By Injection Of Brine With Different Salinity For Reservoir Cores; Society of Petroleum Engineers (SPE), 2007. https://doi.org/10.2118/109849-ms

CHAPTER 4

Challenges, Opportunities, and the Way Ahead in Polymer Flooding

KRISHNA RAGHAV CHATURVEDI[1], RAMESH NARUKULLA[1], RISHI RAJ GOSWAMI[2], and TUSHAR SHARMA[1]

[1]*Enhanced Oil Recovery Laboratory, Rajiv Gandhi Institute of Petroleum Technology, Jais, India*

[2]*Oil India Limited, Assam, India*

ABSTRACT

This chapter explores the application of polymer flooding as a dynamic and effective method in enhanced oil recovery (EOR). Examining the mechanisms behind polymer-induced viscosity modification, reservoir sweep improvement, and the impact on fluid mobility, the chapter aims to provide a comprehensive overview of the enhanced oil recovery potential of polymer flooding. This chapter highlights various aspects of polymer flooding in enhancing oil recovery processes, mainly focusing on permeability modification of injected water to address excessive water production during oil recovery. The injection of polymer solutions into reservoirs is explored to adjust water relative permeability, reduce channeling, and increase trapped oil recovery. This chapter also emphasizes the impact of polymer properties, such as branching, on the residual resistance factor (RRF) and oil recovery. Moreover, this chapter discusses the commonly used polymers in EOR, including polyacrylamide (PAM), partially hydrolyzed polyacrylamide (HPAM), and hydrophobically modified associating (HMA) polymers. This discussion covers the adsorption characteristics of these polymers on rock surfaces, their rheological properties, and the considerations for their application in the field. This also briefly explains the various natural

Advancements in Chemical Enhanced Oil Recovery.
Tushar Sharma, Krishna Raghav Chaturvedi, Tarek Ganat, and Imtiaz Ali (Eds.)

polymers used in EOR process, such as guar gum and xanthan gum. It also highlights their limitations due to specialized properties. Additionally, this chapter also examines the rheological behavior of polymer solutions in porous media, addressing phenomena such as shear thickening and shear thinning. The understanding these rheological characteristics is essential for optimizing polymer flooding efficiency. Subsequently, detailed analysis of biopolymers was provided, including their merits and demerits for oilfield applications. The discussion encompasses various types of biopolymers, such as xanthan gum, carboxymethylcellulose, schizophyllan, hydroxyethylcellulose, welan gum, scleroglucan, guar gum, lignin, and cellulose. It also explores the rheological behavior of polymer solutions in porous media and highlights considerations for their practical application in reservoir conditions. Finally, the chapter also emphasizes the challenges before and the significance of polymer flooding as a promising technique to maximize hydrocarbon extraction, contributing to the sustainable and efficient utilization of reservoirs.

4.1 INTRODUCTION

Because the world's energy demands are increasing, oil is likely to remain the world's principal source of energy. To meet the world's energy demands, it is vital to increase oil production through the use of efficient technologies that can produce large amounts of oil at a lower cost than before. More than 70% of the residual oil in typical oil recovery procedures remains in the reservoir after extraction. As a result, it is critical to increase oil output from all oil-producing reservoirs by employing innovative and unique procedures for recovering oil and gas (Pandey et al., 2021). It is widely assumed that enhanced oil recovery (EOR) processes are necessary to recover massive volumes of unrecoverable oil from the vast majority of fields and fulfil the world's energy demands. Chemical enhanced oil recovery (cEOR) is a significant and promising technique in the field of EOR (Shafiai and Gohari, 2020). Surfactants, polymers, alkalis, nanomaterials, and their combinations have all been employed in previous chemical EOR efforts. Polymers operate as a mobility-regulating agent in polymer flooding studies, enhancing displacement and sweep efficiency by increasing the viscosity of the displacing fluid while simultaneously reducing the mobility ratio of oil and water (Seright et al., 2009). It was also revealed that polymers' viscoelastic nature contributes to their microscopic efficiency, which was previously unknown. Oil output might be boosted by injecting the polymer solution, as per previous studies (Kumar and Mandal, 2017). This method may assist oil fields with a saturation of remaining oil greater than the saturation of residual oil. When compared to other methods of increasing oil recovery, this technique is more efficient and straightforward to adopt and maintain. Using this technology has been effective in gathering a significant volume of oil from a variety of reservoirs in the petroleum industry (Chang et al., 2014; Sheng, 2013, 2011).

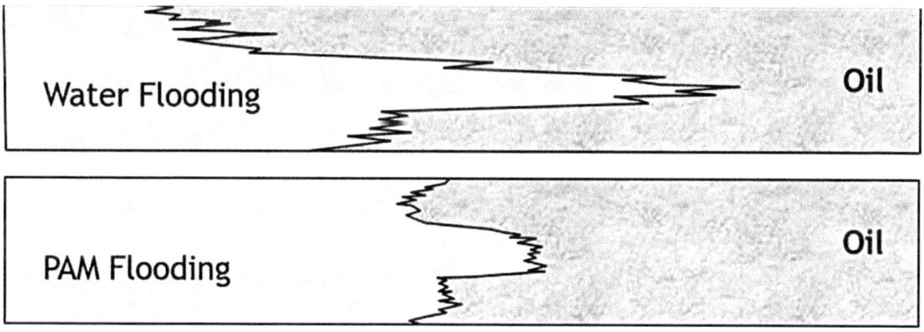

FIGURE 4.1 Comparative analysis between water flooding and polymer flooding showing uniform front formation in polymer flooding.

Source: Reprinted with permission from Sorbie (2005). Also part of previous works (Gbadamosi et al., 2019; Thomas, 2016). http://creativecommons.org/licenses/by/4.0/)

Polymer flooding has proven to be a dependable and successful means of enhancing oil recovery since its inception in the field of enhanced oil recovery more than 40 years ago (Kesarwani et al., 2021; Machale et al., 2020; Sheng, 2011). Oil recovery from water injection is expected to grow by 15–20%, with a total recovery rate of 50%, according to projections (Sheng, 2011). The resistance factor can be enhanced by up to 15% during polymer flooding, while the water cut can be lowered by up to 50%, depending on the material (Singh and Mohanty, 2017). Because it consumes less water and enhances oil recovery while using less water, polymer flooding is more cost-effective than water flooding. There is a wealth of information from numerous fields in China and the United States proving the highest recovery when polymer flooding is used (Chang et al., 2014; Fu et al., 2016). Plastics such as polyacrylamide (PAM), partially hydrolyzed polyacrylamide (HPAM), and the polysaccharide xanthan gum, among others, have been employed in chemical EOR applications (Abidin et al., 2012; Liang et al., 2019). PAMs are synthetic polymers that are commonly used in cEOR. HPAM is a PAM made out of hydrolyzed acrylamide monomers in a straight chain. It belongs to the PAM family. The material's polyelectrolyte nature leads it to have a random coil form, be flexible, and interact with ions. This is due, in part, to its outstanding viscosifying capabilities, low cost, broad molecular weight range, and favorable physiochemical qualities. According to the statistics, HPAM was the most widely used solvent in the field of polymer flooding (Abidin et al., 2012). Because of its propensity to bond to other molecules, xanthan gum is classified as a biopolymer. Xanthan gum is synthesized by microbes from the bacterium *Xanthomonas campestris*, utilizing glucose as a substrate and inorganic nitrogen and proteins as reinforcements. Xanthan gum performs well in high-saline settings and can be used with the majority of surfactants and additives often used during injection in the majority of EOR formulations. In high-salinity conditions, xanthan gum is also beneficial. Numerous review papers on polymer flooding with water-soluble polymers can be found on the internet, with an emphasis on the effectiveness of drag reduction, rheological properties, adsorption on the surfaces of various rocks, including carbonate and sandstone, core flooding experimental results, the salinity/temperature effects, and implementation in various polymers for use in the fields (Le Van and Chon, 2016; Nakhjiri et al., 2018; Raffa et al., 2016; Yousefvand and Jafari, 2015).

Polymers are important in oil recovery because they may be used in several unusual reservoir scenarios, such as high permeability causes or fractures that redirect water forced into the reservoir or channels, or when heavy oils display high flow resistance, among other things (Clarke et al., 2015; Seright et al., 2009). Water-soluble polymers can be used to improve viscosity in certain scenarios, enabling the fluid to flow more freely and improving oil recovery, whereas polymeric gels can be employed to close off high-permeability zones (Urbissinova et al., 2010). More crucially, the chemical characteristics of the reservoir water, reservoir temperature, and reservoir salinity all affect polymer flooding efficiency. Increased salinity or temperature can alter the viscosity of concentrated polymers, resulting in a drop in polymer viscosity as well as a decrease in channeling and sweep efficiency (Azad and Trivedi, 2019). This can be reduced by researching reservoir conditions, the kind of polymer used for injection, and the use of certain additives or copolymers that are resistant to high temperatures and/or a high-saline brine environment (Azad and Trivedi, 2020). As a result, this chapter examines how to control the permeability of injected water, the types of polymers utilized in polymer flooding, the challenges associated with polymer flooding, and prospective polymer flooding uses.

4.2 PERMEABILITY MODIFICATION OF INJECTED WATER

In the petroleum industry, the expansion of a reservoir as a consequence of standard oil recovery processes results in a rise in water production concomitant with oil recovery, which causes significant concerns (Pitts et al., 2004). As a result of this condition, a large amount of surplus water is discharged, which hurts the economics of recovery and contributes to environmental harm. During construction, the permeability of a reservoir to water or brine can be adjusted by injecting a polymer solution into the reservoir at various locations (Azad and Trivedi, 2019; Durucan et al., 2016). The permeabilities of the various strata within the reservoir vary greatly due, in part, to the reservoir's variable structure. Excess water production is possible through the use of channeling and exceptionally pervious reservoir layers. As a result of this phenomenon, a considerable volume of oil becomes trapped in the poor permeability layers, resulting in limited oil recovery during the first two phases of extraction (primary and secondary). As a consequence, by substituting polymer solutions for water injection in heterogeneous reservoirs, it is possible to reduce channeling and water permeability, thereby increasing trapped oil output and lowering water permeability (Wang et al., 2003). Numerous studies have shown that putting polymer solutions into reservoir wells can successfully reduce excessive water flow. When a reservoir contains water, the oil flows within the pores while the water travels along the oil-water contact, between the pore walls, and through the annulus. The polymer is absorbed into the rock's surface or walls after injection, producing a covering on the rock (Chaturvedi and Sharma, 2020). Water is injected into the same reservoir as polymer, causing the polymer coating on the rock's surface to expand and prevent water from flowing through while enabling oil to flow (Telmadarreie and Trivedi, 2016). When water is introduced, the rock swells and prevents water flow while allowing oil to flow. As a result of the lower permeability produced by the polymer injection, a flow barrier forms in the reservoir. Water injection is directed to unswept zones as a result of the higher flow resistance, enhancing output. Allowing the injected polymer to remain in the reservoir for an extended length of time results in a large drop in permeability, resulting in an even bigger fall in permeability than previously reported (Reena et al., 2020).

Lai and colleagues investigated the influence of polymer branching on residual resistance factor (RRF) and oil recovery power in a laboratory setting (Lai et al., 2016). In addition, as part of their research, they created three distinct hyperbranched polymers (HPDAs), each with a different degree of branching. They used these HPDAs to examine changes in a one-dimensional sandpack technique and an RRF made of a high-permeability porous material. This experiment was carried out to gain a better understanding of the method by which these branched polymers alter RRF by modifying the wettability of the rock surface onto which the polymers were placed. SEM and DLS examinations were carried out in addition to the other methods to acquire a better knowledge of the microstructure and diameter distribution of the injected and generated polymer solutions. They discovered that changing the external environment influenced the trends in RRF value variation in all three polymers tested in this study. Because the branching of the polymers produced increased, the RRF finally shifted away from surface adsorption toward mechanical stripping. As the amount of branching in the polymer solution increased, so did the number of chains inside the branches, increasing the hydrodynamic radius of the polymer solution. Furthermore, terminal branches resulted in a stronger structure, indicating that the polymer solution was less degraded when exposed to a high-permeability medium. These observations are highly essential in the synthesis of a newer class of polymers for oilfield applications (Maghzi et al., 2013).

A variety of alternative approaches may be used when polymer flooding is used to increase oil recovery. In general, mixing polymers with water and injecting them into a reservoir thickens and increases the viscosity of the aqueous phase, improving oil recovery efficiency by lowering the mobility ratio and increasing sweep efficiency, among other things. Second, the water permeability of polymer-swept zones can be greatly lowered depending on the type of polymer employed. In the event of polymer flooding, it has a secondary impact on the reservoir, contributing to the enhancement of oil recovery. The lower permeability functions as a sealant for the formation of heterogeneous reservoirs, restoring pressure in the highly permeable zone and enabling the polymer to flow freely. Furthermore, a decrease in permeability acts as a sealant for the formation (Reena et al., 2020; Sun et al., 2020). When these two approaches are combined, the water–oil displacement mobility ratio is reduced by increasing the viscosity of the injected water, which raises the recovery factor (Kargozarfard et al., 2019). Apart from the macroscopic and microscopic components mentioned above, oil displacement is caused by several mechanisms. The macroscopic mechanism, which is concerned with volumetric measurements, determines the efficiency of displacing fluid volumetrically measured in the oil zone, whereas the microscopic mechanism is concerned with the efficacy of displacing fluid on trapped oil mobilization via capillary forces at the pore size when volumetrically measured in the oil zone (Chang et al., 2014; Zhou et al., 2020).

Theoretically, three unique rheological characteristics of polymer solutions across porous surfaces are possible: Newtonian, shear thickening, and shear thinning with shear rate as a function of the shear rate of the solution. As a result of this compression and elongation, polymer molecules are forced down the pore throats and into the pore bodies, making the pores seem smaller. Shear thickening of these elastic polymer chains happens when the flow velocity of polymer molecules in a porous medium is much larger than the flow rate of polymer molecules (high apparent viscosity). However, because of the chemical's nature, it is difficult to restore the missing oil adequately. Shear thickening is a feature that aids in the flow of mobile fluids (Chaturvedi et al., 2021a). This phenomenon has been extensively studied and is widely recognized as one of the most promising ways to boost microscopic effectiveness during polymer flooding due to its vast potential. Shear thickening can be utilized on a larger scale to increase the thickness of sweeps, enhancing their efficacy (Seright et al., 2009). Despite the accuracy with which this effect is crucial in polymer flooding applications, it has considerable limits. Because it is a shear-thinning regime, bulk rheological studies cannot identify this feature, even when shear rates are equal to those used in the research. Recent research has shown that it is possible to make more exact and consistent predictions about the shear thickening regime than was previously thought. All of these forecasts are in terms of apparent viscosity, and their correctness is determined by the effective shear rate. The flow characteristics of xanthan gum and HPAM in porous media were investigated using an empirical viscoelastic model based on empirical viscoelastic theory and similar to Darcy's law (Garrouch and Gharbi, 1999). Polymers have viscoelastic behavior, permeability, and porosity, among other characteristics. The Deborah number (De), which is a dimensionless number, is most often utilized in rheology to characterize the fluidity of materials under specific flow conditions (Azad and Trivedi, 2019). According to the study's findings, the Deborah number alone, as previously thought, may not be sufficient for predicting flow behavior in porous material (Azad and Trivedi, 2020). Shear thickening, according to this theory, does not develop inexorably with Deborah number forever, making it more acceptable. When the shear-thickening area equations and the shear-thinning regime equations were combined, a new model known as the viscoelastic model was created that was applicable over the whole shear rate range. This became known as the viscoelastic model. This model was successfully used to connect previously published laboratory data to the study's conclusions. It was possible to build a library of HPAM polymer solutions with different concentrations, temperatures, hardness, and salinity using this viscoelastic model. Polymer viscosity increases due to elasticity, and the increase is related to particle diameter, porosity, relaxation time, and the square root of tortuosity. It is a tortuosity measure since it is inversely related to the Darcy velocity and the square root of tortuosity. The consistency of experimental and computational data suggests that this equation may detect shear thickening in porous media. Polymer flow in porous media is classified into three steps: extrusion, pore passage, and entry constriction. The first stage is extrusion, which is followed by pore passage and entrance constriction, and so on. This indicates that a combination of viscous and elastic drops can cause pressure drops on their own, allowing the construction of a constitutive model for viscoelastic fluid in porous media that considers both the flow parameters of the polymer solution and the features of the porous medium. It was demonstrated that this unique model is accurate in obtaining the rheological characteristics of polymers in porous media under shear-thickening and shear-thinning regimes by comparing it to actual data from core flooding. Another rheological model takes into account the viscosity of shear thickening and thinning areas. When core flooding experiments were performed on the model, it was discovered that as the shear rate rose, the apparent viscosity fell beyond the shear-thickening zone, which was previously unknown. The mechanical breakdown of the polymer's molecular structure caused by the chain-breaking process is one possible explanation (Azad et al., 2018). This study resulted in the development of a more accurate rheological model multiplied by a degradation factor, which was shown to be capable of precisely forecasting viscosity fluctuations in porous media (Song and Hatzignatiou, 2022).

Also, in most cases, for displacement to occur, the total mobility of the displaced multiphase fluid must be equal to or less than the total mobility of the displacing fluid (Negin et al., 2016). It is difficult to compute the total mobility of the fluids employed in these computations due to unknown saturations of the mobile phase in a few circumstances. Under these conditions, the needed viscosity for polymer flooding may be estimated by starting with the polymer with the least total mobility. A study found that the previously held view was incorrect and identified a new mobility ratio as the ratio of fluid displacement mobility to polymer adsorption mobility in oil (displacing fluid mobility to oil mobility by polymer adsorption) (Mishra et al., 2014; Tekin et al., 2010). HPAM adsorption was detected in both situations in somewhat oil- and water-wet cores, and water's relative permeability was preferentially enhanced relative to oil's relative permeability (Grattoni et al., 2004). Other studies have referred to this phenomenon as adsorption entanglement or bridging adsorption, which are both vernacular phrases for the same phenomenon (Barreau et al., 1997; Tekin et al., 2005). Furthermore, the efficiency of the microscopic displacement is amplified during polymer flooding, which is a beneficial activity (Yekeen et al., 2018).

Longer molecular chain polymers generate normal tension between the polymer solution and the oil as well as shear stress as the elastic fluid flows over the dead ends of the polymer chains. Normal tension and shear stress are not generated by shorter molecular chain polymers (Corredor-Rojas et al., 2018). As a result, the polymer solution applied a great deal of force to the oil droplets, and the droplets were sucked out of the dead ends. This is because the elasticity of the extraction fluid determines how much oil can be recovered from dead ends. This technique for drawing oil droplets out is a well-known technique for creating the pulling effect and is used in a variety of applications. When it comes to viscoelastic fluids (polymers like HPAM), they can both push and pull fluids forward; on the other hand, when it comes to viscoelastic fluids (water), they can drive fluids forward but cannot release the oil from its snag (Berdichevsky and Herman, 2016; Newtonian, 1930). Another experiment looked at the effectiveness of tiny oil displacements using one polymer elastic in combination with another. To achieve their goal, they would need to flow fluids through a dead end with viscoelastic qualities different from the trapped oil in the reservoir. It is necessary to comprehend the link between the elastic modulus (G') and the viscous modulus (V) (G") (Chaturvedi et al., 2021b). Water (a nonelastic fluid) becomes saturated with residual oil when injected over a dead end (Clarke et al., 2015). In the oil-wet reservoir medium, remaining oil gathers on the rock's surface, generating a continuous oil film over the whole surface of the rock (Azad and Trivedi, 2020). It has been observed that elastic fluid's velocity gradient at the capillary wall is significantly bigger than that of Newtonian fluids, showing that the elastic fluid has a greater velocity gradient. As a result, the polymer flow applied greater force than the water flow, resulting in quicker removal of the oil coating from the rock surface and improved oil displacement efficiency. This procedure, known as a stripping mechanism, is performed in a laboratory setting (Urbissinova et al., 2010). The oil stripping operation changed the wettability of the rock from oil-wet to water-wet, resulting in more oil recovery (Azad and Trivedi, 2019; Wu et al., 2014).

Furthermore, oil flow is a system that relies on the elastic properties of the polymer to work effectively. Due to the inherent tension between oil threads, elastic polymer solutions, according to this theory, can be used to stabilize them. Luo et al., discovered oil threads by driving oil into oil columns and interacting with downstream residual oil to produce oil threads (Wei et al., 2014). The normal force pressing on the oil thread's concave surface is frequently smaller than the normal force acting on the convex surface, indicating that the concave surface is more stable. As a result of this occurrence, the natural tension acting on oil threads restricts oil thread deformation, resulting in an improvement in oil displacement efficacy (Gleasure and Phillips, 1990; Hirasaki and Pope, 1974). Because of the increasing Deborah number, the elasticity of polymer solutions in porous media may be described as a dimensionless number, which is referred to as the force contrast in this context. As Deborah's number grows, so does the power disparity between the two characters (Azad et al., 2018). The oil column becomes destabilized as a result of the interfacial tension between the polymer solution and the oil, resulting in the separation of the oil droplets from the column. The elasticity of the polymer solution, on the other hand, can prevent the onset of the instability caused by interfacial tension (Hirasaki and Pope, 1974). Either a thinner film is generated, which degrades into smaller droplets and results in less residual oil saturation, or a longer film is formed, which increases mobility by decomposing the oil column into larger oil drops.

4.3 TYPES OF POLYMERS

PAM is a nonionic, nonhydrolyzed polymer with the structure depicted in Figure 4.2.

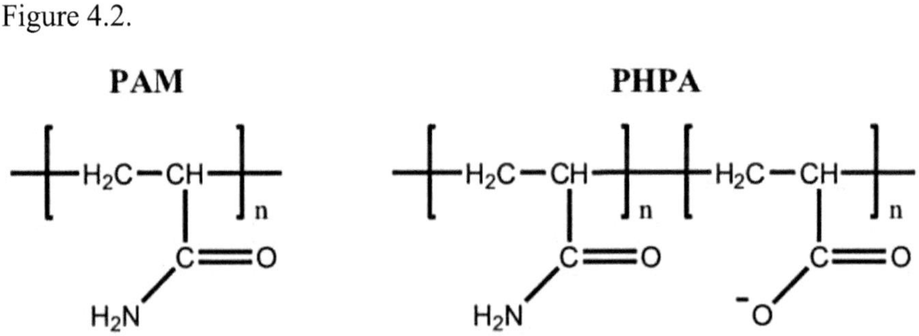

FIGURE 4.2 Chemical structure of polyacrylamide (PAM) and partially hydrolyzed polyacrylamide (PHPA).

Source: Reproduced with permission from (Pinho De Aguiar et al., 2020). http://creativecommons.org/licenses/by/4.0/

Nonionic PAM is generally not recommended for EOR applications due to its proclivity for adsorption on mineral surfaces (Tekin et al., 2010). The majority of current cEOR implementations, however, are based on modified PAM (Kumar and Mandal, 2017; Raffa et al., 2016). These topics are discussed in greater detail in the subsequent sections.

While the polymer PAM does not hydrolyze at ambient temperature, the amide group in the polymer may engage in hydrolysis at elevated temperatures or pH levels. Temperature-induced variations in the pH and viscosity of PAM were also observed when it was hydrolyzed (Ahmed et al., 2017; Majeed et al., 2019). When PAM is hydrolyzed, the resulting carboxylate group facilitates the polymer's conversion to a negatively-charged solution. Because polymer chains are negatively charged, they repel one another, increasing the viscosity of the polymer solution. PAM is only partially hydrolyzed in EOR applications once an alkali is introduced to the polymer solution. Additionally, this partially hydrolyzed PAM is easily prepared by mixing acrylic acid with acrylamide. Adsorption and retention of polymers on reservoir rock surfaces are facilitated by the entrapment of polymer chains and chemical adsorption. In other instances, mechanical trapping of polymer chains on the reservoir rock surface is the major mechanism of PAM retention. PAM adsorption on the surface of the rock is influenced by a variety of parameters, including the molecular weight of the polymer, the functional group distribution, the surface charge, the temperature, and the ionic strength (Barreau et al., 1997; Bera et al., 2013; Uranta et al., 2018). The initial rise in polymer concentration improves surface adsorption, and a plateau is observed at higher polymer concentrations. Although temperature has a significant effect on the adsorption density of nonionic PAM, pH has little effect.

HPAM is a commercially available combination of acrylic acid and acrylamide that is used in cEOR applications, with a structure provided in Figure 4.3.

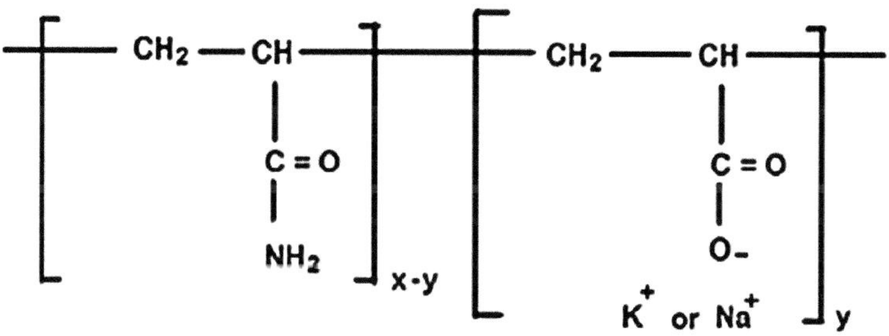

FIGURE 4.3 The structure of partially hydrolyzed polyacrylamide (HPAM).

Source: Reprinted with permission from (Gbadamosi et al., 2019). https://creativecommons.org/licenses/by/4.0

This is often used in EOR applications, and several evaluations have been written on its implementation in a variety of field circumstances (Kang et al., 2019; Sedaghat et al., 2016; Yin et al., 2018). This section discussed HPAM adsorption on rock surfaces. Rheology and thermal stability affect the adsorption isotherm of PAM in the opposite direction. When PAM is heated, it hydrolyzes and becomes ionic, and the hydrogen bonding between the nonionic PAM decreases, resulting in a reduction in adsorption. Electrostatic repulsion rises as a result, whereas adsorption decreases. Calcium improves the adsorption of PAM on sandstone rock surfaces, whereas the addition of a nonionic surfactant enhances the adsorption of nonionic polymers on silica. The percentage of acrylate groups in HPAM dictates the extent of hydrolysis (Zhang and Feng, 2021). The viscosity of the polymer is increased as a result of repulsions between polymer chains generated by the presence of negative groups in HPAM during hydrolysis. As a result of the precipitation that occurs in the presence of divalent cations, a higher degree of hydrolysis is not always desired. In a previous study, the rheological properties of powders and emulsions were compared, and both solutions did not exhibit any plugging in the formation (Han et al., 2021). Temperature and pH are also the primary determinants of hydrolysis rate (Jung et al., 2013; Lewandowska, 2007).

Hydrophobically-modified associating (HMA) polymers have an aqueous backbone modified with hydrophobic monomer fractions, and the hydrophobic groups combine to generate hydrophobic microdomains in aqueous form (Taylor and Nasr-El-Din, 2007). Copolymerization, or grafting, can be used to produce HMA polymers. While hydrophobic molecules can be grafted onto water-soluble chains, hydrophobic monomers can be copolymerized with water-soluble monomers. Hydrophobes are integrated into chains of water-soluble molecules using a variety of different synthetic techniques. Based on the distribution of hydrophobic moieties in HMA polymers, four classifications are possible. Included are associative polymers with a single associating block, multisticker polymers, telechelic polymers, and mixed HMA polymers. The first class of HMA polymers consists of those that include all of the hydrophobes on a single end of the hydrophilic chain. Telechelic HMA polymers are HMA polymers that have a hydrophobic moiety at both ends of their hydrophilic chains. The term "multisticker" refers to HMA polymers that include hydrophobic moieties at several locations throughout their hydrophilic chains. The hydrophobic groups are typically maintained below 5% to ensure the polymer's solubility in water, and coil contraction occurs in dilute solutions due to intramolecular interactions. However, the network structure is formed as a result of intramolecular affiliations that are dominated by intermolecular contacts at concentrations greater than the critical association concentration (CAC). This newly formed network increases the viscosity of the HMA polymer and may be adjusted via a variety of internal and external factors (Wang et al., 2004). A few hydrophobic associating polymers were synthesized in the late 1960s using alkyl vinyl ether and maleic acid. HMA polymers based on acrylamide were developed in the 1980s for use in EOR. There is an abundance of information on the use of HMA polymers in EOR at the laboratory scale.

Another important group of polymers that have been widely explored in field applications includes natural polymers like guar gum, xanthan gum, and some other polymers like PVA. However, on the grand scale of things, these polymers have found limited applications in the field due to their specialized properties. Hence, they have been less discussed in this chapter. However, if the readers are interested to know more about them, a detailed study can be found in the cited works (Pu et al., 2018; Rellegadla et al., 2021; Zhang et al., 2005). A detailed report on the merits and demerits of biopolymers has been provided in Table 4.1.

TABLE 4.1 Commonly Used Biopolymers Along with Advantages and Disadvantages for Oilfield Applications.

Polymer type	Merits	Demerits
Xanthan gum	High resistance to salinity, temperature, and shear	High susceptibility to biodegradation and oxidation
	Possess long-term stability	High risk of plugging of rock pores
Carboxymethylcellulose	The biopolymer is environment friendly	It suffers from oxidative decomposition and thermal degradation
	It has good solubility in water	
Schizophyllan	Good resistance to temperature and salinity	Highly susceptible to biodegradation
	High thickening efficiency	
	Low adsorption of its molecule on rock surface	
	It is a nontoxic material	
Hydroxyethylcellulose	High solubility in water	Biodegradation problem
	Good tolerance to temperature and mechanical shearing	
	Excellent viscosifying effect	

TABLE 4.1 *(Continued)*

Polymer type	Merits	Demerits
Welan gum	Possess long-term stability Excellent viscoelastic effect	Performance dependent on inorganic cations present in reservoir
Scleroglucan	Good resistance to shear and temperature effect	High risk of biodegradation and oxidation
	Excellent viscosifying effect	Poor filterability of the material in rock pores
Guar gum	It is an environment friendly material	High risk of plugging in porous media
	Excellent hydration properties and good compatibility to salts over a wide range of pH	It suffers from poor thermal stability and weak elasticity
Lignin	It is an environment friendly material	Insoluble in water
	Readily available at a low cost	Biodegradation and oxidation effects
Cellulose	Good resistance to temperature and mechanical shearing	Water insolubility and heterogeneous swelling

Source: Adapted from Gbadamosi et al. (2019). https://creativecommons.org/licenses/by/4.0

Pandey and coworkers reported a comprehensive rheological analysis of the natural polymer guar gum for carbon absorption in conjunction with an anionic surfactant and in the presence of salts (Pandey et al., 2021). This work can be used to build a foundation for rheological assessment under high pressure, high temperature, and high salinity conditions, all of which lead to polymer viscosity loss.

It is vital to employ specialized field equipment to monitor and regulate polymer viscosity loss and degradation (Xia et al., 2018). The rheology of polymer solutions on porous surfaces is influenced by microbial activity, salt degradation, and heat degradation. Temperature directly affects the hydrolysis of polymer solutions. Hydrolysis of HPAM typically begins at a temperature of 25°C and escalates to 90°C during polymer flooding. As a result of this process, the viscosity of the polymer solution decreases (Pandey et al., 2021). At elevated temperatures, polymers containing an acrylamide functional group hydrolyze and acquire a negative charge. This results in a decrease in the viscosity of the polymer solution, and injection is not carried out in real time. Additionally, when the polymer solution adsorbs onto the rock surface, the reservoir's permeability decreases, posing challenges during polymer injection. The use of this polymer (HPAM) in high-salinity and high-temperature (HSHT) reservoirs is restricted.

In some polymer flooding applications, xanthan gum has been found to outperform HPAM. Xanthan gum was employed for the first time in this application in the fields of Eddesse-Nord, Germany, and Vorhop-Knesebeck, Norway (Mohsenatabar Firozjaii and Saghafi, 2020). Additionally, biopolymers employed in floods might degrade in conditions of extreme salt and heat. At normal temperatures, the xanthan gum molecules exhibit a double helix configuration; however, when the temperature increases, the shape changes to a chaotic coil. Furthermore, it has been determined that xanthan gum has a maximum temperature of 70°C. When a small amount of salt is added to the xanthan gum solution, charge shielding occurs, resulting in a decrease in the viscosity of the xanthan gum. Xanthan gum is more salt-sensitive than PAM-based polymers. While xanthan gums are more stable than acrylamide polymers at higher temperatures and salinities, they degrade in high salinity, high temperature (HSHT) settings. As a result, this type of bio- and synthetic polymer is incompatible with reservoir conditions characterized by elevated temperatures and salinity (Pandey et al., 2021).

Several new polymers have been created recently that can help in the retention of polymer solution viscosity under high-salinity and high-temperature reservoir conditions (Jouenne, 2020; Mohsenatabar Firozjaii and Saghafi, 2020). A quaternized cationic monomer (a copolymer of acrylamide) is a flocculant that is utilized in the water industry. This copolymer's structure is the result of the combination of two different monomers. Additionally, they observed that the copolymer developed outperformed SAV 37 in terms of mechanical and thermal stability. SNF company developed SAV 37, a terpolymer containing N-vinylpyrrolidone that is stable at elevated temperatures and performs well in brines with a high salinity. A detailed work on synthetic polymers may be found in these cited works (Chang et al., 2014; Seright et al., 2009).

The process of developing a biopolymer capable of proving stability under HSHT conditions is much different from the process of developing the synthetic polymer HPAM (Elyasi Gomari et al., 2021). In comparison to individual polymers, the combination of xanthan and guar displayed increased resistance to salty conditions. Schizophyllan may be the optimal biopolymer for HSHT settings in carbonate and sandstone due to its nonionic (minimal adsorption on the carbonate rock surface) and environmentally friendly characteristics. This polymer may be produced by palm fermentation and agricultural waste products, making it a cost-effective and environmentally beneficial method for polymer flooding applications, especially in conjunction with other chemical additives (Pal et al., 2018). However, this output may be unprofitable under other circumstances. Biopolymer growth is less critical during polymer flooding. This is because of challenges such as microbial and bacterial decay, biomaterial waste on the wellbore wall, biomaterial manufacturing expenses, and injection difficulties. Biological degradation occurs in both biopolymers and synthetic polymers when they are in a water-soluble state; however, biopolymer degradation is more dramatic (Adewole and Muritala, 2019). The biological degradation results in pore obstruction, which damages the formation. However, in many cases, increasing bioplugging has been shown to improve oil recovery by halting flow in highly permeable porous media (Dhanarajan et al., 2017).

4.4 CHALLENGES WITH POLYMER FLOODING

Polymer floods are influenced by the adsorption, precipitation, or mechanical entrapment of polymers in porous media, all of which can happen at the same time, which tends to cause polymer induced formation damage within the reservoir (see Fig. 4.4) (Deng et al., 2006; Greenland, 1963). To be able to perform polymer flooding in the field, one first needs to know about the problems with formation damage (Yuan et al., 2017). The best way to get a full picture of this is to take core flow measurements during polymer flooding (Kumar et al., 2020; Reena et al., 2020). Flooding starts the process of polymer adsorption, which takes up a lot of space in the pore. As a result of not being able to get a lot of oil and the decrease in permeability, the recovery factor is lower. Another problem that comes up when polymers flood is when polymer solutions break down. Chemical, mechanical, or biological decomposition can happen. When ionic ions come into contact and then get oxidized, chemical changes can happen. When macromolecules are sheared at a high speed, mechanical damage may happen. In addition, living things break down biopolymers. Depending on the conditions in the reservoir, some polymers can have a wide range of viscosity and volume ratios; other polymers, on the other hand, don't (Azad et al., 2018; Azad and Trivedi, 2020). Polymer retention on the rock surface makes formation damage more difficult by slowing down flow, increasing external viscosity, and making people worry about formation damage, but it doesn't cause formation damage. As a result, it is more difficult to get the polymer that has been retracted off of the rock surface. Another type of damage to the formation occurs when gelatin is filtered out. Experiments show that raising the differential pressure over and over again doesn't completely remove the filtration effect (Teepakakorn et al., 2018).

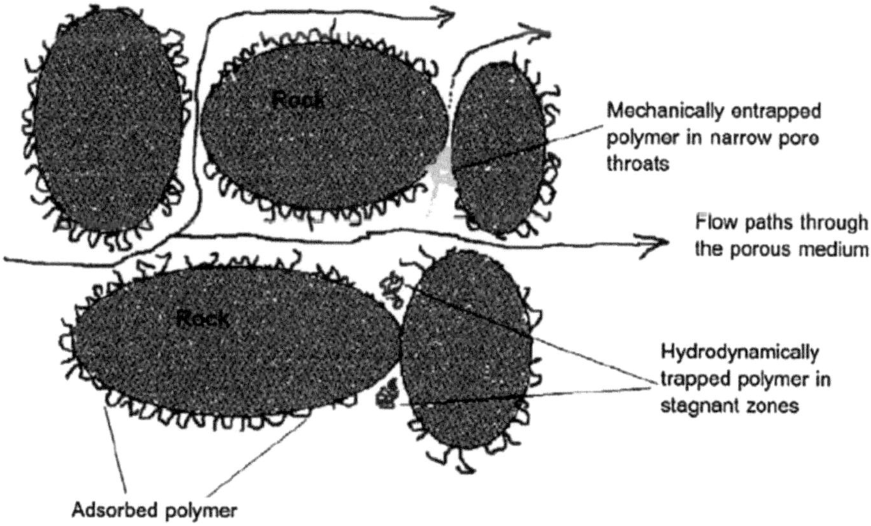

FIGURE 4.4 Polymer flow and retention in porous media, causing formation damage.

Source: Reprinted with permission from (Gbadamosi et al., 2019). (https://creativecommons.org/licenses/by/4.0)

The damage caused by polymer flooding hasn't been talked about as much in the literature, even though polymer adsorption on the surface of the reservoir rock is long-lasting and almost impossible to remove with other methods (Mishra et al., 2014). This is possible because of the irreversible nature of the polymer adsorption process on the surface of the reservoir rock and the near-impossibility of removing it with other methods. This means that the polymer will always get in the way of removal. As a result, the oil that is still in the reservoir after polymer flooding is very hard to get out because the polymer that was deposited permanently reduces the reservoir's permeability. Hydraulic fracturing can sometimes be used to improve injectivity, even though there is a chance that injectivity will be reduced because of higher viscosity and lower permeability. In addition, the polymer that is injected together with water may make it hard to clean because it may form stable emulsions with the oil (Schumi et al., 2020). In a short period of time, it is not possible to get the oil content down to a level that meets the standards for water that can be reinjected (Sheng, 2014, 2011).

4.5 FUTURE SCOPE

Polymers that have recently been synthesized have enhanced resilience to high-salinity and high-temperature conditions. They are appropriate for use in a variety of reservoirs that are subjected to these extreme conditions. To increase their performance, polymers that are resistant to the conditions present in the reservoir are synthesized, or they can be mixed with additional polymers or additives. They are appropriate for use in EOR applications. Advanced synthetic techniques can be used to manufacture a polymer with the properties needed for a specific reservoir (Lai et al., 2016). This can be performed by completing a reservoir analysis prior to flooding (Lv et al., 2015; Sheng, 2011). Reservoirs with variable degrees of permeability require the use of various polymer materials. To improve recovery efficiency and resilience to high salt and high temperatures, polymers can be strengthened with additives such as natural polymers and synthetic or natural surfactants. Furthermore, these compositions change the oil-water mobility ratio, making it easier to find oil in the ground. Combining polymers with synthetic surfactants like TX-100, CTAB, or SDS, as well as bio- and natural surfactants and nanomaterials like silica and titania, may improve oil recovery (Kesarwani et al., 2021; Negin et al., 2016; Raffa et al., 2016). This is being done in both laboratory and field settings (Maghzi et al., 2013).

KEYWORDS

- **polymer flooding**
- **EOR**
- **formation damage**
- **degradation of polymers**
- **polymer retention**
- **polymer adsorption**

REFERENCES

Abidin, A. Z.; Puspasari, T.; Nugroho, W. A. Polymers for Enhanced Oil Recovery Technology. *Procedia Chem.* **2012,** *4*, 11–16. https://doi.org/10.1016/j.proche.2012.06.002

Adewole, J. K.; Muritala, K. B. Some Applications of Natural Polymeric Materials in Oilfield Operations:A Review. *J. Pet. Explor. Prod. Technol.* **2019,** *9*, 2297–2307. https://doi.org/10.1007/S13202-019-0626-9/TABLES/12

Ahmed, S.; Elraies, K. A.; Tan, I. M.; Hashmet, M. R. Experimental Investigation of Associative Polymer Performance for CO2 Foam Enhanced Oil Recovery. *J. Pet. Sci. Eng.* **2017,** *157*, 971–979. https://doi.org/10.1016/j.petrol.2017.08.018

Azad, M. S.; Dalsania, Y.; Trivedi, J. J. Understanding the Flow Behaviour of Copolymer and Associative Polymers in Porous Media Using Extensional Viscosity Characterization: Effect of Hydrophobic Association. *Can. J. Chem. Eng.* **2018,** *96*, 2498–2508. https://doi.org/10.1002/CJCE.23169

Azad, M. S.; Trivedi, J. J. Quantification of the Viscoelastic Effects During Polymer Flooding: A Critical Review. *SPE J.* **2019,** *24*, 2731–2757. https://doi.org/10.2118/195687-PA

Azad, M. S.; Trivedi, J. J. Extensional Effects during Viscoelastic Polymer Flooding: Understanding Unresolved Challenges. *SPE J.* **2020,** *25*, 1827–1841. https://doi.org/10.2118/201112-PA

Barreau, P.; Bertin, H.; Lasseux, D.; Glénat, P.; Zaitoun, A. Water Control in Producing Wells: Influence of an Adsorbed-Polymer Layer on Relative Permeabilities and Capillary Pressure. *SPE Res. Eng.* **1997,** *12*, 234–239. https://doi.org/10.2118/35447-pa

Bera, A.; Kumar, T.; Ojha, K.; Mandal, A. Adsorption of Surfactants on Sand Surface in Enhanced Oil Recovery: Isotherms, Kinetics and Thermodynamic Studies. *Appl. Surf. Sci.* **2013,** *284*, 87–99. https://doi.org/10.1016/j.apsusc.2013.07.029

Berdichevsky, V. L.; Herman, J. N. On Rheology of Cross-Linked Polymers: 1. Slippage of Polymer Chains and its Macroscopic Modeling. *Int. J. Eng. Sci.* **2016,** *100*, 183–198. https://doi.org/10.1016/j.ijengsci.2015.01.003

Chang, H. L.; Zhang, Z. Q.; Wang, Q. M.; Xu, Z. S.; Guo, Z. D.; Sun, H. Q.; Cao, X. L.; Qiao, Q. Advances in Polymer Flooding and Alkaline/Surfactant/Polymer Processes as Developed and Applied in the People's Republic of China. *J. Pet. Technol.* **2014,** *58*, 84–89. https://doi.org/10.2118/89175-jpt

Chaturvedi, K. R.; Fogat, M.; Sharma, T. Low Temperature Rheological Characterization of Single-Step Silica Nanofluids: An Additive in Refrigeration and Gas Hydrate Drilling Applications. *J. Pet. Sci. Eng.* **2021a,** *204*, 108742. https://doi.org/10.1016/j.petrol.2021.108742

Chaturvedi, K. R.; Narukulla, R.; Trivedi, J.; Sharma, T. Effect of Single-Step Silica Nanoparticle on Rheological Characterization of Surfactant Based CO2 Foam for Effective Carbon Utilization in Subsurface Applications. *J. Mol. Liq.* **2021b,** 116905. https://doi.org/10.1016/j.molliq.2021.116905

Chaturvedi, K. R.; Sharma, T. Carbonated Polymeric Nanofluids for Enhanced Oil Recovery from Sandstone Reservoir. *J. Pet. Sci. Eng.* **2020,** *194*, 107499. https://doi.org/10.1016/j.petrol.2020.107499

Clarke, A.; Howe, A. M.; Mitchell, J.; Staniland, J.; Hawkes, L. A. In *How Viscoelastic Polymer Flooding Enhances Displacement Efficiency*, Society of Petroleum Engineers - SPE Asia Pacific Enhanced Oil Recovery Conference, EORC 2015; OnePetro, 2015; pp 926–946. https://doi.org/10.2118/174654-MS

Corredor-Rojas, L. M.; Hemmati-Sarapardeh, A.; Husein, M. M.; Dong, M.; Maini, B. B. Rheological Behavior of Surface Modified Silica Nanoparticles Dispersed in Partially Hydrolyzed Polyacrylamide and Xanthan Gum Solutions: Experimental Measurements, Mechanistic Understanding, and Model Development. *Energy Fuels* **2018,** *32*, 10628–10638. https://doi.org/10.1021/acs.energyfuels.8b02658

Deng, Y.; Dixon, J. B.; White, G. N. Adsorption of Polyacrylamide on Smectite, Illite, and Kaolinite. *Soil Sci. Soc. Am. J.* **2006,** *70*, 297–304. https://doi.org/10.2136/sssaj2005.0200

Dhanarajan, G.; Rangarajan, V.; Bandi, C.; Dixit, A.; Das, S.; Ale, K.; Sen, R. Biosurfactant-Biopolymer Driven Microbial Enhanced Oil Recovery (MEOR) and its Optimization by an ANN-GA Hybrid Technique. *J. Biotechnol.* **2017,** *256*, 46–56. https://doi.org/10.1016/j.jbiotec.2017.05.007

Durucan, S.; Korre, A.; Shi, J. Q.; Govindan, R.; Mosleh, M. H.; Syed, A. The Use of Polymer-Gel Solutions for CO2 Flow Diversion and Mobility Control Within Storage Sites. In *Energy Procedia*; Elsevier Ltd, 2016; pp 450–459. https://doi.org/10.1016/j.egypro.2016.01.046

Elyasi Gomari, K.; Hughes, D.; Islam, M.; Rezaei Gomari, S. Application of Water-Soluble Polymer/Biopolymer Combined with a Biosurfactant in Oil-Wet Fractured Carbonate Reservoirs. *ACS Omega* **2021,** *6*, 15674–15685. https://doi.org/10.1021/ACSOMEGA.1C00855

Fu, L.; Zhang, G.; Ge, J.; Liao, K.; Pei, H.; Jiang, P.; Li, X. Study on Organic Alkali-Surfactant-Polymer Flooding for Enhanced Ordinary Heavy Oil Recovery. *Coll. Surf. A Physicochem. Eng. Asp.* **2016,** *508*, 230–239. https://doi.org/10.1016/j.colsurfa.2016.08.042

Garrouch, A. A.; Gharbi, R. B. An Empirical Investigation of Polymer Flow in Porous Media. *Ind. Eng. Chem. Res.* **1999,** *38*, 3564–3571. https://doi.org/10.1021/ie990121b

Gbadamosi, A. O.; Junin, R.; Manan, M. A.; Agi, A.; Yusuff, A. S. An Overview of Chemical Enhanced Oil Recovery: Recent Advances and Prospects. *Int. Nano Lett.* **2019,** *9*, 171–202. https://doi.org/10.1007/s40089-019-0272-8

Gleasure, R.W.; Phillips, C. R. An Experimental Study of Non-newtonian Polymer Rheology Effects on Oil Recovery and Injectivity. *SPE Res. Eng.* **1990,** *5*, 481–486. https://doi.org/10.2118/17648-pa

Grattoni, C. A.; Luckham, P. F.; Jing, X. D.; Norman, L.; Zimmerman, R. W. Polymers as Relative Permeability Modifiers: Adsorption and the Dynamic Formation of Thick Polyacrylamide Layers. *J. Pet. Sci. Eng.* **2004,** *45*, 233–245. https://doi.org/10.1016/J.PETROL.2004.07.002

Greenland, D. J. Adsorption of Polyvinyl Alcohols by Montmorillonite. *J. Colloid Sci.* **1963,** *18*, 647–664. https://doi.org/10.1016/0095-8522(63)90058-8

Han, X.; Li, C.; Pan, F.; Li, Y.; Feng, Y. A Comparative Study on Enhancing Oil Recovery with Partially Hydrolyzed Polyacrylamide: Emulsion Versus Powder. *Can. J. Chem. Eng.* **2021**. https://doi.org/10.1002/CJCE.24300

Hirasaki, G. J.; Pope, G. A. Analysis of Factors Influencing Mobility and Adsorption in the Flow of Polymer Solution Through Porous Media. *SPE J.* **1974,** *14*, 337–346. https://doi.org/10.2118/4026-pa

Jouenne, S. Polymer Flooding in High Temperature, High Salinity Conditions: Selection of Polymer Type and Polymer Chemistry, Thermal Stability. *J. Pet. Sci. Eng.* **2020,** *195*, 107545. https://doi.org/10.1016/J.PETROL.2020.107545

Jung, J. C.; Zhang, K.; Chon, B. H.; Choi, H. J. Rheology and Polymer Flooding Characteristics of Partially Hydrolyzed Polyacrylamide for Enhanced Heavy Oil Recovery. *J. Appl. Polym. Sci.* **2013,** *127*, 4833–4839. https://doi.org/10.1002/APP.38070

Kang, W.; Cao, C.; Guo, S.; Tang, X.; Lashari, Z. A.; Gao, Y.; Zhang, X.; Iqbal, M. W.; Yang, H. Mechanism of Silica Nanoparticles' Better-Thickening Effect on Amphiphilic Polymers in High Salinity Condition. *J. Mol. Liq.* **2019,** *277*, 254–260. https://doi.org/10.1016/j.molliq.2018.12.092

Kargozarfard, Z.; Riazi, M.; Ayatollahi, S. Viscous Fingering and its Effect on Areal Sweep Efficiency During Waterflooding: An Experimental Study. *Pet. Sci.* **2019,** *16*, 105–116. https://doi.org/10.1007/s12182-018-0258-6

Kesarwani, H.; Saxena, A.; Mandal, A.; Sharma, S. Anionic/Nonionic Surfactant Mixture for Enhanced Oil Recovery through the Investigation of Adsorption, Interfacial, Rheological, and Rock Wetting Characteristics. *Energy Fuels* **2021,** *4*, 3065–3078. https://doi.org/10.1021/acs.energyfuels.0c03767

Kumar, R. S.; Chaturvedi, K. R.; Iglauer, S.; Trivedi, J.; Sharma, T. Impact of Anionic Surfactant on Stability, Viscoelastic Moduli, and Oil Recovery of Silica Nanofluid in Saline Environment. *J. Pet. Sci. Eng.* **2020,** *195*, 107634. https://doi.org/10.1016/j.petrol.2020.107634

Kumar, S.; Mandal, A. A Comprehensive Review on Chemically Enhanced Water Alternating Gas/CO2 (CEWAG) Injection for Enhanced Oil Recovery. *J. Pet. Sci. Eng.* **2017,** *157*, 696–715. https://doi.org/10.1016/j.petrol.2017.07.066

Lai, N.; Zhang, Y.; Xu, Q.; Zhou, N.; Wang, H.; Ye, Z. A Water-Soluble Hyperbranched Copolymer Based on a Dendritic Structure for Low-to-Moderate Permeability Reservoirs. *RSC Adv.* **2016,** *6*, 32586–32597. https://doi.org/10.1039/C6RA06397G

Le Van, S.; Chon, B. H. Chemical Flooding in Heavy-Oil Reservoirs: From Technical Investigation to Optimization Using Response Surface Methodology. *Energies* **2016,** *9*. https://doi.org/10.3390/en9090711

Lewandowska, K. Comparative Studies of Rheological Properties of Polyacrylamide and Partially Hydrolyzed Polyacrylamide Solutions. *J. Appl. Polym. Sci.* **2007,** *103*, 2235–2241. https://doi.org/10.1002/APP.25247

Liang, K.; Han, P.; Chen, Q.; Su, X.; Feng, Y. Comparative Study on Enhancing Oil Recovery under High Temperature and High Salinity: Polysaccharides Versus Synthetic Polymer. *ACS Omega* **2019,** *4*, 10620–10628. https://doi.org/10.1021/acsomega.9b00717

Lv, G.; Li, Q.; Wang, S.; Li, X. Key Techniques of Reservoir Engineering and Injection-Production Process for CO2 Flooding in China's SINOPEC Shengli Oilfield. *J. CO2 Util.* **2015,** *11*, 31–40. https://doi.org/10.1016/j.jcou.2014.12.007

Machale, J.; Majumder, S. K.; Ghosh, P.; Sen, T. K. Role of Chemical Additives and Their Rheological Properties in Enhanced Oil Recovery. *Rev. Chem. Eng.* **2020,** *36*, 789–830. https://doi.org/10.1515/REVCE-2018-0033/MACHINEREADABLECITATION/RIS

Maghzi, A.; Mohebbi, A.; Kharrat, R.; Ghazanfari, M. H. An Experimental Investigation of Silica Nanoparticles Effect on the Rheological Behavior of Polyacrylamide Solution to Enhance Heavy Oil Recovery. *Pet. Sci. Technol.* **2013,** *31*, 500–508. https://doi.org/10.1080/10916466.2010.518191

Majeed, T.; Kamal, M. S.; Solling, T.; Mahboob, A.; Zhou, X. In *Performance Evaluation of Novel Polymers for CO2 Foam Enhanced Oil Recovery*, Abu Dhabi International Petroleum Exhibition and Conference; Abu Dhabi, 2019.

Mishra, S.; Bera, A.; Mandal, A. Effect of Polymer Adsorption on Permeability Reduction in Enhanced Oil Recovery. *J. Pet. Eng.* **2014,** *2014*, 1–9. https://doi.org/10.1155/2014/395857

Mohsenatabar Firozjaii, A.; Saghafi, H. R. Review on Chemical Enhanced Oil Recovery Using Polymer Flooding: Fundamentals, Experimental and Numerical Simulation. *Petroleum* **2020,** *6*, 115–122. https://doi.org/10.1016/J.PETLM.2019.09.003

Nakhjiri, M. T.; Marandi, G. B.; Kurdtabar, M. Poly(AA-co-VPA) Hydrogel Cross-Linked with N-maleyl Chitosan as Dye Adsorbent: Isotherms, Kinetics and Thermodynamic

Investigation. *Int. J. Biol. Macromol.* **2018,** *117*, 152–166. https://doi.org/10.1016/J.IJBIOMAC.2018.05.140

Negin, C.; Ali, S.; Xie, Q. Application of Nanotechnology for Enhancing Oil Recovery – A Review. *Petroleum* **2016,** *2*, 324–333. https://doi.org/10.1016/j.petlm.2016.10.002

Newtonian, N. An Introduction to the Rheology of Polymeric. *Surfactants Polym. Aqueous Solut.* **1930,** 317–335. https://doi.org/10.1002/0470856424.ch15

Pal, S.; Mushtaq, M.; Banat, F.; Al Sumaiti, A. M. Review of Surfactant-Assisted Chemical Enhanced Oil Recovery for Carbonate Reservoirs: Challenges and Future Perspectives. *Pet. Sci.* **2018,** *15*, 77–102. https://doi.org/10.1007/S12182-017-0198-6

Pandey, A.; Sinha, A. S. K.; Chaturvedi, K. R.; Sharma, T. Experimental Investigation on Effect of Reservoir Conditions on Stability and Rheology of Carbon Dioxide Foams of Nonionic Surfactant and Polymer: Implications of Carbon Geo-Storage. *Energy* **2021,** *235*, 121445. https://doi.org/10.1016/J.ENERGY.2021.121445

Pinho De Aguiar, K. L. N.; Frias De Oliveira, P.; Elias Mansur, C. R. A Comprehensive Review of In Situ Polymer Hydrogels for Conformance Control of Oil Reservoirs. *Oil Gas Sci. Technol. Rev. d'IFP Energies Nouv.* **2020,** *75*, 8. https://doi.org/10.2516/OGST/2019067

Pitts, M. J.; Wyatt, K.; Surkalo, H. In Alkaline-Polymer Flooding of the David Pool, Lloydminster Alberta, Proc–eedings of the SPE Symposium on Improved Oil RecoverY; April 1–6, 2004. https://doi.org/10.2118/89386-ms

Polychronopoulos, N. D.; Vlachopoulos, J. Polymer Processing and Rheology. In *Functional Polymers*; Springer: Cham, 2019; pp 133–180. https://doi.org/10.1007/978-3-319-95987-0_4

Pu, W.; Shen, C.; Wei, B.; Yang, Y.; Li, Y. A Comprehensive Review of Polysaccharide Biopolymers for Enhanced Oil Recovery (EOR) from Flask to Field. *J. Ind. Eng. Chem.* **2018,** *61*, 1–11. https://doi.org/10.1016/J.JIEC.2017.12.034

Raffa, P.; Broekhuis, A. A.; Picchioni, F. Polymeric Surfactants for Enhanced Oil Recovery: A Review. *J. Pet. Sci. Eng.* **2016,** *145*, 723–733. https://doi.org/10.1016/j.petrol.2016.07.007

Reena; Kumar, A.; Mahto, V.; Choubey, A. K. Synthesis and Characterization of Cross-Linked Hydrogels Using Polyvinyl Alcohol and Polyvinyl Pyrrolidone and their Blend for Water Shut-Off Treatments. *J. Mol. Liq.* **2020,** *301*, 112472. https://doi.org/10.1016/j.molliq.2020.112472

Rellegadla, S.; Jain, S.; Agrawal, A. A Holistic Approach to Determine the Enhanced Oil Recovery Potential of Hydroxyethylcellulose, Tragacanth Gum and Carboxymethylcellulose. *J. Mol. Liq.* **2021,** *341*. https://doi.org/10.1016/j.molliq.2021.117334

Schumi, B.; Clemens, T.; Wegner, J.; Ganzer, L.; Kaiser, A.; Hincapie, R. E.; Leitenmüller, V. Alkali/Cosolvent/Polymer Flooding of High-TAN Oil: Using Phase Experiments, Micromodels, and Corefloods for Injection-Agent Selection. *SPE Res. Eval. Eng.* **2020,** *23*, 463–478. https://doi.org/10.2118/195504-PA

Sedaghat, M. H.; Mohammadi, H.; Razmi, R. Application of SiO2 and TiO2 Nano Particles to Enhance the Efficiency of Polymer-Surfactant Floods. *Energy Sources A Recovery Util. Environ. Eff.* **2016,** *38*, 22–28. https://doi.org/10.1080/15567036.2012.740552

Seright, R. S.; Seheult, M.; Talashek, T. Injectivity Characteristics of EOR Polymers. *SPE Res. Eval. Eng.* **2009,** *12*, 783–792. https://doi.org/10.2118/115142-PA

Shafiai, S. H.; Gohari, A. Conventional and Electrical EOR Review: The Development Trend of Ultrasonic Application in EOR. *J. Pet. Explor. Prod. Technol.* **2020,** *10*, 2923–2945. https://doi.org/10.1007/s13202-020-00929-x

Sheng, J. J. In *Modern Chemical Enhanced Oil Recovery*; Elsevier Inc., 2011. https://doi.org/10.1016/C2009-0-20241-8

Sheng, J. In *Enhanced Oil Recovery Field Case Studies*; Elsevier Inc., 2013. https://doi.org/10.1016/C2010-0-67974-0

Sheng, J. J. Critical Review of Low-Salinity Waterflooding. *J. Pet. Sci. Eng.* **2014,** *120*, 216–224. https://doi.org/10.1016/j.petrol.2014.05.026

Singh, R.; Mohanty, K. K. Foam Flow in a Layered, Heterogeneous Porous Medium: A Visualization Study. *Fuel* **2017,** *197*, 58–69. https://doi.org/10.1016/j.fuel.2017.02.019

Song, W.; Hatzignatiou, D. G. On the Reduction of the Residual Oil Saturation Through the Injection of Polymer and Nanoparticle Solutions. *J. Pet. Sci. Eng.* **2022,** *208*, 109430. https://doi.org/10.1016/J.PETROL.2021.109430

Sun, F.; Lin, M.; Dong, Z.; Zhang, G. Delayed-Crosslink Hydrogel for Improving Oil Recovery in Differential Heterogeneous Reservoirs. *ACS Omega* **2020,** *5*, 228–235. https://doi.org/10.1021/acsomega.9b02545

Taylor, K. C.; Nasr-El-Din, H. A. In *Hydrophobically Associating Polymers for Oil Field Applications*, Canadian International Petroleum Conference 2007, CIPC 2007; OnePetro, 2007. https://doi.org/10.2118/2007-016

Teepakakorn, A. P.; Hayakawa, T.; Bureekaew, S.; Ogawa, M. Removal of Water-Soluble Polymers from an Aqueous Solution by Adsorption onto an Acidic Clay. *Clays Clay Miner.* **2018,** *66*, 96–103. https://doi.org/10.1346/CCMN.2018.064079

Tekin, N.; Demirbaş, Ö.; Alkan, M. Adsorption of Cationic Polyacrylamide onto Kaolinite. *Microporous Mesoporous Mater.* **2005,** *85*, 340–350. https://doi.org/10.1016/J.MICROMESO.2005.07.004

Tekin, N.; Dinçer, A.; Demirbaş, Ö.; Alkan, M. Adsorption of Cationic Polyacrylamide (C-PAM) on Expanded Perlite. *Appl. Clay Sci.* **2010,** *50*, 125–129. https://doi.org/10.1016/J.CLAY.2010.07.014

Telmadarreie, A.; Trivedi, J. J. Post-Surfactant CO 2 Foam/Polymer-Enhanced Foam Flooding for Heavy Oil Recovery: Pore-Scale Visualization in Fractured Micromodel. *Transp. Porous Media* **2016,** *113*, 717–733. https://doi.org/10.1007/s11242-016-0721-z

Thomas, A. Polymer Flooding. In *Chemical Enhanced Oil Recovery (CEOR) - A Practical Overview*; IntechOpen, 2016. https://doi.org/10.5772/64623

Uranta, K. G.; Rezaei-Gomari, S.; Russell, P.; Hamad, F. Studying the Effectiveness of Polyacrylamide (PAM) Application in Hydrocarbon Reservoirs at Different Operational Conditions. *Energies* **2018,** *11*. https://doi.org/10.3390/en11092201

Urbissinova, T. S.; Trivedi, J. J.; Kuru, E. Effect of Elasticity During Viscoelastic Polymer Flooding: A Possible Mechanism of Increasing the Sweep Efficiency. *J. Can. Pet. Technol.* **2010,** *49*, 49–56. https://doi.org/10.2118/133471-PA

Wang, W.; Liu, Y.; Gu, Y. Application of a Novel Polymer System in Chemical Enchanced Oil Recovery (EOR). *Colloid Polym. Sci.* **2003,** *281*, 1046–1054. https://doi.org/10.1007/s00396-003-0873-6

Wang, Y. Y.; Dai, Y. H.; Zhang, L.; Luo, L.; Chu, Y. P.; Zhao, S.; Li, M. Z.; Wang, E. J.; Yu, J. Y. Hydrophobically Modified Associating Polyacrylamide Solutions: Relaxation Processes and Dilational Properties at the Oil−Water Interface. *Macromolecules* **2004,** *37*, 2930–2937. https://doi.org/10.1021/MA049923V

Wei, B.; Romero-Zerón, L.; Rodrigue, D. Oil Displacement Mechanisms of Viscoelastic Polymers in Enhanced Oil Recovery (EOR): A Review. *J. Pet. Explor. Prod. Technol.* **2014,** *4*, 113–121. https://doi.org/10.1007/S13202-013-0087-5/FIGURES/3

Wu, Z.; Yue, X.; Cheng, T.; Yu, J.; Yang, H. Effect of Viscosity and Interfacial Tension of Surfactant-Polymer Flooding on Oil Recovery in High-Temperature and High-Salinity

Reservoirs. *J. Pet. Explor. Prod. Technol.* **2014,** *4*, 9–16. https://doi.org/10.1007/s13202-013-0078-6

Xia, Q.; Xiao, H.; Pan, Y.; Wang, L. Microrheology, Advances in Methods and Insights. *Adv. Colloid Interface Sci.* **2018,** *257*, 71–85. https://doi.org/10.1016/J.CIS.2018.04.008

Yekeen, N.; Manan, M. A.; Idris, A. K.; Padmanabhan, E.; Junin, R.; Samin, A. M.; Gbadamosi, A. O.; Oguamah, I. A Comprehensive Review of Experimental Studies of Nanoparticles-Stabilized Foam for Enhanced Oil Recovery. *J. Pet. Sci. Eng.* **2018**. https://doi.org/10.1016/j.petrol.2018.01.035

Yin, X.; Kang, W.; Song, S.; Huang, Z.; Hou, X.; Yang, H. Stabilization Mechanism of CO2 Foam Reinforced by Regenerated Cellulose. Colloids Surfaces A Physicochem. *Eng. Asp.* **2018,** *555*, 754–764. https://doi.org/10.1016/j.colsurfa.2018.07.042

Yousefvand, H.; Jafari, A. Enhanced Oil Recovery Using Polymer/Nanosilica. *Procedia Mater. Sci.* **2015,** *11*, 565–570. https://doi.org/10.1016/j.mspro.2015.11.068

Yuan, B.; Wang, W.; Moghanloo, R. G.; Su, Y.; Wang, K.; Jiang, M. Permeability Reduction of Berea Cores Owing to Nanoparticle Adsorption onto the Pore Surface: Mechanistic Modeling and Experimental Work. *Energy Fuels* **2017,** *31*, 795–804. https://doi.org/10.1021/acs.energyfuels.6b02108

Zhang, H.; Feng, Y. Dependence of Intrinsic Viscosity and Molecular Size on Molecular Weight of Partially Hydrolyzed Polyacrylamide. *J. Appl. Polym. Sci.* **2021,** *138*. https://doi.org/10.1002/APP.50850

Zhang, L. M.; Zhou, J. F.; Hui, P. S. A Comparative Study on Viscosity Behavior of Water-Soluble Chemically Modified Guar Gum Derivatives with Different Functional Lateral Groups. *J. Sci. Food Agric.* **2005,** *85*, 2638–2644. https://doi.org/10.1002/jsfa.2308

Zhou, W.; Xin, C.; Chen, S.; Yu, Q.; Wang, K. Polymer-Enhanced Foam Flooding for Improving Heavy Oil Recovery in Thin Reservoirs. *Energy Fuels* **2020,** *34*, 4116–4128. https://doi.org/10.1021/acs.energyfuels.9b04298

CHAPTER 5

Surfactants in EOR

RAMESH NARUKULLA, ALPANA SINGH,
KRISHNA RAGHAV CHATURVEDI, and TUSHAR SHARMA

Enhanced Oil Recovery Laboratory, Rajiv Gandhi Institute of Petroleum Technology, Jais, India

ABSTRACT

From past few decades, there has been a huge interest in exploring various techniques in chemical enhanced oil recovery (CEOR) and it is due to the tremendous increase in energy demand across the globe. Surfactant flooding has marked its footprints as a highly established method among the CEOR methods. Surfactants/surface active agents are the organic compounds comprising of both hydrophobic tail and a hydrophilic head and has the ability to adsorb on fluid–fluid interface in order to alter the rock properties. This chapter delves into the strategic application of surfactant flooding as a pivotal method in CEOR. Investigating the role of surfactants in reducing interfacial tension, altering wettability, and improving displacement efficiency within reservoirs, the chapter provides a comprehensive understanding of the mechanisms driving enhanced recovery. This chapter also provides the detailed understanding of the categories of surfactants, that is, cationic, anionic, nonionic, zwitterionic, amphoteric and cationic-anionic along with their applications in different reservoirs. This is followed by the elaboration of synergy and suitability of different types of surfactants with rock system. A new class of surfactants, that is, natural surfactants (derived from natural origin) along with the application in field of EOR has been highlighted in this chapter. Along with listing the various types of surfactants used in EOR, this chapter also underscores, through extensive literature survey, the potential of surfactant flooding as a valuable tool for optimizing hydrocarbon extraction,

Advancements in Chemical Enhanced Oil Recovery.
Tushar Sharma, Krishna Raghav Chaturvedi, Tarek Ganat, and Imtiaz Ali (Eds.)

fostering reservoir management practices that align with economic and environmental sustainability goals. In spite of the huge acceptance of surfactant flooding as a well-establish method in CEOR, there are some challenges faced during the execution which include the excess adsorption of surfactant on rock surface and the harsh reservoir conditions that make the surfactant unstable. This not only impacts the oil recovery but also reduces the economic strength of the ongoing projects. This chapter also entails about the proper selection of surfactant, depending upon the condition of reservoir as well as the rock type.

5.1 INTRODUCTION

A three-stage technique is used to collect oil from the ground: the primary, secondary, and tertiary stages (enhanced oil recovery [EOR]). Only the first two processes were successful in recovering one-third of the OOIP (original oil in situ); the other two-thirds of the oil in the reservoir is due to trapped oil (residual oil) or bypassed oil due to variations in the mobility of water and oil or heterogeneity in the reservoir (Mungray and Kumar, 2009; Souza et al., 2014). To recover this oil, a tertiary (EOR) method is necessary. Surfactant flooding is a type of chemical flooding that falls within the chemical flooding category. Displacing immiscible oil, which will not mix with water or brine, is a complicated technique that must be completed in a controlled environment. This situation can be changed, on the other hand, by surfactants. Using surfactants in reservoirs to improve oil recovery was first proposed in the early 1990s and has progressed through several stages to optimize its application while taking into consideration a variety of factors such as reservoir condition and cost as well as environmental concerns (Amirianshoja et al., 2013). Surfactants have played an increasingly important role in EOR over the last few years. Surfactants, also known as surface-active agents, are amphiphilic organic compounds that have a hydrophobic tail that is oil-soluble and a hydrophilic head that is water-soluble (water-soluble) (Pal et al., 2018). When used between liquid and liquid, liquid and solid, gas and liquid, or gas and solid surfaces, surfactants can reduce interfacial tension (IFT) and affect the wettability of the surfaces (Sheng, 2013). Because of these characteristics, surfactant injection has sparked a great deal of interest in the subject of increased oil recovery and has been widely used. Surfactants aid in oil recovery by increasing the wettability of the rock, making it more water-soluble, and lowering the interfacial tension (IFT) (between the surfactant and the crude oil), allowing trapped oil to be retrieved from the reservoir. Surfactants are used in a variety of applications, including drilling, drilling mud, and drilling fluid (Sagir et al., 2020).

Although they are ineffective, when the proper surfactant for the reservoir state is used, they can boost oil recovery by as much as 50%. Surfactant selection that is not optimal leads to unfavorable changes in wettability, as well as blockage of reservoir pores and excessive surfactant adsorption on the rock surface, on the other side. The introduction of an adequate amount of surfactant causes the formation of microemulsions with residual oil and reduces the IFT between oil and water, allowing for the conversion of rock wettability from oil to water, overcoming high capillary pressure, and assisting in the mobilization of residual oil from rock pores (oil remaining after water flooding) (see Fig. 5.1). As a result, residual oil saturation decreases, whereas recovery increases as a result of this (Belhaj et al., 2020). A surfactant's dissociation properties in water are classified into five categories: cationic, anionic, nonionic, zwitterionic, amphoteric, and cationic–anionic. Each type of surfactant has distinct constraints depending on the reservoir in which it is used. Especially in the case of sandstone reservoirs, anionic surfactants are beneficial in the extraction of oil from the ground (Paulo et

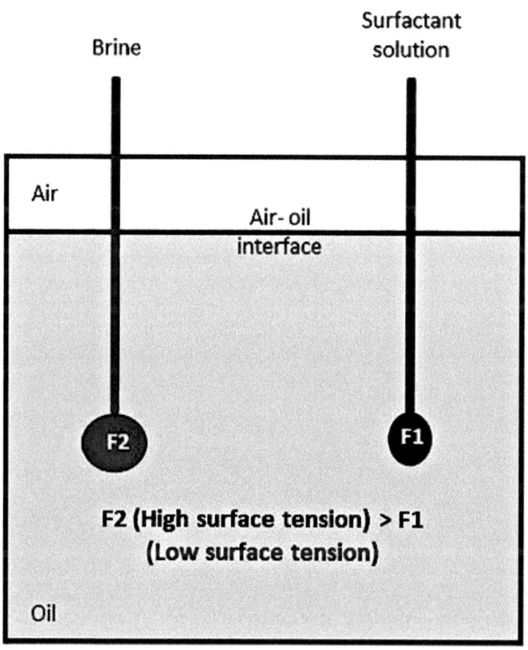

FIGURE 5.1 IFT mechanism of (a) brine and (b) surfactant/brine in oil. Measurements from brine droplet showed that water forms a large spherical droplet inside the oil phase associated with the strong hydrogen bonding, while the surfactant solution forms small oval-shaped droplets with the weaker hydrogen bonds.

al., 2013). Different situations involving carbonate reservoirs revealed that either cationic, nonionic, or a mix of the two worked better than the other types. To pick an appropriate surfactant that will improve oil recovery, it is necessary to understand how wettability fluctuates when surfactant flooding is performed. Multiple ways for improving the wettability of rock surfaces have been discussed in the literature; however, only a few of the research provided an adequate explanation for how the surface wettability was improved by the strategy (Negin et al., 2017). Previous studies have reported the use of cationic surfactants on the wettability of carbonate rock during wet conditions and enhanced oil recovery (EOR), including the effect of cationic surfactants on the wettability of dolomite surface during wet conditions and EOR (Alvarado and Manrique, 2010; Ravera et al., 2006). Scientists asserted that the cationic surfactant C12TAB interacts with dolomite by ionic contact, desorbs stearic acid from its surface in an irreversible manner, and alters its wettability in favor of a more water-wetted condition. The adsorption of TritonX-100 (a nonionic surfactant) on the dolomite surface is accomplished by electron polarization and ion exchange. In response to the hydrophobic interaction, the released stearic acid created a new layer on the rock surface, resulting in a decrease in surface wettability and the formation of a weak water-wet state (Kumar et al., 2021). The anionic surfactant (SDS) rendered the dolomite surface wet and neutral, allowing it to absorb more water. Because of the hydrophobic interaction between the adsorbed acid and the tail of the SDS surfactant, the change in dolomite surface wettability was regulated by a different mechanism. This demonstrates that the ionic nature of the surfactant has an impact on the change in the wettability of the solution (Franco-Aguirre et al., 2018). Most recently, some studies have explored the wettability of oil-wet sandstone by utilizing surfactant combinations that were both cationic and nonionic. They investigated the effect of the synergy of CTAB and TX-100 on the wettability alteration of oil-wet sandstone and discovered that the surfactant mixture reduced the CMC when compared to a single surfactant and successfully altered the wettability to water-wet when compared to CTAB alone, which was previously thought to be impossible (Kesarwani et al., 2021). Extensive work has been performed by Pal and co-workers to investigate the microscopic process by which the wettability of an oil-wet sandstone surface is altered using a synthetic Gemini surfactant (GABEO) (Kumar and Mandal, 2017; Pal et al., 2018; Pal and Mandal, 2020; Samanta et al., 2011; Saxena et al., 2019). In this chapter, we will discuss many surfactant applications in EOR, as well as the microscopic mechanics of surfactants at the atomic level.

5.2 ANIONIC SURFACTANTS

In sandstone reservoirs, anionic surfactants with a negatively charged polar head are good for improving oil recovery since they are non-ionic (EOR) (Fig. 5.2). Since the vast majority of EOR activities are carried out in sandstone reservoirs, which are best suited to anionic surfactants, anionic surfactants are the most often employed surfactants in the industry (Paulo et al., 2013). Sulfonates, sulfates, and carboxylates, to name a few of the more popular anionic surfactants, are available in several different forms. In the presence of high salinity and divalent ions, sulfonate-derived surfactants tend to precipitate, even though they are typically thermally stable. Those containing sulfate, on the other hand, are more resistant to high salinity but are also more unstable when exposed to high temperatures (Karakashev et al., 2011; Negin et al., 2017). The reaction of benzene in the presence of propylene tetramer leads to the formation of alkyl aryl groups, which are then subjected to a further sulfonation method to create alkyl aryl sulfonates. Benzene is also used in the manufacturing of alkyl aryl sulfonates. Salinity effects on EOR were investigated using alkyl ortho-xylene surfactants from the C9, C12, and C15 classes, all of which are members of this family. The highest recovery was found at the salinity appropriate for each surfactant by continuously infusing each surfactant at the salinity appropriate for each surfactant (Hirasaki et al., 2011). These surfactants have hydrophobic qualities, and the longer alkyl chains (C14–C30) form bonds with crude oil, but the shorter alky chains do not. These surfactants, on the other hand, are unstable under high saline conditions or in the presence of divalent ions because they

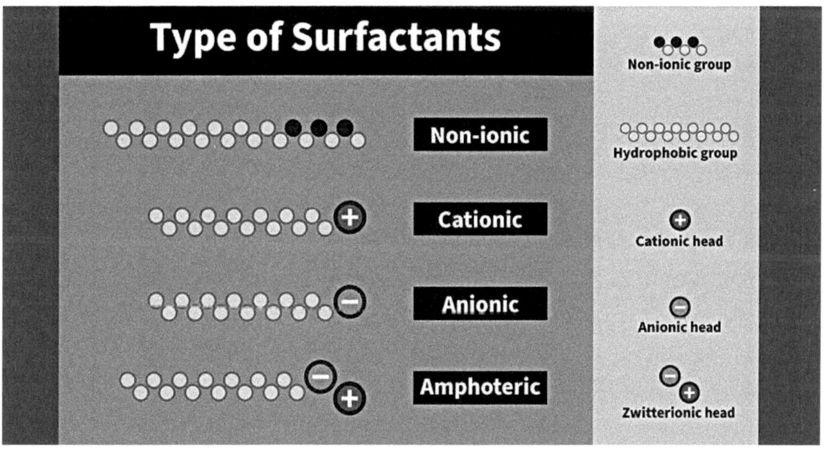

FIGURE 5.2 Various types of surfactants used in enhanced oil recovery investigations.

can be manufactured from low-cost basic ingredients and may be adjusted to match the unique needs of a certain reservoir.

Linear alkylbenzene sulfonate is a member of the alkyl aryl sulfonate family with a negative charge and a hydrophilic group, which comes from benzene sulfonate. It belongs to the alkyl aryl sulfonate family of chemicals. These surfactants are not only detrimental to the environment, but their branched alkyl side chains are also slow biodegradation, making them less efficient. Even though the alkyl groups in these surfactants are branchless, they are hazardous to the environment. It is estimated that 10–14 carbon atoms are present in the alkyl side chain of these surfactants. These surfactants are suitable for use in EOR applications (Liu et al., 2008; Sakthivel et al., 2017). They also observed that increasing the number of carbon atoms in the surfactant's alkyl group led to a decrease in the IFT value between crude oil and water, with better results achieved when the number of carbon atoms in the alkyl group was between 7 and 10. The addition of the sulfonate group in this surfactant boosts the surfactant's temperature resistance by a factor of two. The toxicity of the water increases as the salinity of the water increases due to the surfactant contained in the water's composition while the discovery of increasing surfactant concentration significantly reduced contact angle and IFT value to the smallest possible values up to the CMC value, after which contact angle and IFT became almost constant and increased capillary number with increasing crude oil mobility through porous media. It was feasible to recover up to 70% of the total quantity of original crude oil with this surfactant. Throughout the adsorption process, both linear and branched surfactants adsorbed onto a wet $CaCO_3$ surface; however, linear surfactant did not exhibit any structural alterations, but branched surfactant did show structural variations and form a rearranged layer structure (Sagir et al., 2020). When comparing oil-wet surfaces to branched alkyl surfactants, the interaction of oil-wet surfaces exhibited higher oil desorption with linear surfactant and a considerable change in wettability toward water-wet surfaces. As a result, it is envisaged that these surfactants would perform better in EOR applications. Even though the vast majority of reservoirs operate in high-temperature and high-saline settings, most anionic surfactants are unstable at higher temperatures, disintegrate, and do not work well in high-saline situations. Because of this, the development of anionic surfactants for use in high temperature and high salt environments that are environmentally sustainable is the main priority for improving EOR applications (Sheng, 2011).

5.3 CATIONIC SURFACTANTS

These surfactants have a positively charged hydrophilic head and a hydrophobic alkyl chain. Because of the positive charge on their heads, they may easily attract negatively charged clays, making reservoir rocks wettable. Cationic surfactants have been discovered to successfully recover stranded oil in carbonate reservoirs due to their charge on the surface and little adsorption on calcite mineral surfaces and other carbonate mineral surfaces. In reservoirs, these surfactants are utilized alone or in conjunction with anionic surfactants for EOR applications. Catalytic surfactants for EOR applications include quaternary ammonium salts such as cetyl trimethyl ammonium bromide (CTAB), coco alkyl trimethyl ammonium chloride, cationic Gemini surfactants, dodecyl trimethyl ammonium bromide (DTAB), cetylpyridinium chloride, and ethoxylated alkyl amine (Table 5.1). CTAB is a cationic surfactant also known as hexadecyltrimethylammonium bromide and cetrimonium

TABLE 5.1 Name, Molecular Formula, and Structure of Cationic Surfactants.

Name and molecular formula	Structure
Benzyltrimethyl ammonium bromide (BTAB) $C_{10}H_{16}BrN$	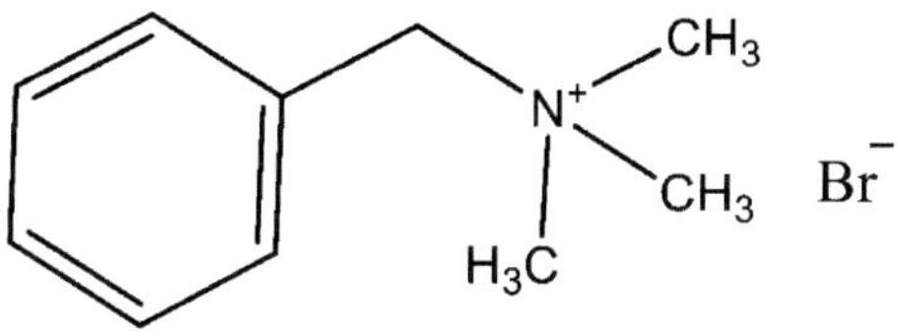
Dodecyltrimethyl ammonium bromide (DTAB) $C_{15}H_{34}BrN$	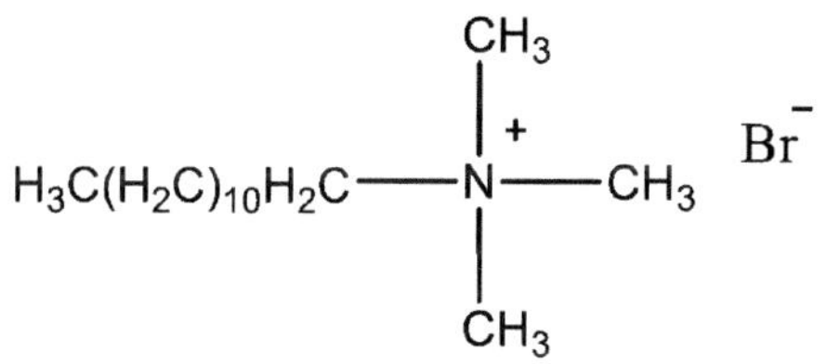
Tetrabutyl ammonium bromide (TBAB) $C_{16}H_{36}BrN$	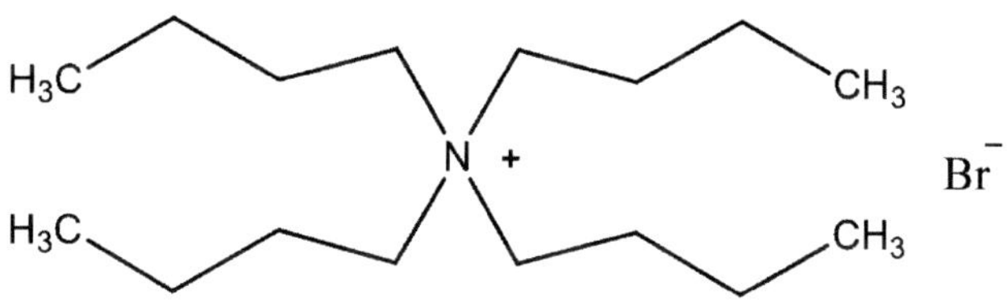
Hexadecyltrimethyl ammonium bromide (CTAB) $C_{19}H_{42}BrN$	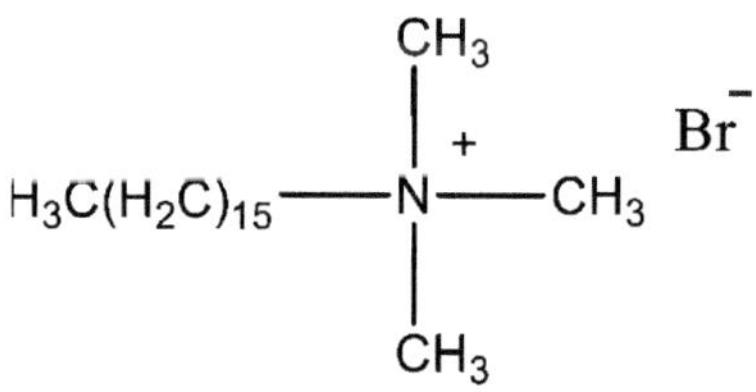

bromide. The IFT decreased with an increase in temperature and increased after reaching a certain point. Increasing the IFT value, on the other hand, does not affect oil recovery, showing that the wettability of the reservoir rock is the key predictor of oil recovery. Studies have also discovered that the presence of salts increased surfactant adsorption and decreased contact angle and that C15TAB affected wettability the fastest of the four surfactants (Fig. 5.3).

Instead of utilizing a single surfactant (anionic or cationic) for wettability modification and oil recovery, a mixture of anionic and cationic surfactants can produce superior outcomes at optimal surfactant concentrations and salinity. Due to the presence of carbonate minerals, clays, and quartz in most reservoirs, a blend of anionic and cationic surfactant compositions may aid in reservoir oil recovery.

5.4 NON-IONIC, GEMINI, AND ZWITTERION SURFACTANTS

Nonionic surfactants are a critical component material in a wide number of applications, ranging from personal care goods to a wide variety of industrial

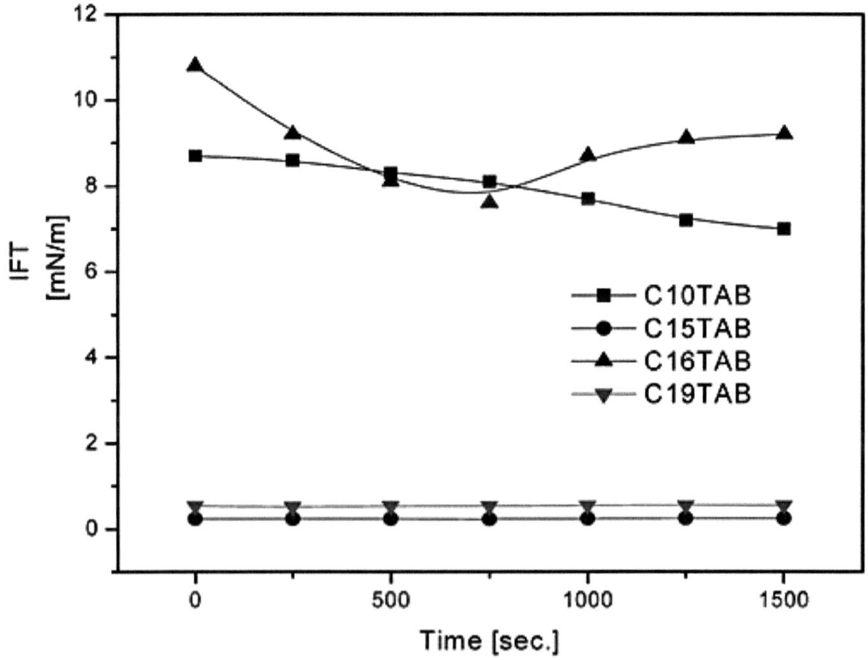

FIGURE 5.3 IFT of different surfactants with an elapse of time.

Source: Reprinted with permission from Kumar et al. (2016). Copyright 2016. American Chemical Society uses (Pandey et al., 2021). They are constituted of uncharged hydrophilic and hydrophobic groups and act as wetting and spreading agents, as well as emulsifiers and foaming agents (Lee et al., 2018; Turro et al., 2002). The hydrophilic component of nonionic surfactants is mostly composed of poly(ethylene oxide), which is generated from petroleum (EO). The manufacturing costs of these commodity chemicals are inextricably linked to the volatility of crude oil prices, offering even another reason to pursue alternative energy sources for production. Due to the increasing demand for EO-based nonionic surfactants, it is anticipated that the future replacement of the present market would be especially beneficial for the high-value utilization of biomass sugars. On a large scale, EO-based surfactants may be difficult to replace, and a market barrier may prevent a new generation of surfactants from establishing a stronger footing in the marketplace. However, we feel the time has come to focus our efforts on high-end applications such as medication emulsification and pharmaceutical formulations that require these technologies and are willing to pay a temporary price premium to achieve them. To summarize, petroleum-based products must be phased out and replaced with novel nonionic surfactants that are highly biodegradable, non-irritating, low in toxicity, and sourced from renewable resources. Existing and innovative sugar-based surfactants will gain popularity in future "green" businesses as their performance qualities improve, experts predict (Negin et al., 2017).

Zwitterionic or amphoteric surfactants are those that have both a positive and a negative charge. As previously stated, these charges may be permanent or conditional on the pH value (Endo et al., 2004; Wu et al., 2020). Typically, the cationic component is an amine or a quaternary ammonium cation, whereas the anionic component is carboxylic, sulfuric, or phosphoric acid, depending on the scenario (or esters thereof). A detailed list of zwitterion chemical surfactants has been provided as under (Rapp, 2016):

- Chemically, carboxylic acid/quaternary ammonium is a compound. Cocamidopropyl betaine or amidosulfobetaine-16, which has a longer hydrophobic tail than cocamidopropyl betaine, or the C80 detergent, which has an alkylphenyl hydrophobic tail, are all examples of beneficial carboxylic acid and quaternary ammonium ion pairings in cleaning solutions. Numerous other carboxylic acid/quaternary ammonium surfactants are used, including lauryl-N,N-(dimethylammonio)butyrate and lauryl-N,N-(dimethyl)-glycinebetaine, both of which are produced from lauryl-N,N-(dimethylammonio)butyrate.

- Sulfuric acid/quaternary ammonium is a sulfuric acid/quaternary ammonium mixture. Three-[(3-cholamidopropyl)dimethylammonio]-1-propanesulfonate is a well-known example of a reaction between sulfuric acid and the quaternary ammonium ion; it is an organic chemical belonging to the hydroxysultaines class. The 3-([3-cholamidopropyl]dimethylammonio)-2-hydroxy-1-propanesulfonate is a slightly modified form of this molecule that is somewhat more hydrophilic in nature than the original 3-[(3-cholamidopropyl)dimethylammonio]-2-hydroxy-1-propane-sulfonate-1-propanesulfonate. Additionally, there is a sulfuric acid equivalent to lauryl-N,N-(dimethylammonio)butyrate known as lauryl-N,N-(dimethyl)-propanesulfonate, which is generated from lauryl-N,N-(dimethylammonio)butyrate. The aromatic quaternary ammonium surfactants 3-(4-tert-butyl-1-pyridinio)-1-propanesulfonate, 3-(1-pyridinio)-1-propanesulfonate, and 3-(benzyldimethylammonio)propanesulfonate are examples. Numerous zwitterionic surfactants, such as surfactin, are commercially marketed.
- Phosphoric acid/quaternary ammonium is a phosphoric acid/quaternary ammonium solution. Numerous phosphoric acid/quaternary ammonium surfactants are being used in addition to carboxylic and sulfuric acids, such as hexadecyl phosphocholine (also known as miltefosine).
- Miscellaneous. In some situations, a quaternary ammonium molecule with an attached hydroxy group can be used as a surfactant to produce zwitterionic surfactants. When attached directly to a nitrogen atom, the hydroxy group becomes significantly more acidic and sensitive to deprotonation. Lauryldimethylamine N-oxide is an excellent representative of this class of surfactants.
- Phospholipids. Phospholipids are a prominent class of chemicals in the realm of zwitterionic surfactants. They consist of a quaternary ammonium head connected to two hydrophobic fatty acids through a phosphate group and glycerol. To be more precise, these specific types of surfactants are responsible for the formation of cell membranes and are thus among the most important surfactants studied in cell biology.

Gemini surfactants represent a new class of surfactants that have been found to have high applicability in field applications (Pal et al., 2019). There are two hydrophilic groups at the top of the list of Gemini surfactants. There are two hydrophobic groups at the bottom of the structure. There is a spacer in front of the head groups or immediately next to them that links them to one another. If all of the hydrophobic portions and hydrophilic groups are the same, then the Gemini surfactant is in the same location and has the same structure. Heterogeminis and symmetric dimeric surfactants are not the same things since they are not symmetric dimeric surfactants. The hydrophobic groups on the two sides are distinct (or the same) on either side. They have a significant impact on how they behave in water as well as the applications for which they are suitable. They have a significant role in this. Hydrophobic and hydrophilic substituents can be found in a wide variety of various substances and situations. The majority of the time, nitrogen atoms in quaternary forms do not have a cyclic structure. Numerous Geminis, on the other hand, have nitrogen present in both the saturated and unsatisfied rings, meaning that they have nitrogen in both locations (Pal et al., 2020; Sekhon, 2013).

5.5 NATURAL SURFACTANTS

Mainly the surfactants used in industries for various purposes are synthetic, that is, they have a chemically based origin. Though synthetic surfactants being chemical based are extensively used in industries but they have some disadvantages, which include their toxicity, non-biodegradable in nature, being associated with high cost and not being environmentally friendly. The evergreen issue of environmental hazards associated with the usage of synthetic surfactants has led researchers to discover a new class of surfactants that comes from a natural origin, and are said to be natural surfactants. Natural surfactants are mainly originated from natural origin. Surfactants are mainly amphiphilic compounds that comprise a hydrophilic head and hydrophobic tail and due to this, they have great solubility in both water and organic solvents. The main role of surfactants is to reduce the interfacial tensions and form emulsions (Saha et al., 2019). In recent times, green synthesis of surfactants has gained huge attention from researchers as the procedure is quite simple, free from harsh chemicals, and apart from all these natural surfactants are known to be proven as a great alternative to chemical-based synthetic surfactants. Major advantages of natural surfactants over synthetic surfactants are as follows:

- Natural surfactants are derived from natural origin, mainly from biomass which also includes various agricultural wastes whereas synthetic surfactants are prepared from various chemicals which are quite hazardous to the environment.
- Natural surfactants being derived from biomass are biodegradable as compared to synthetic surfactants.

- While extracting natural surfactants from natural plant-based products, the use of chemicals is almost diminished that is why natural surfactants are non-toxic.
- As natural surfactants are derived from nature itself, therefore, they are also environmentally friendly.
- As natural surfactants have a natural origin they are abundantly and easily available in nature.
- As compared to synthetic surfactants, natural surfactants are quite cost-effective due to the elimination of the chemical during extraction techniques.

Natural surfactants are broadly classified into two categories, that is, plant-based surfactants and biosurfactants. Plant-based surfactants are the surfactants which are derived from plants whereas biosurfactants are the natural surfactants which are prepared by fermenting the natural substrate, for instance, alkane, sugar, oil, and other waste in the presence of microbes like yeast or bacteria. With the recent advancement in the field of biotechnology, researchers have developed a new alternative to chemical-based synthetic surfactants by giving rise to biosurfactants that are prepared from microorganisms.

Biosurfactants have a wide variety of applications in the food processing industry, cosmetics industry, and microbial-EOR (Md, 2012). Apart from this, biosurfactants have applications in the medicinal field which is due to their anti-microbial, anti-cancer, anti-viral activity, and anti-adhesive agents (Mukherjee et al., 2006). Various types of biosurfactants are produced by the specified microorganism, which includes:

➢ Glycolipids: Glycolipids are very well known biosurfactants that comprise monosaccharides, disaccharides, trisaccharide, and tetra saccharides which contains mannose, glucose, glucuronic acid, galactose, rhamnose, and galactose sulfate.

➢ Rhamnolipids: Rhamnolipids are the crystalline acids that comprise β-hydroxy fatty acid coupled with carboxyl end to rhamnose sugar molecule. They are mainly produced by Pseudomonas aeruginosa which is further categorized as mono-rhamnolipids and di-rhamnolipids. These are quite effective surfactants that are used for eliminating hydrophobic compound from the contaminated soil.

➢ Lipopolysaccharides: Lipopolysaccharides are the main component of bacteria cell which comprise a long chain of sugar groups that are strongly bonded to lipids. They have a great solubility to water and have a higher molecular mass.

- Phospholipids: Phospholipids are the special class of lipids that comprise a glycerol backbone which includes a hydrophilic head carrying a phosphate group and two hydrophobic tails obtained from the chain of fatty acids.

In the field of EOR, surfactants play a major role by reducing interfacial tension at the oil–water interface and recovering more oil to meet the rising demands. In drilling mud, the application of surfactants is quite popular as the surfactants are quite good at emulsifying and wetting (Quintero, 2002) and to avoid various concerns of hazardous environmental impact associated with synthetic surfactants, natural surfactants are gaining enormous attention in the field of EOR and it is due to economic feasibility and a viable option (Singh et al., 2021). A wide variety of natural surfactants have been reported by researchers for their great applicability in the field of EOR by reducing interfacial tension at the oil–water interface, by reducing contact angle and by altering wettability and a combined list of all is mentioned in Table 5.2. Chhetri et al. (2009) prepared a natural surfactant derived from soapnut pericarp which is further used for reducing interfacial tension at the oil–water interface from 19 mN/m to 2.5 mN/m. Ahmadi et al. (2012) developed a novel surfactant derived from the leaves of Glycyrrhiza Glabra and reported that IFT reduction was around 69% in the presence of surfactant solution and kerosene oil. Shahri et al. (2012) formulated a natural surfactant from the leaves of the tree named Zyziphus Spina Christi and reported the reduction of IFT from 48 dynes/cm to 9 dyne/cm with kerosene oil. Ghahfaroki et al. (2015) synthesized the natural surfactants from three new plants that are cationic, derived from leaves of olives, spistan, and prosopis. He reported the IFT reduction from 36.5 to 14 mN/m for olives leaf surfactant, 36.5–20.5 mN/m for spistan leaf surfactant, and 36.5–15.11 mN/m for prosopis leaf surfactant at distilled water and kerosene oil interface. Zhang et al. (2015) highlighted the novel zwitterionic surfactant derived from castor oil which helps in reducing interfacial tension to a very low value of 5.4×10^{-3} mN/m at crude oil and water interface. Rahmati et al. (2015) introduce two new classes of cationic surfactants that are derived from the leaves of mulberry and henna. Both the natural surfactants showed great results in reducing IFT, that is, mulberry leaves surfactant helps to lower from 43.9 to 4.01 mN/m while henna leaves surfactants help to reduce from 43.9 to 3.05 mN/m at distilled water and kerosene oil interface. Saxena et al. (Saxena et al., 2017) developed a new anionic surfactant derived from palm oil which results in

TABLE 5.2 List of Natural Surfactants for EOR Studies.

References	Surfactant	EOR studies			
		IFT range (mN/m)	**Oil recovery**	**Contact angle reduction**	**Reservoir rocks**
Chhetri et al. (2009)	Soapnut pericarp	19–2.5	-----	-----	-----
Ahmadi et al. (2012)	Glycyrrhiza Glabra	33–9	-----	-----	Carbonate
Shahri et al. (2012)	Zyziphus Spina Christi	48–9	-----	-----	Carbonate
Ghahfaroki et al. (2015)	Leaves of olives	36.5–14	-----	-----	-----
Ghahfaroki et al. (2015)	Leaves of Prosopis	36.5–15.11	-----	-----	-----
Ghahfaroki et al. (2015)	Leaves of Spistan	36.5–20.15	-----	-----	-----
Zhang et al. (2015)	Castor oil	5.4×10^{-3}	-----	-----	Quartz sand
Rahmati et al. (2015)	Mulberry leaves	43.9–4.01	-----	62.5?–48.5?	Sandstone
Rahmati et al. (2015)	Henna leaves	43.9–3.05	-----	66?– 37?	Sandstone
Saxena et al. (2017)	Palm oil	1×10^{-3}	Additional 25.8% OOIP	-----	Sandstone
Isari et al. (2018)	Leaves of Eucalyptus	35.2–10.5	-----	140.6?–60.2?	Carbonates
Saxena et al. (2019)	Mahua oil	1×10^{-2}	Additional 20% OOIP	-----	Sandstone

decreases in the IFT to the ultra-low value of 1×10^{-3} mN/m at crude oil and distilled water interface at optimal salinity of 2 wt.% of NaCl. Also, additional oil recovery was reported, that is, 25.8% of original oil in place (OOIP) after conventional water flooding. Isari et al. (Isari et al., 2018) extracted a nonionic surfactant derived from eucalyptus leaves which has shown results by reducing the value of IFT from 35.2 to 10.5 mN/m at oil–water interface and contact angle reduction from 140.6o to 60.2o. Saxena et al. (Saxena et al., 2019) synthesized a new anionic surfactant derived from mahua oil and has resulted in lowering the value of IFT to 10^{-2} mN/m at crude oil–water interface with an optimal salinity of 2% of NaCl. Also, additional oil recovery of 20% OOIP was observed with the surfactant polymer flooding where partially hydrolyzed polyacrylamide (PHPA) was used as polymer after conventional water flooding.

5.6 SHORTCOMINGS OF SURFACTANT EOR

Flooding in an oil reservoir is especially challenging and necessitates the use of more powerful surfactants, the majority of which are expensive. Surfactant loss happens often throughout this flooding process due to the several processes involved, reducing the quantity of surfactant available to mobilize the trapped reservoir oil. Surfactant loss is primarily caused by adsorption on the rock surface, which also involves phase separation and surfactant precipitation. Surfactant adsorption is recognized as a key challenge in EOR due to the high cost of the surfactants employed in the process (Bera et al., 2013). Determine the feasibility of the procedure by calculating the total amount of surfactant required for the whole flooding operation. Another mechanism for surfactant loss is phase entrapment. The process through which surfactants migrate to the oil phase, also known as the microemulsion phase, is referred to as "phase entrapment." Because of the high salinity, temperature, and concentration of divalent ions in the reservoir, these surfactants may interact with the oil phase, resulting in surfactant loss and an inability to attain the required low IFT values. Surfactant effectiveness in fractured reservoirs with dual porosity is limited due to a lack of real-world experience (Sheng, 2017).

Polymers and surfactants are used in a type of chemical-EOR method. It helps to get more oil out of the ground while reducing the amount of oil that comes out with it (SP) (Machale et al., 2020; Raffa et al., 2016). It is possible for surfactants, which have strong capillary forces, to get rid of the oil that has been stuck in the pores of the reservoir. In addition, it lowers the IFT between the displacement oil and water. Polymer, on the other hand, increases the viscosity of the displacing fluid's aqueous phase, which slows down the mobility ratio and delays fingering development caused by flooding with freshwater (Zhang et al., 2006). This polymer-surfactant method is not a stand-alone operation in the reservoir. Instead, it is a synergistic interaction of polymer and surfactant in the reservoir pores that improves oil recovery from blocked reservoir pores. These two requirements must work together, which can be found out by the injection process. Polymer injection order is a term for the order in which the polymer or surfactant is put in. There is no way to avoid excessive polymer adsorption because the polymer is injected first, and the surfactant is injected after. However, water fingering can't be avoided when the surfactant is injected first because of its low viscosity (Chaturvedi et al., 2021). As a result, when polymer surfactant flooding is used, it is important to deal with these problems during the injection process. Because of the processes of diffusion and dispersion, it is important to look into the interaction of polymer and surfactant in porous media to see how they work together. Mixing polymer with a surfactant before injecting it can help reduce water fingering and improve sweep efficiency. However, the salinity of a reservoir affects both the viscosity of the polymer and the concentration of the salt and the type of surfactant used. All of these things affect the sweep efficiency (Wang et al., 2003).

A polymer and a surfactant can work together to help reduce IFT and remove oil from areas in the reservoir where they come into contact with each other (Raffa et al., 2016). A microemulsion is formed when surfactant, water, and oil mix during a flood. In this case, the saltiness of the reservoir affects how it acts. Instabilities can happen when polymer surfactants are used to flood heavy oil reservoirs with polymer surfactants. The behavior of the microemulsion phase, the IFT, and the viscosity difference between the aqueous, oil, and microemulsion phases all play a role in how these instabilities show up. Because the salinity and wettability of the reservoir affect the development of viscous fingers, it is important to know all about viscous instability to flood more polymer surfactants into the ground. Polymer and surfactant could be used to make it easier to sweep in both homogeneous and heterogeneous situations (Sheng, 2011). This shows that, when used with standard recovery methods, polymer surfactant flooding can significantly improve the efficiency of recovery. Many models and tests on the EOR recovery results at the start of the operation have shown that it is very important to figure out where the polymer-surfactant flooding should start at the start of the operation. Theoretically, chemical flooding happens when there is a lot of oil in the water. But this isn't possible because water flooding is cheaper and more important to eliminate uncertainties in porous media; chemical flooding may take longer than the time needed for lab study and preparation, chemical flooding requires technical facilities, is costly for operators, and is difficult to get approval because of environmental concerns. This is why it isn't possible to do this. Water flooding is also cheaper and more important to get rid of uncertainties in porous media.

KEYWORDS

- **anionic**
- **cationic**
- **zwitterion**
- **Gemini**
- **surfactants**

REFERENCES

Ahmadi, M. A.; Zendehboudi, S.; Shafiei, A.; James, L. Nonionic Surfactant for Enhanced Oil Recovery from Carbonates: Adsorption Kinetics and Equilibrium. *Ind. Eng. Chem. Res. 2012, 51*, 9894–9905. https://doi.org/10.1021/ie300269c

Alvarado, V.; Manrique, E. Enhanced Oil Recovery: An Update Review. *Energies* **2010**. https://doi.org/10.3390/en3091529

Amirianshoja, T.; Junin, R.; Kamal Idris, A.; Rahmani, O. A Comparative Study of Surfactant Adsorption by Clay Minerals. *J. Pet. Sci. Eng.* **2013,** *101*, 21–27. https://doi.org/10.1016/j.petrol.2012.10.002

Belhaj, A. F.; Elraies, K. A.; Mahmood, S. M.; Zulkifli, N. N.; Akbari, S.; Hussien, O. S. E. The Effect of Surfactant Concentration, Salinity, Temperature, and pH on Surfactant Adsorption for Chemical Enhanced Oil Recovery: A Review. *J. Pet. Explor. Prod. Technol.* **2020,** *10*, 125–137. https://doi.org/10.1007/s13202-019-0685-y

Bera, A.; Kumar, T.; Ojha, K.; Mandal, A. Adsorption of Surfactants on Sand Surface in Enhanced Oil Recovery: Isotherms, Kinetics and Thermodynamic Studies. *Appl. Surf. Sci.* **2013,** *284*, 87–99. https://doi.org/10.1016/j.apsusc.2013.07.029

Chaturvedi, K. R.; Narukulla, R.; Sharma, T. A Comparative Study of Clay Enriched Polymer Solutions for Effective Carbon Storage and Utilization (CSU) in Saline Reservoirs. *Coll. Polym. Sci.* **2021,** *2021*, 1–13. https://doi.org/10.1007/S00396-021-04868-9

Chhetri, A. B.; Watts, K. C.; Rahman, M. S.; Islam, M. R. Soapnut Extract as a natural surfactant for enhanced oil recovery. *Energy Sources A Recovery Util. Environ. Eff.* **2009,** *31*, 1893–1903. https://doi.org/10.1080/15567030802462622

Endo, T.; Nagai, D.; Monma, T.; Yamaguchi, H.; Ochiai, B. A Novel Construction of a Reversible Fixation–Release System of Carbon Dioxide by Amidines and Their Polymers. *Macromolecules* **2004,** *37*, 2007–2009. https://doi.org/10.1021/MA0305479

Franco-Aguirre, M.; Zabala, R. D.; Lopera, S. H.; Franco, C. A.; Cortés, F. B. Interaction of Anionic Surfactant-Nanoparticles for Gas - Wettability Alteration of Sandstone in Tight Gas-Condensate Reservoirs. *J. Nat. Gas Sci. Eng.* **2018,** *51*, 53–64. https://doi.org/10.1016/j.jngse.2017.12.027

Ghahfarokhi, A. K.; Dadashi, A.; Daryasafar, A.; Moghadasi, J. Feasibility Study of New Natural Leaf-Derived Surfactants on the IFT in an Oil–Aqueous System: Experimental Investigation. *J. Pet. Explor. Prod. Technol.* **2015,** *5*, 375–382. https://doi.org/10.1007/s13202-015-0158-x

Hirasaki, G. J.; Miller, C. A.; Puerto, M. Recent Advances in Surfactant EOR. *SPE J.* **2011,** *16* (04), 889–907. https://doi.org/10.2118/115386-PA

Isari, A. A.; Eslahati, M.; Moradi, S.; Eslahati, A. Feasibility Investigation of A Novel Natural Surfactant Extracted from Eucalyptus Leaves for Enhanced Oil Recovery of Carbonates: Experimental Study. *Pet. Chem. Ind. Int.* **2018,** *1*, 1–5. https://doi.org/10.33140/pcii/00010

Karakashev, S. I.; Ozdemir, O.; Hampton, M. A.; Nguyen, A. V. Formation and Stability of Foams Stabilized by Fine Particles with Similar Size, Contact Angle and Different Shapes. *Coll. Surfaces A Physicochem. Eng. Asp.* **2011,** *382*, 132–138. https://doi.org/10.1016/J.COLSURFA.2010.09.023

Kesarwani, H.; Saxena, A.; Mandal, A.; Sharma, S. Anionic/Nonionic Surfactant Mixture for Enhanced Oil Recovery through the Investigation of Adsorption, Interfacial, Rheological, and Rock Wetting Characteristics. *Energy Fuels* **2021,** *4*, 3065–3078. https://doi.org/10.1021/acs.energyfuels.0c03767

Kumar, R. S.; Goswami, R.; Chaturvedi, K. R.; Sharma, T. Effect of Nanoparticle on Rheological Properties of Surfactant-Based Nanofluid for Effective Carbon Utilization: Capturing and Storage Prospects. *Environ. Sci. Pollut. Res.* **2021,** *2021*, 1–16. https://doi.org/10.1007/S11356-021-14570-6

Kumar, S.; Mandal, A. A Comprehensive Review on Chemically Enhanced Water Alternating Gas/CO2 (CEWAG) Injection for Enhanced Oil Recovery. *J. Pet. Sci. Eng.* **2017,** *157*, 696–715. https://doi.org/10.1016/j.petrol.2017.07.066

Kumar, S.; Panigrahi, P.; Saw, R. K.; Mandal, A. Interfacial Interaction of Cationic Surfactants and Its Effect on Wettability Alteration of Oil-Wet Carbonate Rock. *Energy Fuels* **2016,** *30*, 2846–2857. https://doi.org/10.1021/acs.energyfuels.6b00152

Lee, J.; Zhou, Z. L.; Behrens, S. H. Interfaces Charged by a Nonionic Surfactant. *J. Phys. Chem. B* **2018,** *122*, 6101–6106. https://doi.org/10.1021/ACS.JPCB.8B02853/SUPPL_FILE/JP8B02853_SI_001.PDF

Liu, S.; Zhang, D. L.; Yan, W.; Puerto, M.; Hirasaki, G. J.; Miller, C. A. Favorable Attributes Alkaline-Surfactant-Polymer Flooding. *SPE J.* **2008,** *13*, 5–16. https://doi.org/10.2118/99744-PA

Machale, J.; Majumder, S. K.; Ghosh, P.; Sen, T. K. Role of Chemical Additives and Their Rheological Properties in Enhanced Oil Recovery. *Rev. Chem. Eng.* **2020,** *36*, 789–830. https://doi.org/10.1515/REVCE-2018-0033/MACHINEREADABLECITATION/RIS

Md, F. Biosurfactant: Production and Application. *J. Pet. Environ. Biotechnol.* **2012,** *03*. https://doi.org/10.4172/2157-7463.1000124

Mukherjee, S.; Das, P.; Sen, R. Towards Commercial Production of Microbial Surfactants. *Trends Biotechnol.* **2006,** *24*, 509–515. https://doi.org/10.1016/j.tibtech.2006.09.005

Mungray, A. K.; Kumar, P. Fate of Linear Alkylbenzene Sulfonates in the Environment: A Review. *Int. Biodeterior. Biodegrad.* 2009. https://doi.org/10.1016/j.ibiod.2009.03.012

Negin, C.; Ali, S.; Xie, Q. Most Common Surfactants Employed in Chemical Enhanced Oil Recovery. *Petroleum* **2017,** *3*, 197–211. https://doi.org/10.1016/J.PETLM.2016.11.007

Pal, N.; Kumar, N.; Mandal, A. Stabilization of Dispersed Oil Droplets in Nanoemulsions by Synergistic Effects of the Gemini Surfactant, PHPA Polymer, and Silica Nanoparticle. *Langmuir* , **2019,** *35*, 2655–2667. https://doi.org/10.1021/acs.langmuir.8b03364

Pal, N.; Mandal, A. Enhanced Oil Recovery Performance of Gemini Surfactant-Stabilized Nanoemulsions Functionalized with Partially Hydrolyzed Polymer/Silica Nanoparticles. *Chem. Eng. Sci.* **2020,** *226*, 115887. https://doi.org/10.1016/J.CES.2020.115887

Pal, N.; Saxena, N.; Mandal, A. Characterization of Alkali-Surfactant-Polymer Slugs Using Synthesized Gemini Surfactant for Potential Application in Enhanced Oil Recovery. *J. Pet. Sci. Eng.* **2018,** *168*, 283–300. https://doi.org/10.1016/j.petrol.2018.05.026

Pal, N.; Verma, A.; Ojha, K.; Mandal, A. Nanoparticle-Modified Gemini Surfactant Foams as Efficient Displacing Fluids for Enhanced Oil Recovery. *J. Mol. Liq.* **2020,** *310*, 113193. https://doi.org/10.1016/j.molliq.2020.113193

Pandey, A.; Sinha, A. S. K.; Chaturvedi, K. R.; Sharma, T. Experimental Investigation on Effect of Reservoir Conditions on Stability and Rheology of Carbon Dioxide Foams of Nonionic Surfactant and Polymer: Implications of Carbon Geo-storage. *Energy* **2021,** *235*, 121445. https://doi.org/10.1016/J.ENERGY.2021.121445

Paulo, A. M. S.; Plugge, C. M.; García-Encina, P. A.; Stams, A. J. M. Anaerobic Degradation of Sodium Dodecyl Sulfate (SDS) by Denitrifying Bacteria. *Int. Biodeterior. Biodegrad.* **2013,** *84*, 14–20. https://doi.org/10.1016/j.ibiod.2013.05.027

Quintero, L. An Overview of Surfactant Applications in Drilling Fluids for the Petroleum Industry. *J. Dispers. Sci. Technol.* **2002,** *23*, 393–404. https://doi.org/10.1081/DIS-120003327

Raffa, P.; Broekhuis, A. A.; Picchioni, F. Polymeric Surfactants for Enhanced Oil Recovery: A Review. *J. Pet. Sci. Eng.* **2016,** *145*, 723–733. https://doi.org/10.1016/j.petrol.2016.07.007

Rahmati, M.; Mashayekhi, M.; Songolzadeh, R.; Daryasafar, A. Effect of Natural Leaf-Derived Surfactants on Wettability Alteration and Interfacial Tension Reduction in Water-Oil System: EOR Application. *J. Japan Pet. Inst.* **2015,** *58*, 245–251. https://doi.org/10.1627/jpi.58.245

Rapp, B. E. In *Microfluidics: Modeling, Mechanics and Mathematics*; Elsevier Inc., 2016. https://doi.org/10.1016/C2012-0-02230-2

Ravera, F.; Santini, E.; Loglio, G.; Ferrari, M.; Liggieri, L. Effect of Nanoparticles on the Interfacial Properties of Liquid/Liquid and Liquid/Air Surface Layers. *J. Phys. Chem. B* **2006,** *110*, 19543–19551. https://doi.org/10.1021/jp0636468

Sagir, M.; Mushtaq, M.; Tahir, M. S.; Tahir, M. B.; Shaik, A. R. Surfactants for Enhanced Oil Recovery Applications, In *Surfactants for Enhanced Oil Recovery Applications*, 2020; pp 1–129. https://doi.org/10.1007/978-3-030-18785-9

Saha, R.; Uppaluri, R. V. S.;, Tiwari, P. Impact of Natural Surfactant (Reetha), Polymer (Xanthan Gum), and Silica Nanoparticles to Enhance Heavy Crude Oil Recovery. *Energy Fuels* **2019,** *33*, 4225–4236. https://doi.org/10.1021/acs.energyfuels.9b00790

Sakthivel, S.; Velusamy, S.; Nair, V. C.; Sharma, T.; Sangwai, J. S. Interfacial Tension of Crude Oil-Water System with Imidazolium and Lactam-Based Ionic Liquids and Their Evaluation for Enhanced Oil Recovery Under High Saline Environment. *Fuel* **2017,** *191*, 239–250. https://doi.org/10.1016/J.FUEL.2016.11.064

Samanta, A.; Ojha, K.; Mandal, A. The Characterization of Natural Surfactant and Polymer and Their Use in Enhanced Recovery of Oil. *Pet. Sci. Technol.* **2011,** *29*, 765–777. https://doi.org/10.1080/10916460903485819

Saxena, N.; Pal, N.; Dey, S.; Mandal, A. Characterizations of Surfactant Synthesized from Palm Oil and its Application in Enhanced Oil Recovery. *J. Taiwan Inst. Chem. Eng.* **2017,** *81*, 343–355. https://doi.org/10.1016/j.jtice.2017.09.014

Saxena, N.; Saxena, A.; Mandal, A. Synthesis, Characterization and Enhanced Oil Recovery Potential Analysis Through Simulation of a Natural Anionic Surfactant. *J. Mol. Liq.* **2019,** *282*, 545–556. https://doi.org/10.1016/j.molliq.2019.03.056

Sekhon, B. S. Surfactants: Pharmaceutical and Medicinal Aspects. *J. Pharm. Technol. Res. Manag.* **2013,** *1*, 43–68. https://doi.org/10.15415/jptrm.2013.11004

Shahri, M. P.; Shadizadeh, S. R.; Jamialahmadi, M. Applicability Test of New Surfactant Produced from Zizyphus Spina-Christi Leaves for Enhanced Oil Recovery in Carbonate Reservoirs. *J. Japan Pet. Inst.* 2012, *55*, 27–32. https://doi.org/10.1627/jpi.55.27

Sheng, J. J. In *Modern Chemical Enhanced Oil Recovery*; Elsevier Inc., 2011. https://doi.org/10.1016/C2009-0-20241-8

Sheng, J. In *Enhanced Oil Recovery Field Case Studies*; Elsevier Inc., 2013. https://doi.org/10.1016/C2010-0-67974-0

Sheng, J. J. Critical Review of Alkaline-Polymer Flooding. *J. Pet. Explor. Prod. Technol.* **2017,** *7*, 147–153. https://doi.org/10.1007/s13202-016-0239-5

Singh, A.; Chaturvedi, K. R.; Sharma, T. Natural Surfactant for Sustainable Carbon Utilization in Cleaner Production of Fossil Fuels: Extraction, Characterization and Application Studies. *J. Environ. Chem. Eng.* **2021,** *9*, 106231. https://doi.org/10.1016/j.jece.2021.106231

Souza, E. C.; Vessoni-Penna, T. C.; De Souza Oliveira, R. P. Biosurfactant-Enhanced Hydrocarbon Bioremediation: An Overview. *Int. Biodeterior. Biodegrad.* **2014**. https://doi.org/10.1016/j.ibiod.2014.01.007

Turro, N. J.; Kuo, P. L.; Somasundaran, P.; Wong, K. Surface and Bulk Interactions of Ionic and Nonionic Surfactants. *J. Phys. Chem.* **2002,** *90*, 288–291. https://doi.org/10.1021/J100274A017

Wang, W.; Liu, Y.; Gu, Y. Application of a Novel Polymer System in Chemical Enchanced Oil Recovery (EOR). *Coll. Polym. Sci.* **2003,** *281*, 1046–1054. https://doi.org/10.1007/s00396-003-0873-6

Wu, S.; Liang, F.; Hu, D.; Li, H.; Yang, W.; Zhu, Q. Determining the Critical Micelle Concentration of Surfactants by a Simple and Fast Titration Method. *Anal. Chem.* **2020,** *92*, 4259–4265. https://doi.org/10.1021/acs.analchem.9b04638

Zhang, Q. Q.; Cai, B. X.; Xu, W. J.; Gang, H. Z.; Liu, J. F.; Yang, S. Z.; Mu, B. Z. Novel Zwitterionic Surfactant Derived from Castor Oil and its Performance Evaluation for Oil Recovery. *Coll. Surfaces A Physicochem. Eng. Asp.* **2015,** *483*, 87–95. https://doi.org/10.1016/j.colsurfa.2015.05.060

Zhang, L.; Zhang, D.; Jiang, B. The Rheological Behavior of Salt Tolerant Polyacrylamide Solutions. *Chem. Eng. Technol.* **2006,** *29*, 395–400. https://doi.org/10.1002/ceat.200500306

CHAPTER 6

Binary Combinations for Chemical EOR: Role of Alkalis

TANDRIMA BANERJEE[1] and ABHIJIT SAMANTA[2]

[1]*Department of Chemical Sciences, Indian Institute of Science Education and Research (IISER) Kolkata, West Bengal, India*

[2]*School of Engineering and Applied Sciences, The Neotia University, West Bengal, India*

ABSTRACT

Over time, the energy dependency on petroleum oil has been increased to meet the domestic and industrial necessities. After conventional oil recovery process, a large amount of residual oil is still trapped in the pore spaces of the reservoir rocks due to low permeability, low porosity, and complex geological structures. In this regard, enhanced oil recovery (EOR) techniques have apprehended immense growth as EOR is not only applied for improving the oil recovery but also for enhancing the sweep efficiency. Numerous methods have been used to increase the oil recovery over the past decades with cost-effectiveness. This chapter extensively discussed on the latest breakthroughs in alkali flooding, providing a comprehensive analysis of their mechanisms and effectiveness in enhancing oil recovery. This chapter also entails the applications and effectiveness of different binary mixtures of alkali-surfactant, alkali-polymer and alkali-surfactant-polymer in chemical EOR. Moreover, this study provides a holistic view of the hybrid modes of chemical flooding and also helps to increase the understanding of the current worldwide applications of chemical EOR and future developments will be outlined.

Advancements in Chemical Enhanced Oil Recovery.
Tushar Sharma, Krishna Raghav Chaturvedi, Tarek Ganat, and Imtiaz Ali (Eds.)

6.1 INTRODUCTION

Energy from various fossil fuels like oil, coal, natural gas, etc., has become an important concerning factor of the modern world and global intensive economic growth for past few centuries and petroleum remains the main power source (Okoliegbe and Agarry, 2012; Silva et al., 2014; Najafi-Marghmaleki et al., 2018). The advanced and continuous research on enhanced oil recovery (EOR) started since from the 1980s. Among all EOR methods, the research on alkali–surfactant–polymer (ASP) techniques has been started by Reed and Healy (1977), Pope (1980), and Hirasaki (1981). The effect of surface tension reduction on water flooding has been first implemented by Uren and Fahmy (1927). This thought of oil recovery was not put forward till 1960s due to the limitation of technology. However, the dependency of the increase in displacement efficiency with the reduction of interfacial tension (IFT) has been proved by Wagner and Leach in the year 1966. Subsequently, it has been observed that zero oil saturation has approached if IFT of oil/water reduced to 10^{-3} mN/m (Healy and Reed, 1977). Alkali can also reduce the IFT by forming in situ surfactants by reacting with the acidic components of crude oil. Several studies have been reported about the effects of alkali on reduction of IFT, wettability alteration, mobility of injected chemical slug through the pores of reservoirs, and phase behavior analysis of chemical EOR process. Coreflood experiments have also reported the effects of alkali on the salinity of injected fluid on the recovery efficiency (Nasr-El-Din and Taylor, 1993, 1998; Nasr-El-Din et al., 1992; Torrealba and Johns, 2017; Torrealba et al., 2018a, 2018b). Use of combined alkali–surfactant (AS), alkali–polymer (AP), and alkali/surfactant/polymer (ASP) flooding are effective methods for enhancing conventional oil recovery.

6.2 MECHANISMS OF USE OF ALKALI IN EOR

In EOR methods, the ultra-low interfacial tension between the trapped oil and the injection fluid has achieved through use of a very low concentration of the surfactant. The ultra-low interfacial tension can also attain due to the presence of alkali in the injection fluid which can easily penetrate deeply into the trapped oil globules. Alkali saponifies the naphthenic acids in the crude oil to form in situ surfactant (sodium naphthenate), continuously providing ultra-low IFT and freeing the trapped oil. Sometimes, formation of in situ surfactant may reduce the quantity of injected surfactant during surfactant flooding due to addition of alkali. Though surfactant formation is the key mechanism, the reduction in surfactant adsorption due to high pH and improvement of behavior of injected synthetic and in situ surfactant are possibly even more important mechanisms. A large injection volume of high concentrations solutions or low concentrations solution of a weak alkali (pH < 9) is required for effective oil recovery.

In chemical EOR technique, polymer is used to increase the viscosity of the injection fluid, to minimize channeling, and provide mobility control. A schematic diagram of ASP flooding process is shown in Figure 6.1, which clearly shows the different steps of the process, namely: a preflush of brine to lower the salinity of the reservoir, an alkaline–surfactant solution used to reduce the interfacial tension between the aqueous and oleic phases, a polymer solution to perform a uniform sweep of the oil and the previous slugs, and then freshwater buffer to finally drive the oil and the chemicals to the producer wells. Finally, drive fluid (water) is used to move the front toward the producing well.

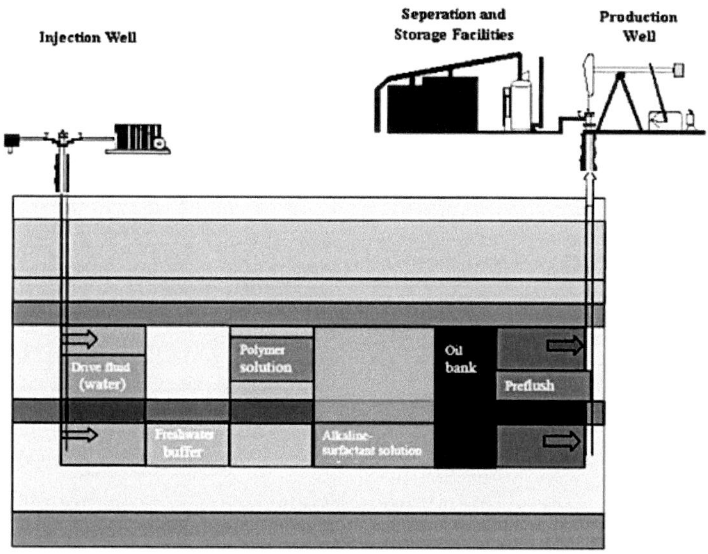

FIGURE 6.1 Schematic diagram of ASP flooding technique.

The oil recovery (N_P) in cEOR can be represented by the following eq 6.1:

$$N_P = E_D E_A E_{VI} \left(\frac{S_O V_P}{B_O} \right) \tag{6.1}$$

Where E_D is the pore to pore displacement efficiency (microscopic efficiency) improved with alkali and surfactant; E_A is the areal displacement efficiency improved with polymer (Kamal et al., 2013, 2015; Saboorian-Jooybari et al., 2016); E_{VI} is the vertical displacement efficiency improved with polymer; both the E_A and E_{VI} are known as macroscopic efficiencies (Samanta et al., 2010; Wang et al., 2010; Samanta et al., 2012; Bera et al., 2013, 2014; Olajire 2014; Zhu et al., 2014) and $\left(\frac{S_O V_P}{B_O}\right)$ is the stock tank oil saturation.

Oil that comes into contact with water during water flooding (secondary recovery) is not possible to displace because of trapping of oil by capillary forces. The capillary forces and the viscous forces are relates by a dimensionless capillary number eq 6.2 (Chatzis and Morrow, 1984; Ayirala and Rao, 2004).

$$\text{Capillary number} = \frac{\mu v}{\gamma \cos\theta} \tag{6.2}$$

Where μ = the viscosity of the aqueous phase, v = the velocity, γ = the IFT between oil and water, and θ = the contact angle.

The pore displacement efficiency can be improved by reducing the capillary forces and the oil/water IFT. The oil recovery and residual oil are directly dependent on capillary number (from 10^{-7} to 10^{-6} for conventional brine injection). It has been reported that if the capillary number increases between 10^{-4} and 10^{-3}, the oil saturation reduces to 90% (Howe et al., 2015), and when the capillary number reaches 10^{-2}, the residual oil saturation approaches to zero (Hou et al., 2005). Application of surfactants with brine during cEOR process can reduce the IFT from value of 20–30 mN/m to values in the range of 10^{-2} to 10^{-3} to achieve sufficient oil recovery (Spildo et al., 2012; Song et al., 2016; Pal et al., 2016). Figure 6.2 shows a schematic diagram of microscopic and macroscopic sweep efficiencies of trapped the oil droplet inside the reservoir pore.

6.3 ALKALI FLOODING

Alkali compounds are basic and ionic salts of alkaline metal or alkaline earth metal element. It produces hydroxide ions (OH^-) in water or alcohol. The alkalies are considered for EOR which can generate high pH. Commonly used alkali by oil industry includes sodium hydroxide, sodium carbonate, sodium silicate, sodium phosphate, ammonium hydroxide, etc. These alkalis

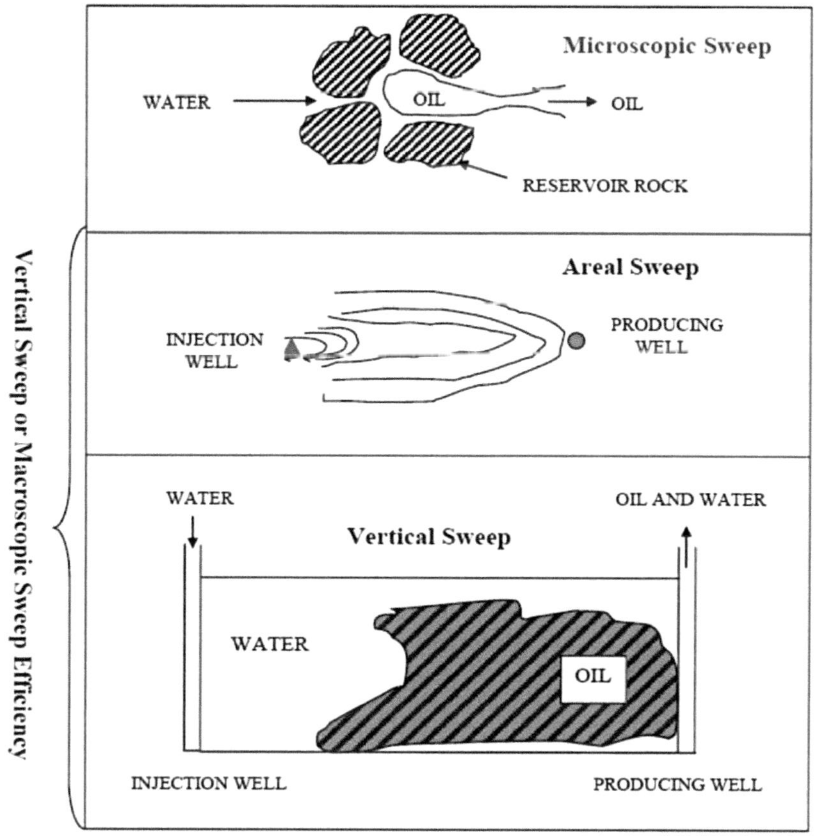

FIGURE 6.2 Simplistic diagram of microscopic and macroscopic sweep efficiencies (Olajire, 2014). Copyright © 2014 Elsevier Ltd.

Source: Reprinted with permission from Olajire, 2014. Copyright © 2014 Elsevier Ltd

react with the acidic groups of the oil and form in situ surfactant. Normally, 0.2–5% concentrations of alkali and 10.5% of pore volume slug sizes are injected during alkali flooding in EOR (Lyons and Plisga, 2011). The weaker base like sodium carbonates and silicates can improve the waterflooding performance by adding synthetic surfactants in alkaline solution.

Oil recovery mechanisms in alkali flooding are complicated and there is a divergence of opinion on the principles. At least eight postulated recovery mechanisms are reported in the literature (deZabala et al., 1982; Ramakrishnan and Wasan, 1983). These include emulsification with coalesence, alteration of wettability, swelling of oil-phase, low interfacial tensions, and disruption of rigid films. The existence of diverse mechanisms should be featured by the crude oil character and the reservoir rock properties. The crude oils from different reservoirs can lead to extensively dissimilar behaviors under disparate environmental conditions such as pH, temperature, salinity, and hardness. Though, all the researchers agree on the fact that the acidic components in the crude oil are most important factor for alkali flooding.

The alkali flooding can be differentiated from other recovery techniques on the basis of in situ saponification of crude oil. Soap production from the acidic part of a crude oil and the injected alkali is the principal mechanism of alkali flooding. Formation of surface active materials extensively depends on the presence of amount of carboxylic acids and to a lesser amount as asphaltene or porphyrins and carboxyphenols fractions (Gao and Towler, 2011). The acidic components of oil are measured and referred to as acid number. The acid number of petroleum oil is one of the most important factors for alkali flooding and it characterizes by the amount of generated natural soap with alkali addition. Acid number is defined as the amount of potassium hydroxide (KOH) in milligrams is required to neutralize the acids present in one gram of crude oil (deZabala et al., 1982). However, the acid number may not be a good indicator of the acid components present in crude oil that generate soap. Alkali also reduces the amount of surfactant needed in two ways: (1) alkali reduces surfactant adsorption by making it preferentially water-wet and (2) alkali reacts with the acids in the crude oils to produce in situ soaps.

Some new protocols have been discussed for measurements of acid number which reflect the soap production in a better way (Green and Willhite, 1998; Satter et al., 2008). The soap formation is modeled by partitioning of acidic part of the crude oil (HA_O) to water (HA_W) according to the following solubility as:

$$HA_O \overset{K_D}{\Leftrightarrow} HA_W \tag{6.3}$$

Where, K_D is the partition coefficient of HA between oil and water.

The acid is then dissociated in the aqueous phase to produce soluble anionic surfactant (A^-), which is referred as soap as express below:

$$HA_W \overset{K_a}{\Leftrightarrow} H^+ + A^- \tag{6.4}$$

Where, K_a is the reaction constant following the equilibrium relationship:

$$K_a = \frac{[H^+][A^-]}{[HA_W]} \tag{6.5}$$

This reaction is one of the sources of alkali consumption since the alkali uses OH^- to generate soap by the following reaction:

$$HA_w + OH^- \overset{K_b}{\Leftrightarrow} H_2O + A^- \tag{6.6}$$

According to eq 6.6, this result increases with $[A^-]$. The net effect of all the reaction is shown in Figure 6.3.

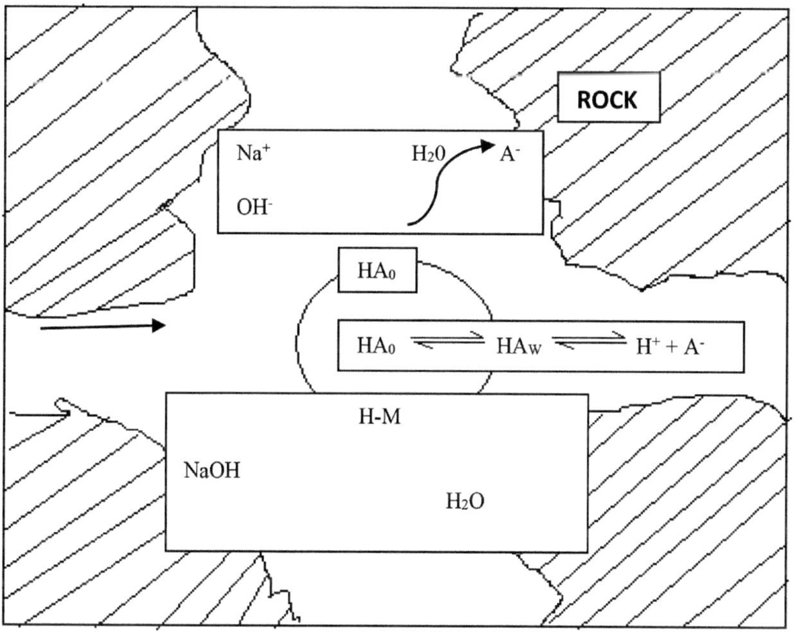

FIGURE 6.3 Simplistic diagram of interaction between alkali and petroleum acid.

Source: Reprinted with permission from Samanta et al., 2011). Copyright © 2010 American Chemical Society

Alkali can interact with reservoir rock in several ways depending on the rock mineralogy, which may include surface exchange and hydrolysis, congruent and incongruent dissolution reactions, and insoluble salt formation by reaction with hardness ions in the fluid and those exchanged from rock surface. The reversible exchange between sodium and hydrogen-base is a very important mechanism of alkali consumption among alkali–rock (clay) interactions and cannot be neglected.

$$MH + Na^+ + OH^- \leftrightarrow MNa + H_2O \tag{6.7}$$

Where M denotes a mineral-base exchange site.

A typical IR spectrum of the crude oil and alkali treated crude oil are shown in Figure 6.4. From the figure, it has been illustrated that the strong absorption peaks at 2923 and 2852 cm^{-1} are due to C–H stretching and the peak at 1460 cm^{-1} and 1376 cm^{-1} are due to C–H deformation and C–H symmetric deformation, respectively. The absorption peak of long chain alkyl groups, $(CH_2)_n$, $n > 4$ is observed at 722 cm^{-1}. The region between 1800 and 1640 cm^{-1} indicates the presence of carbonyl groups, such as carboxylic acids. The vibration of the carboxylic acid group presents in the crude oil corresponds to 1643 cm^{-1}. The absorption at wave number range 3700–3100 cm^{-1} indicates the presence of small amount of phenolic functional groups. The presence of C=C ring stretching corresponds to is weak absorption bands at 1590–1470 cm^{-1}. It has been observed from the figure that the IR spectrum of the alkali-treated crude oil gives almost same peak as that of crude oil (Samanta et al., 2011). It may be concluded that the presence of long-chain carboxylic acids is responsible for formation of in situ surfactants.

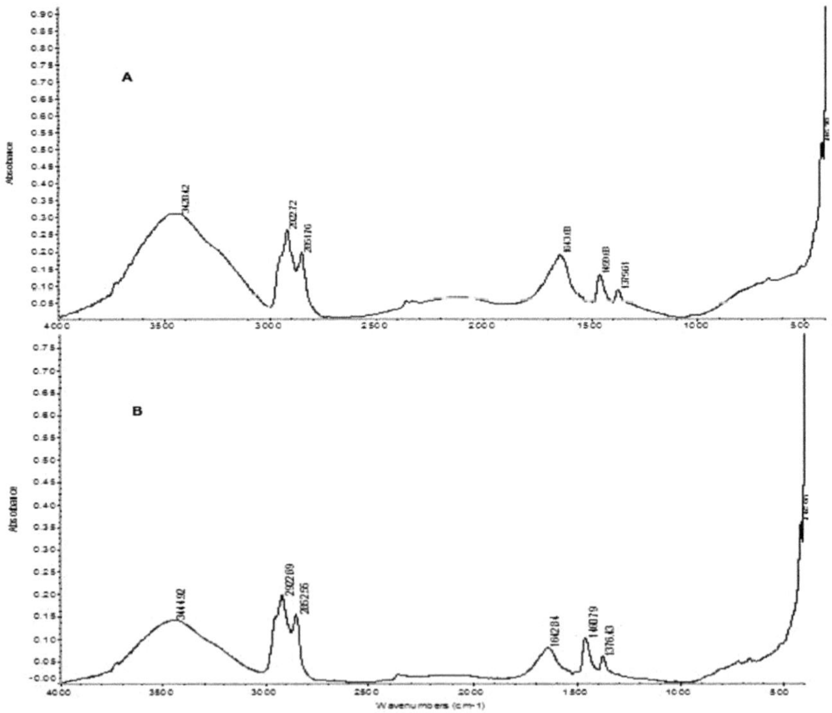

FIGURE 6.4 FTIR spectrum of (A) Crude Oil and (B) after interaction with 1.0% alkali.

Source: Reprinted with permission from Samanta et al., 2011). Copyright © 2010 American Chemical Society

Application of alkali flooding as an EOR process has been investigated in various laboratory experiments by many investigators. In the early 1920's, the injection of alkali was proposed for oil recovery (Nutting, 1927). Later in the year 1927, Atkinson has proposed that residual oil is trapped within the pores of sands grains by capillary forces and adhesion in conjunction with oil viscosity and these oil droplets can be released by injection of alkali solutions (Atkinson, 1927). Mungan (1966) came out with experimental observations that the wettability and caustic flooding have effects on the additional oil recovery. Later on McAuliffe (1973a, 1973b) has also provided evidence that injection of oil-in-water emulsions can improve primary oil recovery in very much the same way as emulsions generated in situ. Cooke et al. (1974) reported a third mechanism by which caustic NaOH could advance the oil recovery after waterflooding. Some crude oils and porous media are converted from water-wet to oil-wet under proper conditions of pH, salinity, and temperature. A fourth mechanism was proposed by Jennings et al. (1974) in which caustic injection can improve oil recovery. Their laboratory experiments showed that if interfacial tension were low enough, residual oil in a preferentially water-wet core could be emulsified in situ. Subsequently, it could move downstream with the flowing caustic, and could be entrapped again by pore throats that are too small for the oil emulsion droplets to penetrate. A laboratory experiment on caustic floods of a viscous Lloyminster crude oil has observed that stable oil-in-water emulsions were produced in situ and that there was evidence for reduced water mobility during displacement (Dranchuk et al., 1974). Lieu et al. (1985) studied the chemical reactions with reservoir matrix in the recovery of petroleum by alkali flooding. They observed that an important initial combination of rapid reversible adsorption and chemical reactions followed by slower non-reversible chemical reactions that can later assume great importance because of the long contact time that may prevail in flooding a petroleum reservoir. The effectiveness of alkali, such as NaOH, sodium orthosilicate, and Na_2CO_3 has been studied to recover residual oil during hot waterfloods (Mehdizadeh and Handy, 1989). The caustic flooding experiments for heavy oil recovery have been performed using NaOH as the alkaline agent, and Wainright (408 mPa.s) and Horsfely (18 mPa.s) as crude oils (Symonds et al., 1991). The effect of injection rate and loss of caustic to the porous medium found that performance in all cases was sensitive to injection rate, and loss of caustic to the rock and fluids were significant. Hornof et al. (1994) worked on the effect of flooding sequence on the displacement of acidic oil by alkaline solutions. Paraffin oil doped with linoleic acid served as acidic oil while NaOH solutions of different concentrations were used as the aqueous flooding agents. They tried two different flooding strategies; firstly the injection of 5 PV alkali flooding followed by injection of 1 PV water and secondly the continuous alkali flooding. They observed that the oil recovery increased with increasing NaOH concentration in the first strategy while in the second one, the oil recovery decreased with increasing NaOH concentration. The feasibility of employing various alkaline agents (NaOH, sodium orthosilicate, and sodium metasilicate) has been investigated to enhance the recovery of specific Saskatchewan heavy oil from the Kindersley region (Chiwetelu et al., 1994). The dynamic IFT of the alkali solutions with the crude oil at different concentrations and temperatures is an important parameter to measure the formulations active surfactants and their efficiencies in oil recovery. Charterjee et al. (1998) measured ultralow dynamic IFT with surfactant-enhanced alkaline systems by using a modified technique parallel to the sessile drop principal and found an ultralow IFT in alkali/acidic oil systems at an optimal pH. The intermittent alkali flooding can significantly enhance the oil recovery in oil-wet carbonate reservoirs by reducing the interfacial tension between the reservoir fluids and by reversing the wettability to a more favorable condition (Bortolotti et al., 2009). A simulation technique has been developed to simulate and match the experimentally measured pressure drop and cumulative oil production during the alkali flooding processes by incorporating both the measured W/O emulsion viscosity and relative permeability (Arhuoma et al., 2009). $NaBO_2$ is also used as alkali instead of Na_2CO_3 at pH values of about 11 at 1 wt% alkali concentration to produce soap for acidic crude oils. This alkali can tolerate divalent cations; hence it can be used in carbine reservoir. Sodium metaborate is also applied in carbonate rocks. Another alkali $NaHCO_3$ is preferred in reservoir containing clays, although most commonly used alkali is Na_2CO_3 because it is economical and transport better in porous media.

An experimental study shows that the cumulative oil recovery by alkali flooding is higher in case of higher concentration of alkali solution due to significant changes in physicochemical and petro-physical properties (Fig. 6.5). The additional recovery of oil over water flooding is 13.88%, 14.77%, and 15.27% when 0.5%, 0.7%, and 1.0% alkali slug was used, respectively (Samanta et al., 2011).

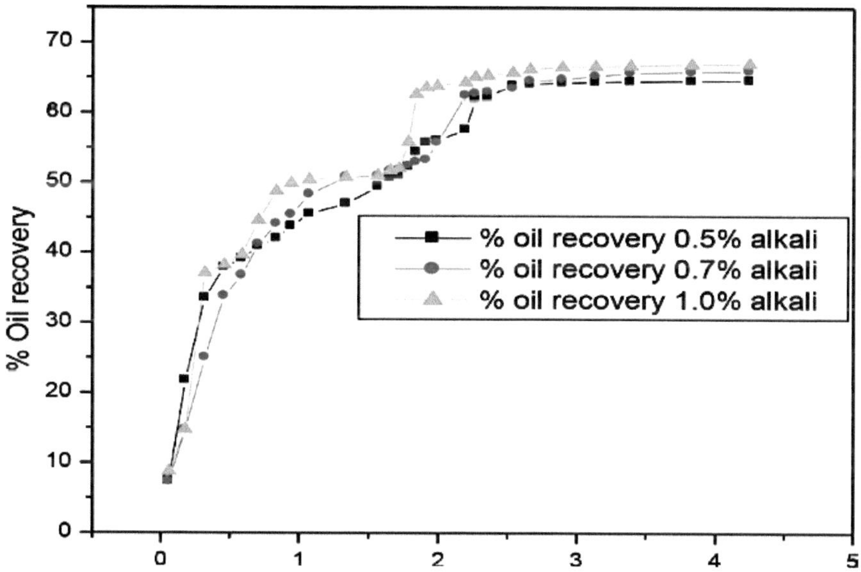

FIGURE 6.5 Cumulative oil recovery of NaOH flooding.

Source: Reprinted with permission from Samanta et al., 2011). Copyright © 2010 American Chemical Society

6.4 ALKALI–SURFACTANT FLOODING

Alkali–surfactant (AS) flooding is already an established technique in conventional oil reservoirs, whereby EOR is a result of reduced traping oil due to the lowered oil/water IFT. In addition, the injection of these chemicals may lead to the formation of emulsions. In the 1950's, the combination of alkali and surfactant injection was first discussed (Doscher, 1956) and since then alkali and/or surfactant injection has become an accepted method over many conventional recovery process in EOR.

Surfactant has both hydrophilic and hydrophobic group. Usually, an injected surfactant is more hydrophilic in nature than a soap. Addition of alkali in surfactant solution can also act as an electrolyte in the binary mixture. The interaction between AS and provided salinity is very complicated. The relationship of optimum salinity of mixing is given by eq 6.8.

$$\ln(C_{se,\mathrm{m}}^{opt}) = X_1 \ln(C_{se,1}^{opt}) + X_2 \ln(C_{se,2}^{opt}) \tag{6.8}$$

Where $C_{se,\mathrm{m}}^{opt}$, $C_{se,1}^{opt}$ and $C_{se,2}^{opt}$ are the optimum salinities of the mixture and surfactant components 1 and 2, respectively; X_1 and X_2 are the mole fractions of surfactants 1 and 2, respectively.

The optimum solubilization ratios both logarithmic and linear mixing is expressed as (Mohammadi, 2008):

$$R^{opt}_{\text{sol,m}} = X_1 R^{opt}_{\text{sol,1}} + X_2 R^{opt}_{\text{sol,2}}$$

or

$$\ln(R^{opt}_{\text{sol,m}}) = X_1 \ln(R^{opt}_{\text{sol,1}}) + X_2 \ln(R^{opt}_{\text{sol,2}}) \tag{6.9}$$

Where $R^{opt}_{sol,\text{m}}$, $R^{opt}_{sol,1}$ and $R^{opt}_{sol,2}$ are the optimum solubilization ratio of the mixture and surfactant components 1 and 2, respectively.

However, a very high concentration of alkali is required to saponify the naphthenic acid in the crude oil, but a system containing higher oil can generate more soap. Therefore, the phase behavior of these surfactants system depends on the oil/water ratio (WOR). Hence, the plot the optimum salinity versus the mole ratio of the soap to the surfactant is independent of WOR (Zhang et al., 2006).

Several studies have been observed that low IFT between a crude oil and an alkali solution is similar to conventional oil (Thomas et al., 2001; Pitts et al., 2004). On the other hand, the formation of water-in-oil (W/O) emulsions has also been recommended as a promising recovery method (Liu et al., 2006; Mai et al., 2006; Bryan and Kantzas, 2007). Therefore, AS system has synergetic effect on the phase behavior, IFT, surfactant adsorption, emulsion, and wettability.

An optimum salinity of AS system of an alkali-generated soap solution and surfactant is required to achieve lowest IFT. To get minimum IFT, the required alkali concentration of an AS mixture becomes higher than that of the initial optimum salinity of the soap solution, according to a mixing rule eq 6.1, because the optimum salinity of a synthetic surfactant is generally higher than that of a soap. Addition of a synthetic surfactant, such as NEODOL 25-3S to 1.55% $Na_2O.SiO_2$ alkali solution with the Gulf Coast crude oil can lead to the optimum type III microemulsion as shown in Figure 6.6 (Nelson et al., 1984). Microemulsion Type II and I regions are shown in the above and below the shaded areas, respectively. It has been found that the addition of NEODOL 25-3S to the alkali-oil system, the optimum salinity of a new system becomes above the 1.55% $Na_2O.SiO_2$ solution due to the higher optimum salinity of the surfactant. To make this system at an optimum condition (type III), addition of more salt like NaCl is required to the system. Therefore, the phase behavior of original alkali-oil system can be improved by addition of synthetic surfactant. Sometimes addition of nonionic surfactant, for example, BD-185, can improve the salinity-resistant potential of anionic surfactant petroleum sulfonate in AS system (Liu et al., 1995; Zhou et al., 2009).

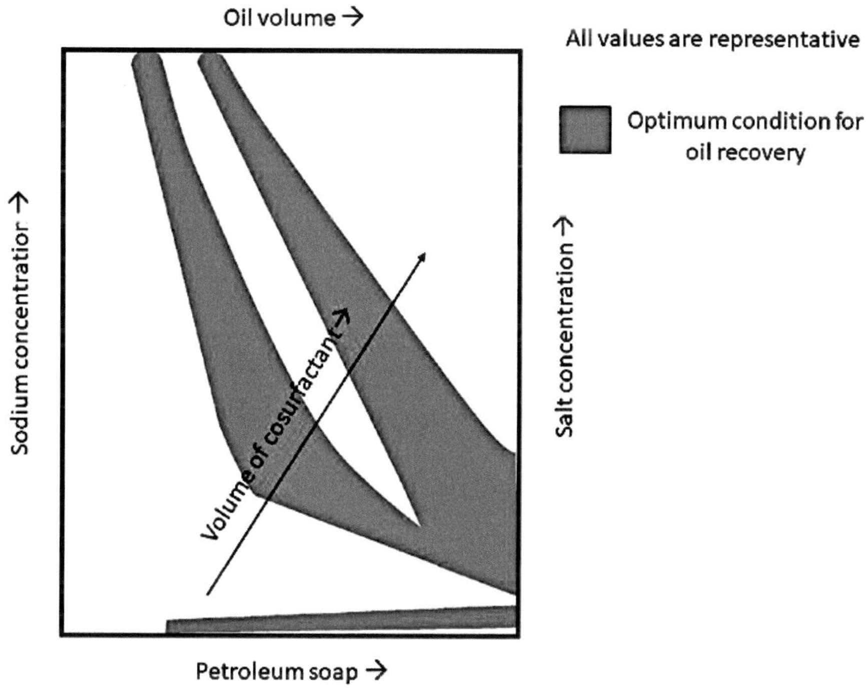

FIGURE 6.6 Activity maps for a AS system with crude oil

Source: Reprinted with permission from (Nelson et al., 1984)

Lowering of IFT is an example of soap-surfactant synergy. Figure 6.7 shows the IFT between crude oil (Yates oil) and the microemulsion from 0.2% mixture of NEODOL 67-7PO sulfate and internal olefin sulfonate 15–18 (4:1 by weight) (Liu et al., 2007). It has been observed that the salt concentration range for the low IFT region ($<10^{-2}$ mN/m) becomes much wider when 1% sodium carbonate was added to a surfactant system. This soap-surfactant synergy reveals the AS system more robust in useful applications in oil recovery. Addition of sodium carbonate can form microemulsion quickly and such synergy effect can enhance with a combination of NaOH and Na_2CO_3 to surfactant solution (Liu et al., 2006; Jackson 2006; Pei et al., 2012). The IFTs between Bohai oil and different alkalis such as NaOH, ethylene-diamine, and Na_2CO_3 solutions are shown in Figure 6.8 to compare their effectiveness in the reduction of oil/water IFT. All of these can reduce the IFT of oil/water interface to lower than 10^{-1} mN/m (Chen et al., 2015). The characteristics of these curves are; first decrease, and then increase, finally tends to be balanced with time. This pattern of the curve is due to the adsorption of *in situ* surfactant on the oil/water interface increases with time at the starting point and leads to reduction of IFT. Then diffusions of in situ surfactants occur to water or oil phase, which reduce the quantity of adsorbed surfactant on the oil/water interface and hence increase IFT with time.

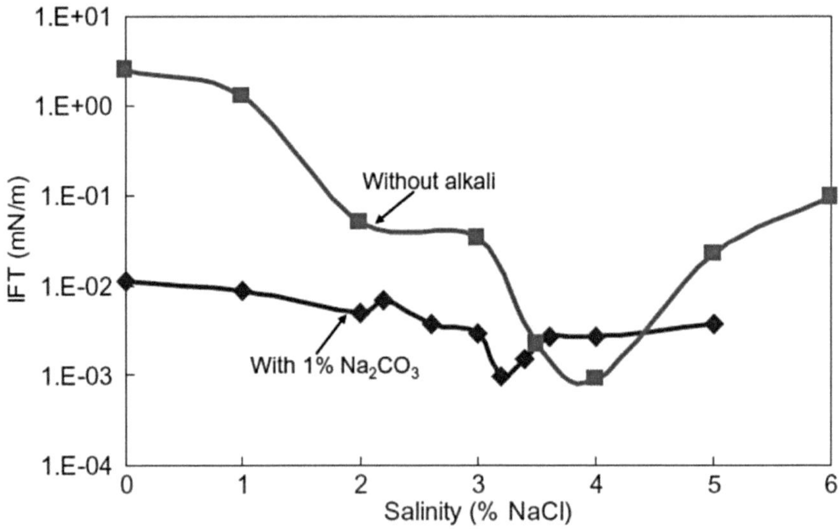

FIGURE 6.7 IFT between Yates oil and the microemulsion.

Source: Reprinted with permission from Liu et al., 2007. Copyright © 2006 Elsevier B.V.

Anionic surfactants are adsorbed on sandstone and carbonate rock due to the ionic attraction between surfactant anion and mineral sites (Zhang and Somasundaran, 2006). Mineral sites become more negatively charged with increasing pH; hence, the adsorption of anionic surfactant on reservoir rock is retarded. Sometimes the surfactant retention may be reduced with alkali addition due to removal of multivalent ions. The carbonate ions (CO_3^{2-}) can reduce the adsorption of surfactant on calcite because of change in the surface charge (Hirasaki and Zhang, 2004). Calcite or dolomite is the carbonate formations and sandstone cementing materials which consist of an isoelectric point (or point of zero charge (PZC)) at pH = 9. Here, the calcium and magnesium ions become more significant factor to determine

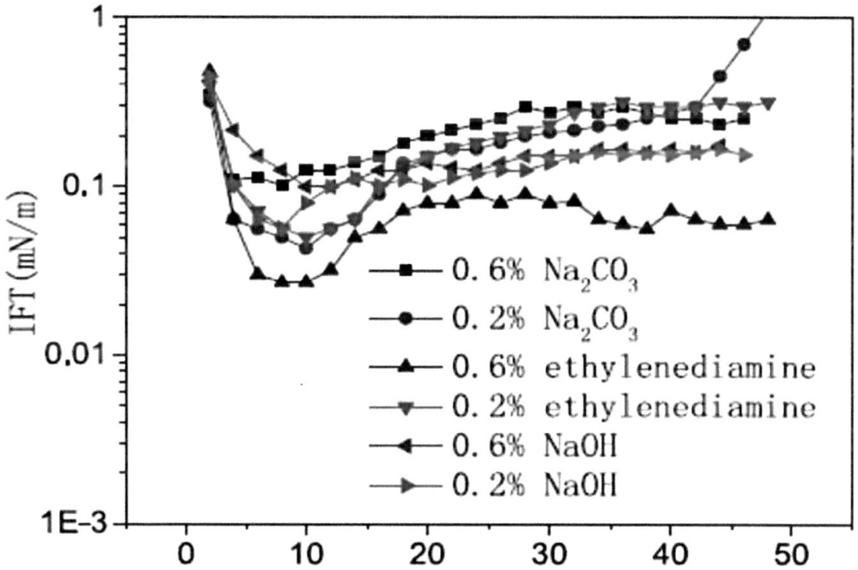

FIGURE 6.8 Dynamic IFTs of crude oil in presence of different alkaline solutions.
Source: Reprinted with permission from (Chen et al., 2015). Copyright © 2015 Elsevier B.V.

the potential with carbonate ion (Hirasaki et al., 2008). Sulfate ions are also adsorbed in the fractured chalk surfaces during injection of seawater and alter the surface potential as well as change the potential (Austad et al., 2005). At high alkaline concentration, the flocculation of surfactant micelles may occur due to increase in ionic strength. However, the adsorption of cationic surfactant increases with increasing pH. pH affects the results of chemical EOR software (UTCHEM version 9) as the adsorption of surfactant depends on pH. AS solution containing anionic surfactant changes the wettability of water-wet sand slightly while AS solution contain a cationic surfactant changes a water-wet rock to oil-wet permanently. This indicates that the application of anionic surfactant is more favorable for AS flooding.

Several research groups have conducted the laboratory studies on AS flooding in heavy oil reservoirs (Bryan and Kantzas, 2009; Kumar et al., 2010). But some have targeted fields studies directly, e.g., study for the Hepler field in Kansas (French, 1994). However, not many pilots were conducted in fields, partly owing to the low oil price. Probably more is owing to the fact that mobility control using polymer is essential in AS flooding.

The Big Sinking field in the Appalachian Basin of Eastern Kentucky and the White Castle Field in Louisiana are another example of AS flooding application. In Big Sinking field, the AS solution contain 1.5% sodium hydroxide and 0.1% (active concentration) ORS-62HF surfactant was injected after 1500 bbls of fresh water preflush. The compositions of AS solutions applied in White Castle Field are presented in Table 6.1. The Upper Assam Basin in India is a good example for implementing chemical EOR (Gogoi and Kakoty, 2017). Previously, BL surfactant from waste of paper mill has showed effect on Upper Assam Basin, but IFT reduction was not efficient enough (Gogoi, 2010, 2014; Gogoi and Das, 2012; Gogoi and Gogoi, 2005). AS flooding with NaOH as alkali and SLS as surfactant was investigated and also studied the adsorption and desorption phenomenon of surfactants in the porous media (Sarmah et al., 2017; Gogoi, 2011). The surfactant in the form of micellar can decrease the IFT to 10^{-3} mN/m, thus leading to the EOR (Das and Gogoi, 2015a, 2015b). Anionic surfactant, SDS is most effective surfactant for Upper Assam Basin in EOR (Hazarika et al., 2018; Hazarika and Gogoi, 2014). About 13 field projects of AS flooding are documented in literature, most of them in the U.S. onshore except Angsi in Malaysia offshore.

TABLE 6.1 Compositions of AS Solutions (wt.%).

	Carbonate Slug	Silicate Slug
Sodium carbonate	2.53	2.03
Sodium silicate		0.21
Internal olefin sulfonates (IOS)	0.44	0.45
Sodium chloride	0.4	0.4
NEODOL 25-12	0.06	0.06
Sodium thiocyanate	0.05	0.05
Water	96.52	96.8

6.5 ALKALI–POLYMER FLOODING

The field performance of alkali projects shows 1–5% of the incremental oil recovery due to formation of in situ surfactants (Mayer et al., 1983; Sheng et al., 2015). Polymer injection controls the necessary mobility for alkali flooding, but this can't lower the residual oil saturation significantly as the capillary number can't increase by high polymer viscosity. Hence, the

AP flooding offers the synergistic effect of alkaline flooding and polymer flooding.

The interactions between alkali and polymer are very important in AP flooding. Alkali has salt effect to reduce the viscosity of polymer solution due to increase in ionic strength (Kang, 2001; Kazempour et al., 2012). However, alkali hydrolyses polymer which may increase polymer solution viscosity. Normally, the viscosity of polymer solution decreases at high pH. But in this case, the salt effect becomes the predominant factor as a result the viscosity of polymer solution decreases with alkali concentration as shown in Figure 6.9 (Samanta et al., 2010; Sheng, 2011). Actually, the effect of alkali on polymer viscosity is complex, and it depends on the nature and concentration of polymers and alkalis.

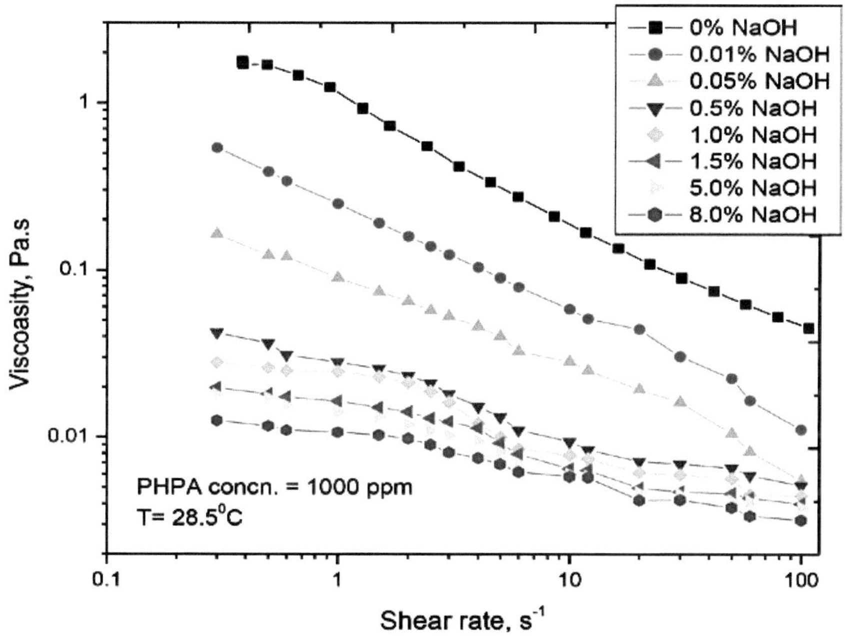

FIGURE 6.9 Viscosity of partial hydrolyzed polyacrymide in presence of different NaOH solution.

Source: Reprinted with permission from (Samanta et al., 2010). Copyright © 2011 American Chemical Society

The residual resistance factor of AP flooding is higher than that of polymer flooding. The overall result is considered during post-AP brine flooding, as the ionic strength reduced considerably and the polymer viscosity is restored. It has been proposed that, the emulsion trapping is one of the reasons of high residual resistance factor during post-AP brine flooding. The decrease in viscosity of AP solution can be overcome by addition of copolymer, for example, P(AAc-st-VP) synthesized from acrylic acid (AA) and N-vinyl-2-pyrrolidinone (NVP) and a proper initiator (Doda et al., 2013). This copolymer is stable in alkali solution. The adsorption of alkali in reservoir rock is lower in case of AP solution than alkali solution alone, as polymer acts as a cover on rock surfaces. Addition of alkali generates high pH in an AP system and alters the charge density with more negative charges, which leads to reduction of polymer adsorption on the rock surface (Krumrine and Falcone, 1987; Chen et al., 1999).

The residual oil recovery (%) versus pore volume (PV) injected of different combinations and sequences of injection of polymer and orthosilicate have been shown in Figure 6.10 (Sheng, 2017). In this figure, A and P indicates only alkali and polymer solution, A/P represents the alkali injection followed by polymer injection, P/A is the polymer injection followed by alkali injection and A + P means AP flooding. It has been observed that any combination is better than injection of a single component and AP recommends the highest oil recovery which indicates the synergy of alkali and polymer. Sheng et al. (1994) and Chen et al. (1999) have reported similar results.

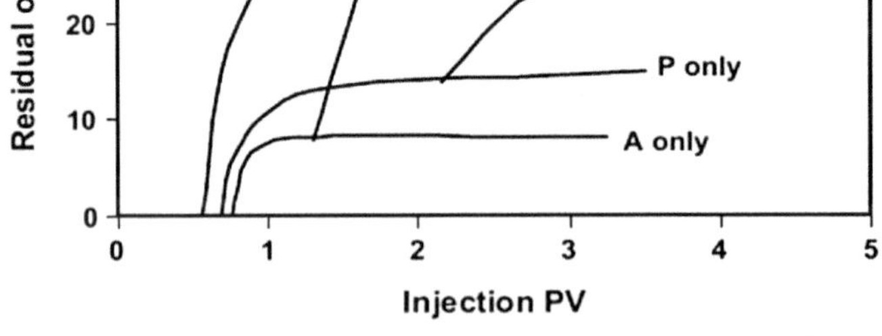

FIGURE 6.10 Evaluation of residual oil recovery factors of AP flooding containig polyacrylamide and orthosilicate.

Source: Reprinted with permission from Sheng, 2017.

For AP flooding experiment, at first the compatibility of alkali and polymer needs to verify and conduct corefloods using the combined solution

(Sheng et al., 2015). It is also required to do a comparative corefloods study to ensure whether the incremental oil recovery is higher for combined solution than alkali solution alone or polymer solution alone. An experimental study of AP flooding in a heavy oil reservoir (oil viscosity 1202 cP and acid number 1.07 mg KOH/g oil) within the high-permeability channels shows an efficient oil recovery of 25–30% of original oil in place (OOIP) (Wu et al., 2011). The optimum formulation was used 0.4% NaOH, 0.2% Na_2CO_3, and 1000 ppm polymer.

Overall only eight AP flooding field projects are reported; four in Canada, two in China, and two in USA (Sheng, 2017). These projects were applied in sandstone onshore reservoirs after high water cuts and high water saturations. Na_2CO_3 is used as alkali with polymers polyacrylamide type in most of fields. The 0.05 PV at 215 ppm of a cationic polymer slug was preflushed to block the high permeability zones in the anionic sandstone reservoirs of Almy Sands (Isenhour Unit) in Wyoming to block (Doll, 1988a, 1988b). This type of the CAT-AN process was also applied in the West Moorcroft AP project (Bala et al., 1992). About 364 AP well injections were reported as "flow-diverting technologies" applications with an incremental oil recovery of little bit over 1000 tons per well (Shandrygin and Lutfullin, 2008). There are no detailed data of AP flooding in Russia or the former Soviet Union available in literature. In some cases where the subsequent water fingering not occurs during AP flooding, polymer post-flush is injected before waterflooding (Sheng et al., 2015). The average EOR factor was 14% and the average water cut decrease about 12.5 % after AP injection.

6.6 ALKALI–SURFACTANT–POLYMER FLOODING

Currently, ASP is considered as the most promising chemical method in EOR of oilfield development because it integrates the advantages of alkali, surfactant, and polymer. The effect of ASP flooding depends on many factors, including reservoir geological features, oil-water interfacial tension, reservoir wettability, and rheology of displacing fluid. The behavior of rheological properties of displacing fluids determines the oil displacement efficiency (Wang et al., 2000; Wang et al., 2001). The oil displacement efficiency of ASP flooding is better than that of individual alkali, surfactant, and polymer as all the different displacement mechanisms are applicable in the former. The ultra-low interfacial tension between oil and water can obtain using specific surfactants. Alkali can also reduce the interfacial tension by forming in situ surfactants by reacting with the acidic components of crude oil. Use of combined alkali and surfactant not only has good synergistic effect on reduction of IFT but also alkali reduces the absorption and loss of costly surfactants. A complete knowledge of rheology of polymers in ASP flooding is essential to examine its ability to improve the mobility control during the displacement process (Xia et al., 1999; Wang and Cheng, 2003). A great progress has been made either in laboratory studies or in pilot tests for alkali/surfactant/polymer (ASP) and surfactant/alkali/polymer (SAP) combination flooding (Song et al., 1995; Han et al., 1996; Gao et al., 1996). ASP flooding is a technique which is developed out on the basis of alkali flooding, surfactant flooding, and polymer flooding (Wang et al., 2007), and oil recovery is enhanced gently by decreasing interfacial tension, increasing capillary number, enhancing microscopic displacing efficiency, improving mobility ration, and increasing macroscopic sweep efficiency (Shen and Yu, 2002).

Substantial research works are being carried out worldwide on alkali, surfactant, and polymer flooding by different researchers (Nasr-El-Din et al., 1994; Al-Hashim et al., 1996; Taylor and Nasr-El-Din, 1996). Hawkins et al. (1994) reported that the simultaneous injection of alkali and polymer is more effective than the same chemicals injected sequentially with no contact between alkali and polymer . Other mechanisms of ASP flooding may be interface producing, bridging between inner-pore and outer-pore and oil–water emulsion (Tong et al., 1998). In a vertical heterogeneous reservoir, ASP flooding increases displacing efficiency by displacing residual oil through decreasing interfacial tension and improving sweep efficiency. The ASP flooding is more effective for oil with high acid value. It has been observed that flooding system's rheology and IFT between flooding system and oil with high acid value were key factors effecting oil recovery. The fluid-flow mechanism of EOR in porous media by ASP flooding can be explained as it displaces not only the residual oil in the high-permeability layer but also the remaining oil in the low and middle permeability layers by increasing both swept volume and displacement efficiency.

The EOR method so-called ASP flooding has been proved to be effective in reducing the oil residual saturation in laboratory experiments and field projects through reduction of interfacial tension and mobility ratio between oil and water phases. A critical step for the optimal design and control of ASP recovery processes is to find the relative contributions of design variables such as, slug size and chemical concentrations, in production performances. In this work, a series of flooding experiments have been performed varying the concentrations and combinations of chemical slugs to find out the optimum formulation and design of ASP flooding for EOR.

Recently, there have been several field pilot tests using ASP (Pratap and Gauma, 2004; Qiao et al., 2000; Li et al., 2003; Yang et al., 2003; Zhao et al., 2006). Oil recovery could be further enhanced by 20% by ASP flooding for heterogeneous reservoirs. The experimental studies of oil recovery performance with polymer and small injection of concentrated slugs of alkali/surfactant (3.75 wt% alkali and 2.5 wt% surfactant, 0.05 PV each) mobilizes as much oil as larger injection volume with more dilute slugs of alkali/surfactant (0.75 wt% alkali plus 0.5 wt% surfactant, 0.25 PV). An ASP flood in a sand reservoir has been reported, where co-injection of alkali, low concentration surfactant, and polymer was applied in the proportion 0.8% Na_2CO_3, 0.1% Petrostep B-100, and 0.105% Pusher 700. The estimated recovery of ASP flooding is 20.7% OOIP over the waterflood recovery. ASP processes are also confirmed to be a promising EOR scheme for carbonate reservoirs (Al-Hashim et al.,

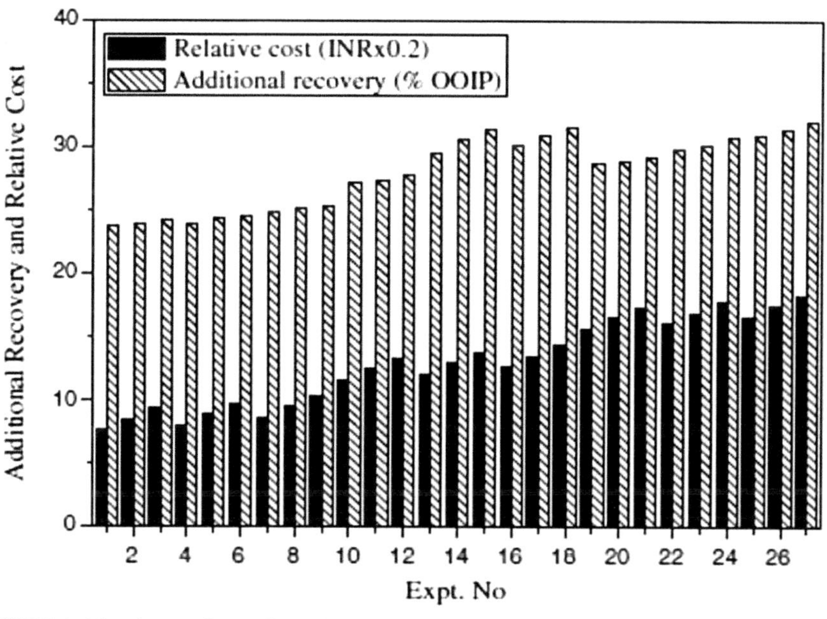

FIGURE 6.11 Comparison of additional oil recovery versus relative cost of chemicals.

Source: Reprinted with permission from Manrique et al., 2007

1996). Compatibility tests showed that the ASP slug has to be prepared by softening the brine and protecting by the pre- and post-flush softening brine slugs. The ASP slugs for core flooding consisted of 0.5% surfactant, 500 ppm polymer, and 1% of alkalis. The additional oil recovery increases with increasing concentration of alkali, surfactant, and polymer at a certain limit. The increase in recovery is only marginal after this optimum concentration of ASP slug; hence, it becomes economically attractive processes which plays significant role in EOR. An analysis has been reported on the relative cost of the ASP slug applied and the corresponding incremental oil recovery as shown in Figure 6.11. But the oil recovery is very much dependent on the reservoir characteristics and crude oil properties, the concentration disparities may vary from field to field.

ASP flooding becomes economically attractive processes which play a significant role in EOR. Numerous ASP flooding projects have been widely reported in the literature during last 25 years. Successful results of pilot tests have been reported from two fields in China. The Daqing oil field represents one of the largest, where oil has high paraffin content and low acidity, two tests, the PO and XF pilot tests, were conducted in 1994 through 1996 (Gao et al., 1996; Wang et al., 1997). It has also been reported that another ASP pilot test in the Daqing field which has been conducted to evaluate the process of applicability in the flooded-out zones (Wang et al., 1998; Pu and Xu, 2009). Other documented examples of Chinese ASP EOR projects are Gudong (Qu et al., 1998), Karamay (Qiao et al., 2000; Gu et al., 1998), Liahoe, and Shengli (Chang et al., 2006) fields. Further ASP flooding processes are reported including Viraj field, India (Pratap and Gauma, 2004) and West Kiehl (Meyers et al., 1992), Sho-Vel-Tum (French, 1999), Cambridge Minnelusa (Vargo et al., 2000), and Tanner (Pitts et al., 2006) fields in the USA during the last decade.

Only one chemical flooding (polymer flooding) application has been recorded for carbonate reservoir in Yates field in the USA (Manrique et al., 2007). ASP flooding has not been implemented yet in carbonate reservoirs, though effective EOR results have been obtained from laboratory studies with combination of an alkali and anionic and non-ionic surfactants (Manrique et al., 2007). The chemical EOR has been implemented in the Kharaib B carbonate reservoir to find out the suitable polymer to improve displacement efficiency and blocked the high adsorption of surfactants in the reservoir rock (Frank et al., 2009).

6.7 CONCLUSION

Alkali reacts with the acidic component of crude oil and forms in situ surfactants. The recovery of residual oil by alkali flooding follows two mechanisms: low interfacial tension and wettability reversal. Though the mechanisms of the AS flooding are still not fully understood, most researchers have the same opinion that key issues are IFT reduction at low surfactant concentration, wettability alteration, and low adsorption of surfactant by alkali and mobility control. The combination of alkali and polymer can reduce the adsorption of polymer as well as consumption of alkali in the reservoir surface. AP solution becomes more viscous with polymer and lowers IFT due to alkali-generated soap that may improve both the sweep efficiency and displacement efficiency which leads to more effective oil recovery than either polymer or alkali alone. ASP flooding embodies the characteristics of alkali flooding, AS flooding, and AP flooding. In ASP flooding, synergistic effect of alkali, surfactant, and polymer makes it more efficient. Simultaneous injection of mixture of alkali, surfactant, and polymer is more successful than the same chemicals injected sequentially with combination of any two chemicals or no contact between the alkali, surfactant, and polymer. The synergism of alkali, surfactant, and polymer results from a combination of improved sweep and mobilization of residual oil due to reduction of interfacial tension, wettability alteration, low adsorption of chemicals, mobility control, etc. These EOR technologies are adaptable in tertiary oil recovery techniques that should be considered for future application in both offshore and onshore to maximize total oil production and meet increasing energy needs. Though there are limited applications of these EOR methods, a large number of ongoing and planned pilot projects execute at present and in near future. However, the EOR methods are already effective and successfully applicated in onshore in pilot and field scales, but the plan and operational experiences are required to develop the efforts to make possible offshore applications to improve global oil production in near future. These technologies need to develop in more cost-effective manner to make it more popular among researchers. Therefore, new challenges of formulation of new effective surfactants or polymers, fluid flow behaviors in porous media, subsurface, and surface challenges also attack the oil researchers to address most cost-effective and environmentally acceptable field applications of onshore and offshore EOR methods.

KEYWORDS

- **interfacial tension,**
- **alkali-surfactant,**
- **polymer,**
- **alkali-surfcatant-polymer,**
- **tertiary oil recovery**

REFERENCES

Adibhatla, B.; Mohanty, K. K. Oil Recovery From Fractured Carbonates by Surfactant-Aided Gravity Drainage: Laboratory Experiments and Mechanistic Simulations. *SPE Reserv. Evaluat. Eng.* **2008,** *11*, 119–130.

Al-Hashim, H. S.; Obiora, V.; Al-Yousef, H. Y.; Fernandez, F.; Nofal, W. In *Alkaline Surfactant Polymer Formulation for Saudi Arabian Carbonate Reservoirs*, SPE/ DOE Improved Oil Recovery Symposium; Tulsa, Oklahoma, Paper SPE/DOE 35353, 1996.

Arhuoma, M.; Yang, D.; Dong, M.; Li, H.; Idem, R. Numerical Simulation of Displacement Mechanisms for Enhancing Heavy Oil Recovery during Alkaline Flooding. *Energy Fuels* **2009,** *23*, 5995–6002.

Atkinson, H. Recovery of petroleum from Oil Bearing Sands. US Patent No. 1,615,311, Nov. 29, 1927.

Austad, T.; Strand, S.; Høgnesen, E. J.; Zhang, P. In *Seawater as IOR Fluid in Fractured Chalk*, Paper SPE 93000 Presented at SPE International Symposium on Oil field Chemistry, Woodlands, TX, Feb 2–4, 2005.

Ayirala, S. C.; Rao, D. N. Multiphase Flow and Wettability Effects of Surfactants in Porous Media. *Colloids Surf. A* **2004,** *241*, 313–322.

Bala, G. A.; Duvall, M. L.; Jackson, J. D.; Larsen, D. C. In *A Flexible Lowcost Approach to Improving Oil Recovery from a (very) Small Minnelusa Sand Reservoir in Crook County*, Wyoming, Paper SPE 24122 Presented at the SPE/DOE Enhanced Oil Recovery Symposium, Apr 22–24, 1992; Tulsa, Oklahoma, USA, 1992.

Bera, A.; Kumar, T.; Ojha, K.; Mandal, A. Adsorption of Surfactants on Sand Surface in Enhanced Oil Recovery: Isotherms, Kinetics and Thermodynamic Studies. *Appl. Surf. Sci.* **2013,** *284*, 87–99.

Bera, A.; Kumar, T.; Ojha, K.; Mandal, A. Screening of Microemulsion Properties for Application in Enhanced Oil Recovery. *Fuel* **2014,** *121*, 198–207.

Bortolotti, V.; Gottardi, G., Manchi, P.; Srrisuriyachai, F. In *Intermittent Alkali Flooding in Vertical Carbonate Reservoir*, Annual Conference and Exhibition; SPE EUROPECE/ EAGE: Amsterdem, Netherlands, 2009; vol. SPE 121832.

Bryan, J.; Kantzas, A. In *Potential for Alkali-Surfactant Flooding in Heavy Oil Reservoirs Through Oil-in-Water Emulsification*, 58th Annual Technical Meeting of Petroleum Society, June 12–14, 2007; Calgary, AB Canada, 2007; CIM Paper 2007-134.

Bryan, J.; Kantzas, A. Potential for Alkali-Surfactant Flooding in Heavy Oil Reservoirs Through Oil-in-Water Emulsification. *JCPT* **2009,** *48*, 37–46.

Chang, L.; Zhang, Z. Q.; Wang, Q. M.; Xu, Z. S.; Guo, Z. D.; Sun, G. Q.; et al. Advances in Polymer Flooding and Alkaline/Surfactant/Polymer Processes as Developed and Applied in the People's Republic of China. *J. Pet. Technol.* **2006,** *58*, 84–91.

Chartterjee, J.; Nikolov, A.; Wasan, D. T. Measurement of Ultralow Interfacial Tension with Application to Surfactant-Enhanced Alkaline Systems. *Ind. Eng. Chem. Res.* **1998,** *37*, 1306–2301.

Chatzis, I.; Morrow, N. Correlation of Capillary Number Relationships for Sandstone. *Soc. Pet. Eng. J.* **1984,** *24*, 555–562.

Chen, Z. Y.; Sun, W.; Yang, H. J. Mechanistic Study of Alkaline– Polymer Flooding in Yangsanmu Field. *J. Northwest Univ. (Nat. Sci. Ed.)* **1999,** *29*, 237–240.

Chen, Z.; Zhao, X.; Wang, Z.; Fu, M. A Comparative Study of Inorganic Alkaline/Polymer Flooding Andorganic Alkaline/Polymer Flooding for Enhanced Heavy Oil Recovery. *Colloids Surf. A Physicochem. Eng. Aspects* **2015,** *469*, 150–157.

Chiwetelu, C. I.; Neale, G. H.; Hornof, V.; George, A. E. Recovery of Saskatchewan Heavy Oil Using Alkaline Solutions. *J. Can. Pet. Technol.* **1994,** *33*, 37–42.

Cooke, C. E. Jr.; Williams, R. E.; Kolodzie, P. A. Oil Recovery by Alkaline Flooding. *J. Pet. Technol.* **1974,***26* (12), 1365–1374.

Das, B. M.; Gogoi, S. B. Relating IFT with Oil Recovery with Special Reference to Bhogpara Porous Media of Upper Assam Basin. *J. Petrol. Eng. Technol.* **2015a,** *5*, 1–8.

Das, B. M.; Gogoi, S. B. Adsorption Study of a Natural Product into the Oil Producing Porous Media of Upper Assam Basin. *Int. J. Res. Eng. Appl. Sci.* **2015b,** *5*, 20–30

deZabala, E. F.; Vislocky, J. M.; Rubin, E.; Radke, C. J. A Chemical Theory for Linear Alkaline Flooding. *Soc. Pet. Eng. J.* **1982,** *22*, 245–258.

Doda, A.; Kotsuchibashi, Y.; Trivedi, J. J.; Narain, R. In *Investigation of Alkali Resistant Polymer for Improved Heavy Oil Recovery*, Paper SPE 165514 Presented at the SPE Heavy Oil Conference-Canada, June 11–13, 2013; Calgary, Alberta, Canada, 2013.

Doll, T. E. In *Performance Data Through 1987 of the Isenhour Unit, Sublette County, Wyoming, Polymer-Augmented Alkaline Flood*, Paper SPE 17801 Presented at the SPE Rocky Mountain Regional Meeting, May 11–13, 1988a; Casper, Wyoming.

Doll, T. E. An Update of the Polymer-Augmented Alkaline Flood at the Isenhour Unit, Sublette County, Wyoming. *SPE Reserv. Eng.* **1988b,** *3*, 604–608.

Doscher, T. M. Oil Recovery From Tar Sands. U.S. Patent No.3279538, 1956.

Dranchuk, P. M.; Scott, J. D.; Flock, D. L. Affect of Addition of Certain Chemicals on Oil Recovery During Waterflooding. *J. Can. Pet. Technol.* **1974,** *13* (03), 27–36.

Frank, S.; van Lingen, P.; Mogensen, K.; Noman, R. In *Screening of EOR Processes for the Kharaib B Reservoir of the Giant Al Shaheen Field, Offshore Qatar*, IPTC 13329, 2009; Presented at the Internatinal Petroleum Technology Cenference (IPTC), Doha, Qatar, Dec 7–9, 2009.

French, T. Evaluation of a Site on Hepler Oil Field for a Surfactant-Enhanced Alkaline Chemical Flooding Project and Supporting Research, Report NIPER/BDM-0073, 1994.

French, T. Evaluation of the Sho-Vel-Tum Alkali-Surfactant-Polymer (ASP) Oil Recovery Project. DOE/SW/45030e; U.S. Department of Energy: Tulsa, OK, USA, 1999.

Gao, S.; Li, H.; Yang, Z.; Pitts, M. J.; Surkalo, H.; Wyatt, K. In *Alkaline-Surfactant-Polymer Pilot Performance of the West Central Saertu, Daqing Oil Field*, SPE/DOE 35383,

Presented at the 1996 SPE/ DOE Tenth Symposium on Improved Oil Recovery, Tulsa, OK, 1996.

Gao, P.; Towler, B. Investigation of Polymer and Surfactant-Polymer Injections in South Slattery Minnelusa Reservoir, Wyoming. *J. Pet Explor. Prod. Technol.* **2011,** *1*, 23–31.

Gogoi, S. B. Characterization of Vesicle for Enhanced Oil Recovery. *Indian J. Chem. Technol.* **2010,** *17*, 282–290.

Gogoi, S. B. Adsorption–Desorption of Surfactant for Enhanced Oil Recovery. *Transp. Porous. Med.* **2011,** *90*, 589–604.

Gogoi, S. B. Effluent as Surfactant for Enhanced Oil Recovery. *J. Innov. Energy Policies* **2014,** *3*, 40–50.

Gogoi, S. B.; Das, B. M. Use of an Effluent for Enhanced Oil Recovery. *Indian J. Chem. Technol.* **2012,** *19*, 366–370.

Gogoi, S. B.; Gogoi, K. D. Emulsion Flow in Bhogpara Core Sample with Emphasis on Enhanced Oil Recovery. *Indian J. Petrol. Geol.* **2005,** *14*, 55–57.

Gogoi, S. B.; Kakoty, M. A Study of CO_2 Flooding on Wave Velocities in the Naharkatiya Oil Reservoir of Upper Assam basin. *Resour. Effic. Technol.* **2017,** *3*, 101–112.

Green, D. W.; Willhite, G. P. In *Enhanced Oil Recovery, SPE Textbook Series*; Henry, L., Doherty Memorial Fund of AIME, Society of Petroleum Engineers: Richardson, Texas, 1998; vol 6.

Gu, H.; Yang, R.; Guo, S.; Guan, W.; Yue, X.; Pan, Q. In *Study on Reservoir Engineering: ASP (Alkali, Surfactant, Polymer) Flooding Pilot Test in Karamay Oilfield*; SPE-50918 Proceedings of 6th SPE Oil & Gas International Conference, Nov 2–6, 1998; Beijing, China, 1998.

Han, W.; Chen, P. Y.; Wang, S. D. An Experimental Study On Formation Damage Due To Alkali in ASP Flooding. *J. Oilfield Chem.* **1996,** *13*, 248–252.

Hawkins, B. F.; Taylor, K. C.; Nasr-El-Din, H. A. Mechanisms of Surfactant and Polymer Enhanced Alkaline Flooding: Application to David Lloyminster and Wainwright Sparky Fields. *J. Can. Pet. Technol.* **1994,** *33*, 52–63.

Hazarika, K.; Gogoi, S. B. Comparative Study of an Enhanced Oil Recovery Process with Various Chemicals for Nahorkatiya Oil Field. *Int. J. Appl. Sci. Biotechnol.* **2014,** *2*, 432–436.

Hazarika, K.; Yadav, R.; Gogoi, S. B.; et al. Characterization of Crude Oil for Enhanced Oil Recovery: A Study with Anionic Surfactant. *Int. J. Ambient Energy* **2018,** *39*, 1–12.

Healy, R. N.; Reed, R. L. In *Improved Oil Recovery by Surfactant and Polymer Flooding*; Academic Press, New York; *Soc. Petrol. Eng. J.* **1977,** *17*, 129–383.

Hirasaki, G. J. Application of the Theory of Multicomponent. Multiphase Displacement to Three-Component, Two-Phase Surfactant Flooding. *Soc. Petrol. Eng. J.* **1981,** *21*, 191–204.

Hirasaki, G. J.; Miller, C. A.; Puerto, M. In *Recent Advances in Surfactant EOR*, Paper SPE 115386 Presented at the SPE Annual Technical Conference and Exhibition, Denver, CO, Sept 21–24; The Revised Version Presented at the International Petroleum Technology Conference, Kuala Lumpur, Malaysia, Dec 3–5, 2008.

Hirasaki, G. J.; Zhang, D. L. Surface Chemistry of Oil Recovery from Fractured, Oil-Wet, Carbonate Formations. *Soc. Pet. Eng. J.* **2004,** 151–162.

Hornof, V.; Neale, G. H.; Yu, A. Effect of Flooding Sequence on the Displacement of Acidic Oil by Alkaline Solutions. *J. Pet. Sci. Eng.* **1994,** *10*, 291–297.

Hou, J.; Liu, Z.; Zhang, S.; Yue, X. A.; Yang, J. The Role of Viscoelasticity of Alkali/Surfactant/Polymer Solutions in Enhanced Oil Recovery. *J. Pet. Sci. Eng.* **2005,** *47*, 219–235.

Howe, A. M.; Clarke, A.; Mitchell, J.; Staniland, J.; Hawkes, L.; Whalan, C. Visualising Surfactant Enhanced Oil Recovery. *Colloids Surf. A* **2015,** *480*, 449–461.

Jackson, A. C. Experimental Study of the Benefits of Sodium Carbonate on Surfactants for Enhanced Oil Recovery. M.S. Thesis, University of Texas at Austin, 2006.

Jennings, H. Y. Jr.; Johnson, C. E. Jr.; McAuliffe, C. D. A Caustic Waterflooding Process for Heavy Oils. *J. Pet. Technol.* **1974,** *26* (12), 1344–1352.

Kamal, M. S.; Hussien, I. A.; Sultan, A. S.; Han, M. Rheological Study on ATBS-AM Copolymer-Surfactant System in High-Temperature and High-Salinity Environment. *J. Chem.* **2013,** *2013*, 801570.

Kamal, M. S.; Sultan, A. S.; Al-Mubaiyedh, U. A.; Hussein, I. A. Review on Polymer Flooding: Rheology, Adsorption, Stability, and Field Applications of Various Polymer Systems. *Polym. Rev.* **2015,** *55*, 491–530.

Kang, W. L. Study of Chemical Interactions and Drive Mechanisms in Daqing ASP Flooding; Petroleum Industry Press, 2001.

Kazempour, M.; Sundstrom, E. A.; Alvarado, V. Effect of Alkalinity on Oil Recovery During Polymer Floods in Sandstone. *SPE Reserv. Eval. Eng.* **2012,** *15*, 195–209.

Krumrine, P. H.; Falcone, J. S. In *Beyond Alkaline Flooding: Design of Complete Chemical Systems*; Paper SPE 16280 Presented at the SPE International Symposium on Oilfield Chemistry, San Antonio, Feb 4–6, 1987.

Kumar, R.; Dao, E.; Mohanty, K. K. In *Emulsion Flooding of Heavy Oil*, Paper SPE 129914 Presented at SPE Improved Oil Recovery Symposium, Tulsa, OK, Apr 24–28, 2010.

Li, H.; Liao, G.; Han, P.; Yang, Z.; Wu, X.; Chen, G.; Xu, D.; Jin, P. Alkaline/Surfactant/Polymer (ASP) Commercial Flooding Test in Central Xing2 Area of Daqing Oilfield; Paper SPE 84896, 2003.

Lieu, V. T.; Miller, S. G.; Miller, S. A Laboratory Study of Chemical Reactions with Reservoir Sand in the Recovery of Petroleum by Alkaline Flooding. *Soc. Pet. Eng. J.* **1985,** *25* (04), 587–856.

Liu, Q.; Dong, M.; Ma, S. In *Alkaline/Surfactant Flood Potential in Western Canadian Heavy Oil Reservoirs*, 2006 SPE/DOE Symposium on Improved Oil Recovery, Apr 22–26, 2006; Tulsa, Okahoma, USA, 2006; vol SPE 99791.

Liu, Q.; Dong, M.; Maa, S. Surfactant Enhanced Alkaline Flooding for Western Canadian Heavy Oil Recovery. *Coll. Surf. A Physiochem. Eng. Aspects* **2007,** *293*, 63–71.

Liu, H.; Zhang, S. F.; Peng, Li X.; Sun, X. L. Synergy of Surfactants in Alkaline Flooding. *Pet. Explor. Dev.* **1995,** *22*, 70–73.

Lyons, W. C.; Plisga, G. J. In *Standard Handbook of Petroleum and Natural Gas Engineering*, 3rd Ed.; Elsevier, 2011.

Mai, A.; Bryan, J.; Goodarzi, N.; Kantzas, A. In *Insights Into Non-Thermal Recovery of Heavy Oil*, 1st World Heavy Oil Conference, Nov 12–15, 2006; Bejing, China, 2006; WHOC Paper 2006-533,

Manrique, E. J.; Muci, V. E.; Gurfinkel, M. E. In *EOR Field Experiences in Carbonate 100943*, Presented at the SPE Asia Pacific Oil and Gas Conference and reservoirs in the United States. *SPE Reserv. Eval. Eng.* **2007,** *10*, 667–886.

Mayer, E. H.; Berg, R. L.; Carmichael, J. D.; Weinbrandt, R. M. Alkaline Injection for Enhanced Oil Recovery—A Status Report. *J. Petrol. Technol.* **1983,** *35*, 209–22.

McAuliffe, C. D. Crude Oil-in-Water Emulsions to Improve Fluid Flow in an Oil Reservoir. *J. Pet. Technol.* **1973a,** *25* (06), 721–726.

McAuliffe, C. D. Oil-in-Water Emulsions and Their Flow Properties in Porous Media. *J. Pet. Technol.* **1973b,** 727–733.

Mehdizadeh, A.; Handy, L. L. Further Investigation of High-Temperature Alkaline Floods. *SPE Reservoir Eng.* **1989,** *4* (02), 171–177.

Meyers, J. J.; Pitts, M. J.; Wyatt, K. Alkaline-Surfactant-Polymer Flood of the West Kiehl Minnelusa Unit. SPE/DOE 24144, 1992.

Mohammadi, H. Mechanistic Modeling, Design and Optimization of Alkaline/Surfactant/Polymer Flood. Ph.D. Dissertation, University of Texas at Austin, 2008.

Mungan, N. Certain Wettability Effects in Laboratory Water Floods. *J. Pet. Technol.* **1966,** *18* (20), 247–252.

Najafi-Marghmaleki, A.; Kord, S.; Hashemi, A.; Motamedi, H. Experimental Investigation of Efficiency of MEOR Process in a Carbonate Oil Reservoir Using *Alcaligenes faecalis*: Impact of Interfacial Tension Reduction and Wettability Alteration Mechanisms. *Fuel* **2018,** *32*, 27–35.

Nasr-El-Din, H. A.; et al. Recovery of Residual Oil Using the Alkali/Surfactant/Polymer Process: Effect of Alkali Concentration. *J. Petrol. Sci. Eng.* **1992,** *6*, 381–401

Nasr-El-Din, H. A.; Green, K. A.; Schramm, L. L. Recovery of Waterflood Residual Oil Using the Alkali/Surfactant/Polymer Process: Effect of Slug Size, Core Length and a Polymer Chase. *Rev. Inst. Fr. Pe't.* **1994,** *49*, 359–377.

Nasr-El-Din, H. A.; Taylor, K. C. Interfacial Tension of Crude Oil/Alkali Systems in the Presence of Partially Hydrolyzed Polyacrylamide. *Colloids Surf.* **1993,** *49*, 273–279.

Nasr-El-Din, H. A.; Taylor, K. C. The Role of Surfactants in Enhanced Oil Recovery. In *Micelles, Microemuslions and Monolayers: Science and Technology*, Chapter 11; Shah, D. O., Eds.; 1998; pp 249–287.

Nelson, R. C.; Lawson, J. B.; Thigpen, D. R.; Stegemeier, G. L. In *Cosurfactant Enhanced Alkaline Flooding*, Paper SPE 12672 Presented at the SPE/DOE Fourth Symposium on Enhanced Oil Recovery Held in Tulsa, OK, Apr 15–18, 1984.

Nutting, P. G. *Oil Gas J.* **1927,** *25*, 106.

Okoliegbe, I. N.; Agarry, O. O. Application of Microbial Surfactant (A Review). *Scholarly J. Biotechnol.* **2012**, *1*, 15–23.

Olajire, A. A. Review of ASP EOR (Alkaline Surfactant Polymer Enhanced Oil Recovery) Technology in the Petroleum Industry: Prospects and Challenges. *Energy* **2014,** *77*, 963–982.

Pal, N.; Babu, K.; Mandal, A. Surface Tension, Dynamic Light Scattering and Rheological Studies of a New Polymeric Surfactant for Application in Enhanced Oil Recovery. *J. Pet. Sci. Eng.* **2016,** *146*, 591–600.

Pei, H.; Zhang, G.; Ge, J.; Ding, L.; Tang, M., et al. In *A Comparative Study of Alkaline Flooding and Alkaline/Surfactant Flooding for Zhuangxi Heavy Oil*, SPE 156852 Presented at the SPE Heavy Oil Conference Canada Alberta, Canada, 2012.

Pitts, M. J.; Dowling, P.; Wyatt, K.; Surkalo, H.; Adams, C. In *Alkalineesurfactantepolymer Flood of the Tanner Field.* SPE 100004 Presented at the SPE/DOE Symposium on Improved Oil Recovery in Tulsa, Oklahoma, 2006.

Pitts, M. J.; Wyatt, K.; Surkalo, H. In *Alkaline-Polymer Flooding of the David Pool, Lioydminster Alberta*, SPE/DOE 14th Symposium on Improved Oil Recovery, Apr 17–21, 2004; Tulsa, Okahoma, USA, 2004; SPE Paper 89386, 2004.

Pope, G. A.; Lake, L. W.; Helfferich, F. G. Cation Exchange in Chemical Flooding: Part 1-Basic Theory Without Dispersion. *Soc. Petrol. Eng. J.* **1978**, *18*, 418–434.

Pope, G. A. The Application of Fractional Flow Theory to Enhanced Oil Recovery. *Soc. Petrol. Eng. J.* **1980,** *20*, 191–205.

Pratap, M.; Gauma, M. S. Field Implementation of Alkaline-Surfactant-Polymer (ASP) Flooding: A Maiden Effort in India. SPE 88455, 2004.

Pu, H.; Xu, Q. An Update and Perspective on Field-Scale Chemical Floods in Daqing Oilfield, China, 2009; SPE 118746.

Qiao, Q.; et al. In *The Pilot Test of ASP Combination Flooding in Karamay Oil Field*, Paper SPE 64726, Presented at the 2000 SPE International Oil and Gas Conference and Exhibition in China, Nov 7, 2000; Beijing, 2000.

Qiao, Q.; Gu, H.; Li, D.; Dong, L. The Pilot Test of ASP Combination Flooding in Karamay Oil Field. Paper SPE 64726, 2000.

Qu, Z.; Zhang, Y.; Zhang, X.; Dai, J. A. In *Successful ASP (Alkaline/Surfactant/Polymer) Flooding Pilot in Gudong Oil Field*, Proceedings of 11th SPE/DOE Improved Oil Recovery Symposium, Apr 19–22, 1998; Tulsa, OK, USA; 1998, SPE-39613.

Ramakrishnan, T. S.; Wasan, D. T. A Model for Interfacial Activity of Acidic Crude Oil/Caustic Systems for Alkaline Flooding. *Soc. Pet. Eng. J.* **1983,** *23*, 602–612.

Saboorian-Jooybari, H.; Dejam, M.; Chen, Z. Heavy Oil Polymer Flooding from Laboratory Core Floods to Pilot Tests and Field Applications: Half-Century Studies. *J. Pet. Sci. Eng.* **2016,***142*, 85−100.

Samanta, A.; Bera, A.; Ojha, K.; Mandal, A. Effects of Alkali, Salts, and Surfactant on Rheological Behavior of Partially Hydrolyzed Polyacrylamide Solutions. *J. Chem. Eng. Data* **2010**, *55*, 4315− 4322.

Samanta, A.; Bera, A.; Ojha, K.; Mandal, A. Comparative Studies on Enhanced Oil Recovery by Alkali−Surfactant and Polymer Flooding. *J. Pet. Explor. Prod. Technol.* **2012,** *2*, 67−74.

Samanta, A.; Ojha, K.; Mandal, A. Interactions Between Acidic Crude Oil and Alkali and Their Effects on Enhanced Oil Recovery. *Energy Fuels* **2011,** *25*, 1642–1649.

Sarmah, S.; Gogoi, S. B.; Barman, J.; et al. In *Study on the Effects of the Rheological Properties for Flow Behaviour of Assam Crude Oil with Various Chemicals*, Proceedings of the 1st GeoMEast International Congress and Exhibition, Egypt 2017 on Sustainable Civil Infrastructures, 2017; pp 148–168.

Satter, A.; Iqbal, G.; Buchwalter, J. In *Practical Enhanced Reservoir Engineering*; PennWell: Tulsa, Oklahoma, 2008.

Shandrygin, A. N.; Lutfullin, A. In *Current Status of Enhanced Recovery Techniques in the Fields of Russia*, Paper SPE 115712 Presented at the SPE Annual Technical Conference and Exhibition, Sept 21–24, 2008; Denver, Colorado, USA, 2008.

Shen, P. P.; Yu, J. Y. Fundamental Research on Enhanced Oil Recovery in Large Scale; Petroleum Industry Publication Company: Beijing, 2002.

Sheng, J. J. In *Modern Chemical Enhanced Oil Recovery: Theory and Practice*; Elsevier, 2011.

Sheng, J. J. Critical Review of Alkaline-Polymer Flooding. *J. Petrol. Explor. Prod. Technol.* **2017,** *7*, 147–153.

Sheng, J. J.; Leonhardt, B.; Azri, N. Status of Polymer-Flooding Technology. *JCPT* **2015**, *54*, 116–126.

Sheng, D. C.; Yang, P. H.; Liu, Y. L. Effect of Alkali-Polymer Interaction on the Solution Properties. *Petrol. Explor. Develop.* **1994,** *21*, 81–85.

Silva, R. C. F. S.; Almeida, D. G.; Rufino, R. D.; Luna, J. M.; Santos, V. A.; Sarubbo, L. A. Applications of Biosurfactants in the Petroleum Industry and There Mediation of Oil Spills. *Int. J. Mol. Sci.* **2014,** *15*, 12523–12542.

Song, B.; Hu, X.; Shui, X.; Cui, Z.; Wang, Z. A New Type of Renewable Surfactants for Enhanced Oil Recovery: Dialkylpolyoxyethylene Ether Methyl Carboxyl betaines. *Colloids Surf. A* **2016,** *489*, 433–440.

Song, W.; Yang, C.; Han, D.; Qu, Z.; Wang, B.; Jia, W. In *Alkaline Surfactant Polymer Combination Flooding For Improving Recovery of the Oil with High Acid Value*, SPE 29905-MS, International Meeting on Petroleum Engineering, Beijing, China, 1995.

Spildo, K.; Johannessen, A. M.; Skauge, A. In *Low Salinity Waterflood at Reduced Capillarity*, SPE Improved Oil Recovery Symposium, Tulsa, Oklahoma, USA, Apr 14–18, 2012;Society of Petroleum Engineers: Richardson, TX, USA, 2012; vol. SPE-154236-MS.

Symonds, R. W. P.; Farouq Ali, S. M.; Thomas, S. Caustic Flooding for Heavy Oil Recovery. *J. Can. Pet. Technol.* **1991,** *30*, 44–49.

Taylor, K.; Nasr-El-Din, H. A. The Effect of Synthetic Surfactants on the Interfacial Behavior of Crude Oil/Alkali/Polymer Systems. *Colloids Surf A* **1996,** *108*, 49–72.

Thomas, S.; Farouq Ali, S. M.; Scouler, J. R.; Verkoczy, B. Chemical Methods for Heavy Oil Recovery. *J. Can. Pet. Tech.* **2001,** *40*, 56.

Tong, Z. S., Yang, C.Z., Wu, G.Q. Yuan, H., Yu, L., Tian, G. In *A Study of Microscopic Flooding Mechanism of Surfactant/Alkali/Polymer*, Paper SPE 39662, Presented at SPE/ DOE Improved Oil Recovery Symposium, Tulsa, Oklahoma, 1998.

Torrealba, V. A.; Johns, R. T. Coupled Interfacial Tension and Phase Behavior Model Based on Micellar Curvatures. *Langmuir* **2017,** *33*, 13604–13614.

Torrealba, V. A., et al. Curvature-Based Equation of State for Microemulsion-Phase Behavior. *Soc. Petrol. Eng. J.* **2018a,** *24* (2), 1–13

Torrealba, V. A., et al. Description of Micellar Radii for Phase Behavior and Viscosity Modeling of Aqueous Surfactant Solutions and Microemulsions. *Langmuir* **2018b,** *34*, 15327–15334.

Uren, L. C.; Fahmy, E. H. Factors Influencing the Recovery of Petroleum from Unconsolidated Sands by Waterflooding. *Trans. AIME* **1927,** 77, 318–335.

Vargo, J.; Turner, J.; Vergnani, B.; Pitts, M. J.; Wyatt, K.; Surkalo, H. Alkaline-Surfactant-Polymer Flooding of the Cambridge Minnelusa Field. *SPE Reserv. Eval. Eng.* **2000,** *3*, 552–558.

Wagner, O. R.; Leach, R. O. Effect of Interfacial Tension on Displacement Efficiency. *Soc. Petrol. Eng. J.* **1966,** *6*, 335–344.

Wang, Y; Cheng, J. The Scaling Charateristics and Adaptability of Mechanical Recovery During ASP Flooding. *J. Daqing Pet. Inst.* **2003,** *27*, 20–22.

Wang, D.; Cheng, J.; Wu, J.g.; Wang, F.; Li, H.; Gong, X. In *An Alkaline/Surfactant/Polymer Field Test in a Reservoir with a Long-term 100% Water Cut*, SPE 49018Presented at the 1998 SPE Annual Technical Conference and Exhibition, Sep 27–30, 1998; New Orleans, Louisiana, 1998.

Wang, D.; Cheng, J.; Yang, Q. Y. Viscous-Elustic Polymer Can Increase Micro-Scale displacement Efficiency In Core. *J. Acta Petroli Sinica* **2000,** *21*, 45–51.

Wang, J.; Yuan, S.; Shen, P.; Zhong, T.; Jia, X. In *Understanding of Fluid Flow Mechanism in Porous Media of EOR by ASP Flooding from Physical Modeling*, International Petroleum Technology Conference, Dubai, ITPC-11257, 2007.

Wang, D.; Xia, H.; Liu, Z.; Yang, Q. In *Study of the Mechanism of Polymer Solution With Visco-Elastic Behavior Increasing Microscopic Oil Displacement Efficiency and the Forming of Steady "Oil Thread" Flow Channels*, SPE 68723, SPE Asia Pacific Oil and Gas Conference and Exhibition, Apr 17–19, 2001; Jakarta, Indonesia, 2001; pp 1–9.

Wang, D.; Zhang, Z.; Cheng, J.; Yang, J.; Gao, S.; Li, L. Pilot Tests of Alakali/Surfactant/Polymer Flooding in Daqing Oil Field. *SPERE* **1997,** *229.*

Wang, Y.; Zhao, F.; Bai, B. In *Optimized Surfactant IFT and Polymer Viscosity for Surfactant-Polymer Flooding in Heterogeneous Formations*, SPE Improved Oil Recovery Symposium, Tulsa, OK, USA, Apr 24–28, 2010;Society of Petroleum Engineers: Richardson, TX, USA, 2010; vol SPE-127391-MS, pp 1–11.

Wu, Y.; Dong, M.; Shirif, E. Study of Alkaline/Polymer Flooding for Heavy-Oil Recovery Using Channeled Sandpacks. *SPE Reser. Eval. Eng.* **2011,** *14*, 310–319.

Xia, H.; Zang,, Y. X.; Zang, Y. L. Rheological Behavior of Alkali-Surfactant-Polymer Solution in Porous Media. *J. Daqing Pet. Inst.* **1999,** *23*, 18–22.

Yang, X.; Liao, G.; Han, P.; Yang, Z.; Yao, Y. In *An Extended Test Study on Alkaline-Surfatant-Polymer Flooding in Beiyidaunxi of Daqing Oilfield*, SPE 80532, SPE Asia Pacific Oil and Gas Conference and Exibition in Jakarta, Indonesia, Apr 15–17, 2003.

Zhao, Z. K.; Bi, C. G.; Li, Z. S.; Qiao, W. H. Interfacial Tension Between Crude Oil and Decylmethylnaphthalene Sulfonate Surfactant Alkalifree Flooding Systems. *Colloids Surfaces A Physicochem. Eng. Asp.* **2006,** *276*, 186.

Zhang, D. L.; Li, S.; Puerto, M.; Miller, C. A.; Hirasaki, G. J., et al. Wettability Alteration and Spontaneous Imbibition in Oil-Wet Carbonate Formations. *J. Pet. Sci. Eng.* **2006,** *52*, 213–226.

Zhang, R.; Somasundaran, P. Advances in Adsorption of Surfactants and Their Mixtures at Solid/Solution Interfaces. *Adv. Colloid Interface Sci.* **2006,** *123–126*, 213–229.

Zhou, Y.; Liu, J.; Yu, T.; Li, Y.; Pan, L. Experimental Study of Alkaline- Surfactant Flooding to Enhance Heavy Oil Recovery. *Adv. Fine Petrochem.* **2009,** *10*, 20–23.

Zhu, D.; Wei, L.; Wang, B.; Feng, Y. Aqueous Hybrids of Silica Nanoparticles and Hydrophobically Associating Hydrolyzed Polyacryla- Mide used for EOR in High-Temperature and High-Salinity Reservoirs. *Energies* **2014,** *7*, 3858–3871.

CHAPTER 7

Foam Flooding for EOR: Viability in Modern Oil Reservoirs

RAVI SHANKAR KUMAR[1,2], MUHAMMAD ARIF[1], and TUSHAR SHARMA[2]

[1]Department of Petroleum Engineering, Khalifa University, Abu Dhabi, United Arab Emirates

[2]Enhanced Oil Recovery Laboratory, Department of Petroleum Engineering, Rajiv Gandhi Institute of Petroleum Technology, Jais, Amethi, India

ABSTRACT

As conventional oil reserves are declining at an alarming rate, a new approach to enhanced oil recovery (EOR) is the need of the era. As the gas injection is associated with the problem of viscous fingering as well as channeling of gas, there is need for such an agent that may be fruitful in reducing the mobility of gas and thereby enhances the volumetric sweep efficiency within porous media. As a result, the foam plays a vital role in EOR. Therefore, this chapter delves into the application of foams for EOR and elucidates the processes involved in their creation. The interfacial tension reduction and mobility control facilitated by foam lead to improved sweep efficiency and capillary desaturation. Stability of foams is highly influenced by lamellae, that is, the thin layer that separates the gas bubbles within foam system. Thinner lamellae are not recommended for EOR as bubble ruptures quickly, as a result, the role of surfactants was explored. Foams, formed by injecting gas into surfactant solutions, exhibit unique properties that enhance oil displacement in reservoirs. Surfactant based foam has the ability to migrate across porous and permeable rock surfaces, which in turns increases the surface tension between immiscible phases. The creation of stable foams

Advancements in Chemical Enhanced Oil Recovery.
Tushar Sharma, Krishna Raghav Chaturvedi, Tarek Ganat, and Imtiaz Ali (Eds.)

involves optimizing surfactant concentration, gas injection rates, and reservoir conditions. This chapter also focuses on the composition of crude oil and its impact on foam stability as well as the stability of foam in depleted reservoir. The study emphasizes the mechanism by which foams generate a favorable mobility contrast, enabling efficient displacement of oil. This was followed by the use of viscoelastic surfactants and polymer in the foam system. Application of nanoparticles (NPs) and viscosifiers in making highly stable foam has also been discussed. NPs stabilizes the foam even at higher salinity and higher temperature by providing a strong barrier which prevents bubble coalescence. Use of viscosifiers enhances the life of foam by enhancing lamellae viscosity and thereby reduced lamellae drainage at elevated temperatures.

7.1 INTRODUCTION

Gas injection is a widely used and effective method for enhanced oil recovery (EOR), and it has recently proven to have promising results in a variety of different areas across the world. Due to the gas's high mobility, the recovery factor achieved during gas injection is frequently lower than it would be otherwise (Liu et al., 2019). Additionally, the effects of viscous fingering and gas channeling via high-permeability zones in porous surfaces reduce the volumetric sweep's efficacy (Xue et al., 2015). By utilizing water alternating gas (WAG) injection, which alternates between gas and water slugs, it has been feasible to significantly alleviate some of the issues associated with continuous gas injection (Moradi et al., 2015). On the contrary, gravity segregation during WAG flooding may occur, resulting in yet another early gas exit from the system. Foaming is another approach to reducing gas mobility that may help improve the volumetric sweep efficiency of the gas or liquid being recovered in oil recovery practices. Foams are frequently used to define a discontinuous gas phase or trapped gas bubbles inside a continuous liquid phase, as well as a discontinuous liquid phase. Additionally, it is vital to understand foam's stability before employing it in EOR. The durability of foam is mostly influenced by the lamellae, a thin aqueous layer that separates gas bubbles within the foam system; thinner lamellae tend to rupture more quickly and are hence not recommended for EOR operations. A foaming agent is used to help in the formation of frothy material. For example, adding a surfactant to the aqueous phase inhibits the coalescence of separated gas bubbles, hence increasing foam stability and decreasing foam instability (Kumar et al., 2021).

Numerous studies indicate that the EOR technique of foam flooding may aid in increasing oil recovery. To mobilize the remaining oil, a mixture of gas and liquid-phase slugs must be injected into the reservoir, followed by the formation of drive foam for mobility control. To maximize the efficiency of the EOR process, the foam is generally employed in conjunction with a chemical EOR approach called foam-assisted chemical flooding. As with other chemical flooding systems, foam-assisted flooding involves injecting foam slug into the remaining oil during the waterflood to aid in oil mobilization and mobility control. Foam is a liquid phase dispersion of a huge volume of metastable gas that occurs when this occurs. A significant amount of gas is present in typical foam (often 60–97 vol%). Bulk foams may be created in the presence of gas by manually spinning a surfactant slug, and they are frequently utilized in oilfield operations to improve fluid flow through porous surfaces (Sydansk et al., n.d.).

CO_2 foams have been lauded as a viable solution for CO_2 mobility control (Chaturvedi et al., 2020) and steam flooding mobility control (Hirasaki et al., 2011). According to some, foams are utilized to prevent movement during gas flooding, which is critical in depleted reservoirs with elevated pressures and temperatures (Hirasaki et al., 2011). Due to the high mobility of CO_2 gas in porous media, viscous fingering occurs. This phenomenon, which may be controlled by modifying the gas's mobility, is caused by the movement of foam inside the porous medium (Chaturvedi et al., 2020). To allow CO_2 to flow freely through the foam liquid phase, the nonwetting phase fills up as many holes as feasible. Viscosity and capillary forces were able to adjust the wettability of these foams in porous media as a consequence of their interaction with the pores in response to these "sequences" of flowing CO_2 foam bubbles coming into contact with the pore surfaces and displaying limits. As surfactant-based foams migrate over porous and permeable surfaces, surface tension between the immiscible phases increases, complicating the process of foam migration. These occurrences impart non-Newtonian features to the flowing foam, giving it the appearance of having a high viscosity or restricted mobility as compared to pure CO_2 and water passing through holes without surfactant or lamellae, respectively (Chaturvedi et al., 2020; Kumar et al., 2021). Intermediate-sized holes can get filled with immobile, trapped gas phase bubbles, decreasing the available pore volume for CO_2 foam passage. At steady state, it is feasible to trap a large amount of the gas contained in sandstone foam within these intermediate-sized pores (Chen et al., 2019). Gas entrapment leads to gas obstruction, further restricting the entrapped gas's movement (Chaturvedi et al., 2020).

7.2 WHY FOAMS?

Foam is better than polymers at sweeping up oil in applications called EOR. This reduces the risk of polymer-caused formation damage. Aside from retaining carbon in underground formations, CO_2 foam has very little water in it, which reduces damage to water-sensitive formations and speeds up the cleanup. There are a few things that make foam less stable, though. For example, when it comes into contact with crude oil, foam stability goes down. This chapter discusses how different foam-making methods work, as well as how the degradation of crude oil affects the stability of CO_2 foam. This chapter, in particular, looks at how nanoparticles (NPs) and viscosifiers can be used to make foams more stable. The chapter also shows how some different NPs, such as silicon dioxide and aluminum oxide, have different effects on the body. They also talk about how viscoelastic surfactants (VESs) and polymers can be used in foam systems in the field. The foam stability and increased oil recovery are the two main things that control these unique systems, as shown in the diagram below (Fig. 7.1). You can see it in Figure 7.1. For example, foams can be used to make a liquid more viscous or to

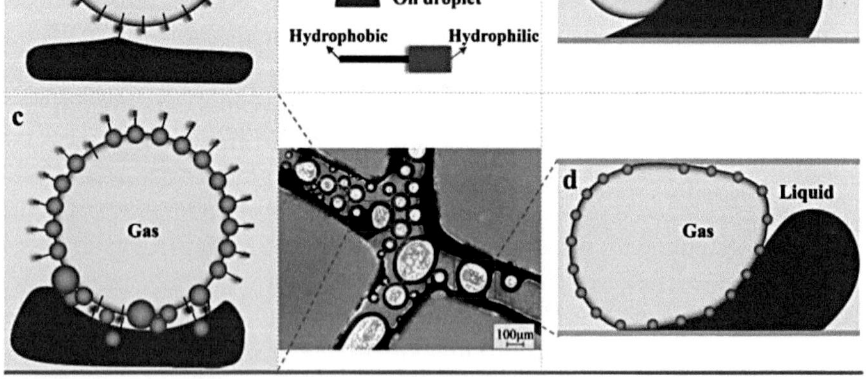

FIGURE 7.1 Schematic showing foam stabilization in the presence of gas, surfactant, NPs, and crude oil.

Source: Reprinted from Sakthivel et al. (2019). Copyright (2019), with permission from ACS.

make a liquid less permeable (Sydansk and Romero-Zerón, n.d.). Foam has a lot of advantages over the chemical EOR method.

- A "mobility control foam" is one in which the foam's mobility is reduced to a level that is about the same as the amount of oil that has been displaced. This is done to stop fingering and channeling. There are usually slugs of surfactant solution and CO_2 that are mixed to fill the reservoir. Once the foam is formed, it is supposed to spread through the formation to keep it from moving too far, which will make the CO_2 flood more efficient (Chaturvedi et al., 2020).
- Conformance control foam is used to build strong, very limited mobility foams in areas that are very porous and have been watered down. Also called injection profile improvement foams or foams that block or divide, these foams are used to improve the profile of the injection. This is done by adding more surfactant to the aqueous solution, which is then pumped with CO_2.

7.3 FOAM STABILITY IN DEPLETED RESERVOIRS

The efficacy of oil displacement is dependent on foam stability. The physical and chemical properties of the surfactant solution influence foam stability in a variety of ways. These qualities include surface tension, surface viscosity, and elasticity. It is crucial to consider disjoining pressure. The disjoining pressure is essentially the pressure difference between the gas and liquid phases in a film, and it is highly dependent on the thickness of the film. According to the Derjaguin, Landau, Verwey, and Overbeek (DLVO) hypothesis, disjoining forces are primarily created by two types of electrostatic potentials: repulsive positive electrostatic potentials and attractive negative van der Waals potentials (Osei-Bonsu et al., 2016). Because of the dipole-induced–dipole interactions that occur between the molecules throughout the film, the molecules are pulled to the van der Waals component. When there is a scarcity of surfactants, these forces take precedence, resulting in negative disjoining pressure and temporary lamellae. When a surfactant is added, it binds to the gas–liquid interface and adsorbs. The repulsive force that holds the lamellae together is provided by the creation of an electrical double layer as a result of surfactant adsorption. As a result, the ionic strength of the aqueous solution, together with other factors, is used to determine the degree of stabilization. One of the most significant challenges with foam in EOR is that oil may make it unstable, making understanding the interplay between foam and oil challenging. The bulk foam experiments show that the literature on the effect of oil on foam durability is split. According to some authors, the presence of oil, particularly lighter hydrocarbons, reduces or delays foam production. Others, on the contrary, have proved that stable foams may be generated in the presence of oil by using a suitable foaming agent. Furthermore, the foam has been shown to develop even when a large amount of heavy oil is present. Other trials have shown that oil can help the foam last longer. Aveyard et al., for example, undertook several studies to identify the link between alkanes and the half-life of a single foam bubble (Aveyard et al., 1994). Their findings show that foams made of shorter-chain hydrocarbons disintegrate quickly. They also observed that foams made of longer-chain hydrocarbons were more stable (Shojaei et al., 2018). Oil is thought to destabilize foam via direct surface interactions between oil and foam, which are impacted by a range of the foam's physicochemical properties. We know that these interactions occur in three unique forms. Oil enters the gas–liquid interface, expands, and creates an unstable bridge over the lamella. When Ibrahim et al. exposed alpha olefin sulfonate (AOS) foam to oil, they noticed that it became unstable and decomposed quickly. The foam evaporated in 30 minutes when it would have taken more than 5 hours without the oil (Ibrahim and Nasr-El-Din, 2018). Because crude oil hastens the decomposition of foam, it is harmful to foam for two reasons. To begin with, oil droplets spread along with the gas–liquid interface, or where gas and liquid meet. Second, oil forms an emulsion within the foam's lamellae. This enhances the instability of the gas–liquid interface, hastening foam lamellae disintegration. Oil droplets congregate in an unstable emulsion and speed up the drainage of foam lamellae, causing the foam to die faster. Within the AOS foam lamellas, there are also layers of crude oil and an emulsion. Oil droplets gather in the unstable emulsion and enhance drainage, progressively weakening the lamellas. As a result, the foam loses stability and degrades faster.

7.4 EFFECT OF CRUDE OIL COMPOSITION

Experiments with the specific crude oil in a porous material are necessary to properly comprehend and explain the behavior of foam in an oil reservoir, and extrapolation from one crude oil to another is not yet conceivable. This technique might take a long time. As a result, modeling the influence of crude oil on foam solely based on its composition is preferable. This would improve the efficiency of reservoir filtering for foam application. No published model predicts the influence of crude oil on foam based on GC, total acid number (TAN), total base number (TBN), or saturated, aromatic, resin, and asphaltene (SARA) components. There are several reasons why creating such a model is difficult. This is due to a variety of factors, including a lack of information on soil composition, a large number of different components in crude oil, and certain compounds that weaken foam while others stabilize it. However, one can prepare "simulated" crude oil using this technique by combining seven pure chemical components (OC). It includes the most prevalent components of crude oil, such as TAN and TBN, as well as the organosulfur concentration seen in "sweet" crude oils.

Jensen et al. conducted steam-foam studies on porous media with four different crude oils and discovered that independent of the crude oil, the pressure drop through their core was solely dependent on oil saturation (Jensen and Friedmann, 1987). However, they were unable to find a clear correlation between the SARA composition of their crude oils and the influence of bulk foam. Others have looked at the effect of crude oil's pure components on foam in bulk and porous media (alkanes, organic acids, alcohols, and aromatics). Tang et al. mixed pure alkanes (C16, C10, C8, and C6) with foam and injected them into sandstone, but none of them produced an apparent viscosity as low as our crude oil (Tang et al., 2019). Furthermore, in terms of their influence on bulk and porous foams, it is unknown how pure oils vary from crude oils or oil combinations. The goal of this study is to see how the presence of OCs (as a distinct phase) in a surfactant solution influences bulk foam production and collapse; in other words, the antifoaming abilities of different pure OCs are investigated. The antifoaming action of an OC may be separate from the instability generated by an OC scattered throughout a formed bulk foam, as previously proven (i.e., defoaming). Antifoaming chemicals cause foam films to rupture in two ways: first, the OC drop comes into contact with the air–water interface, then goes into the foam film and ruptures it. OC drops can affect foam foamability by causing froth bubbles to clump together during foam creation. Furthermore, the anionic surfactant in the foaming solution may react with the natural cationic surfactant in crude oil, generating a bigger aggregate with no hydrophilic head that is typically insoluble in water. High-viscosity emulsions that help in the stabilization of bulk foam may also be made using a surfactant solution and an OC. In a work by Chaturvedi and coworkers, the role of foam degradation as a function of Carbon number on foam degradation was observed and reported. However, the work lacked a detailed understanding of bulk-foam degradation.

However, bulk-foam research has the benefit of providing a clear picture of how an OC affects foam in a variety of ways. In an opaque porous environment, only the pressure gradient and saturation of the several phases of foam can be recognized. While the effluent from porous media can be visually examined for hints regarding the porous medium's foam properties, it does not always accurately represent the porous medium's foam characteristics due to foam creation at the outflow, such as via the capillary end effect. The behavior of foam in bulk and porous media does not always match when an OC is present. For oil-free foam, Jones et al. discovered a relationship between its maximum apparent viscosity in a porous medium and its bulk-foam half-life, but not for oil-containing foam (Jones et al., 2016). However, we feel that looking at both porous media and bulk foam will give us a better understanding of how foam interacts with an OC in a porous medium. This chapter, however, does not address interfacial tensions (IFT) between OC, surfactant solution, and gas. Furthermore, Foam stability prediction using IFT values is erroneous in both bulk and porous media.

7.5 IMPROVING FOAM STABILITY

For EOR applications, the combination of NP and surfactant has also been extensively investigated. Due to the presence of surface-active groups in their constituents, mixing alters the reservoir's characteristics by altering wettability and adsorption at the oil–water interface. When precursor NPs or surfactants alone are unable to form stable emulsions, surfactant nanofluids, also known as nanosurfactants, achieve ultralow IFT and create stable emulsions. Additionally, the surfactant's interaction with the NPs' surface reduces the surfactant's adsorption on rock pores via a competitive adsorption process. On the contrary, the relative concentration ratio between the NP and the surfactant is critical and determines the system's characteristics. When the surfactant-to-NP ratio is low, just a trace amount of surfactant is available to coat the NP's surface. On the contrary, a high concentration ratio results in the formation of a surfactant bilayer on the surface of the NP.

The lamellae's stability dictates the foam's stability in porous environments. Recently, NPs have been proposed as a means of generating stable foam in difficult reservoir conditions, either by incorporating them into the foaming surfactant solution or by altering their surface wettability. NPs performed equally in foams stabilized with NPs and foams stabilized with NP–surfactant combinations. The majority of the challenges associated with surfactant-stabilized foams can be resolved by stabilizing the foam using

NPs. Due to the solid nature of NPs, the foams that they stabilize should be extremely resistant to reservoir conditions such as high salinity and temperature. As with chemically stabilized foams, changes in the polymer chain or surfactant chemical structure have no influence on the stability of the NP-stabilized foam. Three critical processes for NP foam stabilization have been identified: particle detachment energy, maximum capillary pressure of coalescence, and film drainage kinetics. At the gas–liquid interface of foam, NPs may adsorb with very high adhesion energy, resulting in irreversible attachment. This attachment energy is several times more than that of a surfactant molecule, according to research. Surfactants having lower adsorption energy due to smaller particle sizes and less energetic interaction with the fluid-fluid interface generate more stable foam due to the particles being aggressively and persistently adsorbed on the gas–liquid interface. However, at extremely low and high contact angles (30° and 150°, respectively), very little detachment energy is generated, resulting in highly hydrophilic or hydrophobic particles incapable of forming or stabilizing powerful foams. By raising the maximum capillary pressure that the foam lamellae may bear without rupturing, the steric barrier generated by NP adsorption on bubble surfaces increases the endurance of nanoparticle-stabilized foams. Figure 7.1 illustrates how a three-dimensional network structure formed by nonadsorbing NP stratification in the thin layer separating the dispersed phase enhances foam durability. Sun and colleagues investigated the effect of NPs on the production, dispersion, and stability of SiO_2/sodium dodecyl sulfate (SDS)-stabilized foam in micromodels and porous media containing sand packs (Sun et al., 2014). According to their findings, Figures 7.1 and 7.2 demonstrate how silica NPs stabilize SiO_2/SDS foam in the presence of oil. Due to the small amount of microforce exerted on the oil droplet in the case of SDS-stabilized foam, the foam was incapable of changing its form. Coalescence occurs when the bubbles explode, trapping a massive amount of oil in the porous substance. In this case, the enhanced microforce acting on the oil droplet resulted in a significant volume of oil being displaced by the SiO_2/SDS foam. The NPs coupled together increased the viscoelasticity of the bubble surface, resulting in a greater microforce.

7.5.1 INCREASING FOAM RESILIENCE BY THE USE OF NPS

The preceding section demonstrated that when foam systems are exposed to crude oil at elevated temperatures, they become unstable (Fig. 7.3). NPs have been extensively researched as a technique of stabilizing foams used

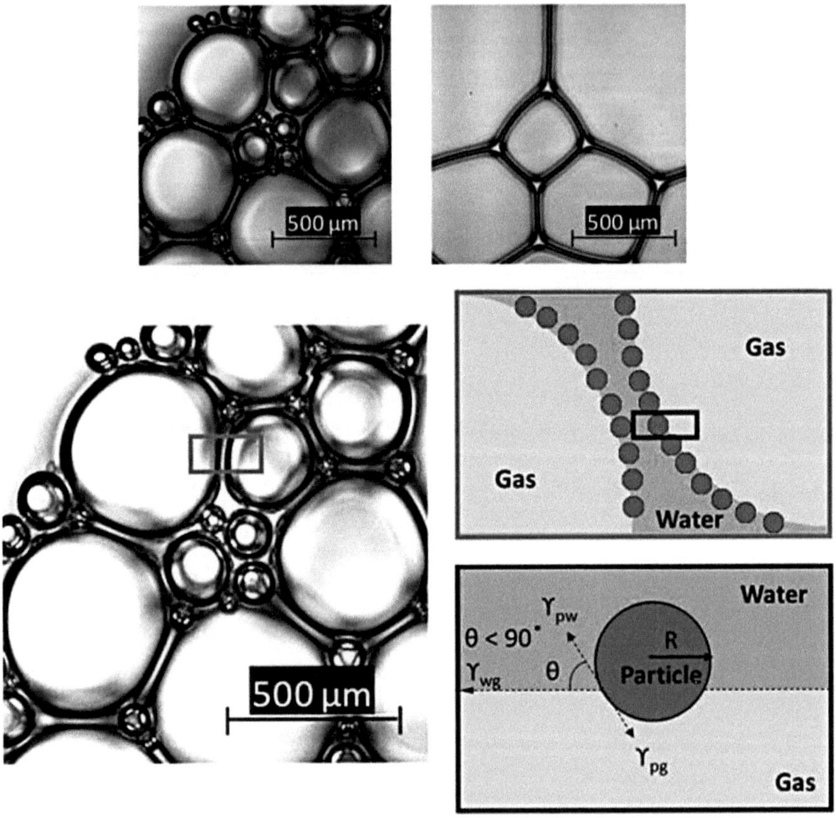

FIGURE 7.2 Mechanism for foam stability using NPs and creation of lamella.

Source: Reprinted from Sakthivel et al. (2019). Copyright (2019), with permission from ACS.

in oil production activities, particularly in settings with high salinity and temperature. NPs can create a strong barrier that prevents bubble coalescence due to their adsorption at the gas–liquid interface. The adsorption of NPs on the surface of the lamella is illustrated. As a result, NP-stabilized foams should be able to withstand extreme reservoir conditions such as high salinity, high temperatures, and crude oil for an extended period of time. For the present being, it is believed that silica NPs are the most effective method of increasing foam stability. The size of the NPs has a considerable effect on the stability of the foam. Numerous research conducted using silica NPs has demonstrated that the smaller the NPs, the more stable the foam. In comparison to bigger NPs, the tiny particle will migrate quickly to the gas–liquid interface. As a result, NP adsorption and concentration on the lamella surface

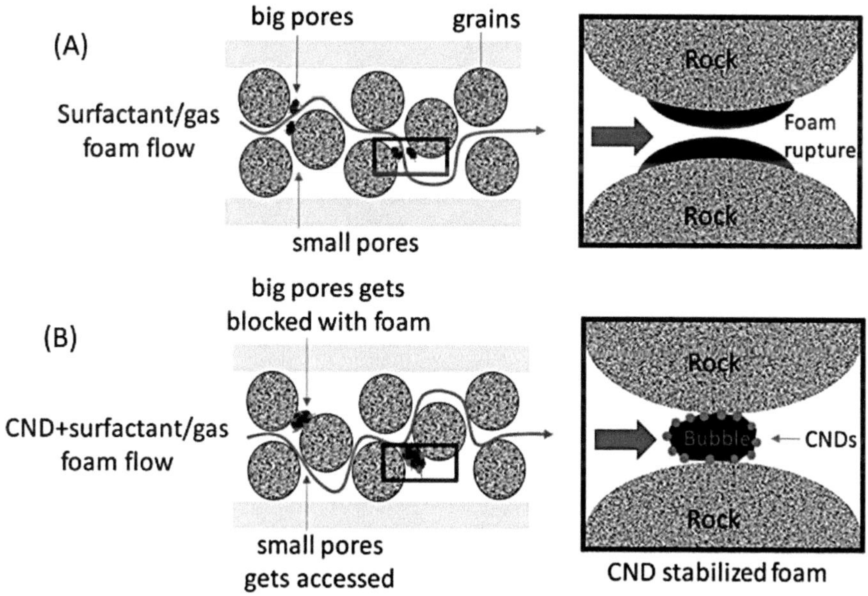

FIGURE 7.3 Foam flow in porous media (A) without NPs (B) with NPs (CND).
Source: Reprinted from Sakthivel et al. (2019). Copyright (2019), with permission from ACS.

increases, resulting in a more stable foam, as seen in Figure 7.2. As the size of the NPs increases, foam quality, salinity, and the hydrophobicity of the NPs all contribute to foam stability. When the foam quality is 70–80%, larger NPs enhance foam stability; whereas, when the foam quality is 50–60%, smaller NPs improve foam stability. Furthermore, in terms of foam stability, 140-nm silica NPs with an 86° contact angle surpassed 100-nm silica NPs with a 54° contact angle.

Emrani and colleagues investigated the effect of silica NPs with a diameter of 140 nm on an AOS foam system (Emrani and Nasr-El-Din, 2017). When NPs are added, a fine-textured foam with a greater apparent viscosity and MRF is produced. Adding NPs to the AOS foam system quadrupled the half-life in a bulk stability test, compared to the initial result without NPs. Additionally, NPs can adsorb at the gas–liquid interface, forming dense, solid coatings that are resistant to film thinning and inter-bubble diffusion. As a result, when crude oil is present, the dispersion of oil droplets along the foam lamellae is reduced, limiting the creation of bridges. The IFT analysis of a silica NP foam system revealed negative spreading and bridging coefficients (0.69, 4.43 at 1200 psi and 150°F, respectively), implying greater foam stability.

The concentration of NPs at the liquid–gas interface is crucial for foam stabilization and must be maintained at a consistent value. The liquid–gas contact is not saturated at low concentrations, resulting in poor foam stability. Foam stability improves as the concentration of NPs rises. At increasing concentrations, NPs agglomerate and form larger particles, limiting foam stability. At concentrations greater than 0.2 wt%, NPs become unstable and agglomerate, as shown by an increase in the zeta potential value. Ibrahim and Nasr-El-Din compared the EOR performance of an AOS foam system to that of a silica–NP system using core flood testing. The oil recovery rate was increased to 60% using water-assisted gas (WAG) technology. Oil recovery increased by 1.8% when AOS was used because it created a weak foam with a similar MRF to the CO_2/water system. Oil recovery improved to 68.2% when silica–NPs were added to the foam system. Suspension's absolute zeta potential decreases in a high-salinity environment. As a result, NPs will have a high predisposition for aggregation formation. To minimize NP instability in a high-salinity environment, surface-modified NPs were employed to promote steric repulsion between particles.

Another critical factor affecting foam stability is the surface wettability of NPs. The surfaces of NPs must be sufficiently hydrophilic to diffuse in the water while being enough hydrophobic to aggregate at the water–gas interface. In comparison to nanoparticles coated with polyethylene glycol (PEG) or NPs coated with a greater SiOH concentration of dichloro dimethyl silane, NPs coated with 50% SiOH dichloro dimethyl silane formed a stable foam at an 8 wt% NaCl salt concentration. Apart from silica, a variety of different types of NPs can be employed to enhance foam stability. 0.1-wt% Fe_2O_3 NPs Fe_2O_3 was shown to be capable of extending the half-life of AOS foam from 1 to 7 hours when used at 75°F and 300 pressure. Due to their enormous surface area, these NPs, On the contrary, prefer to aggregate, as shown by their low absolute zeta potential values. Bayat et al. experimented to determine the foam stability of SiO_2, Al_2O_3, TiO_2, and CuO NPs (Bayat et al., 2016). 0.008 wt% was determined to be the optimal concentration for these NPs.

7.5.2 USING VISCOSIFIERS TO IMPROVE FOAM STABILITY

Figures 7.4 and 7.5 demonstrate how adding thickeners to CO_2 foam, such as polymers or VES, extends the life of the foam by increasing lamella viscosity, which slows lamella drainage when exposed to high temperatures and crude oils. The foam stability graph in Figure 7.5 demonstrates that the

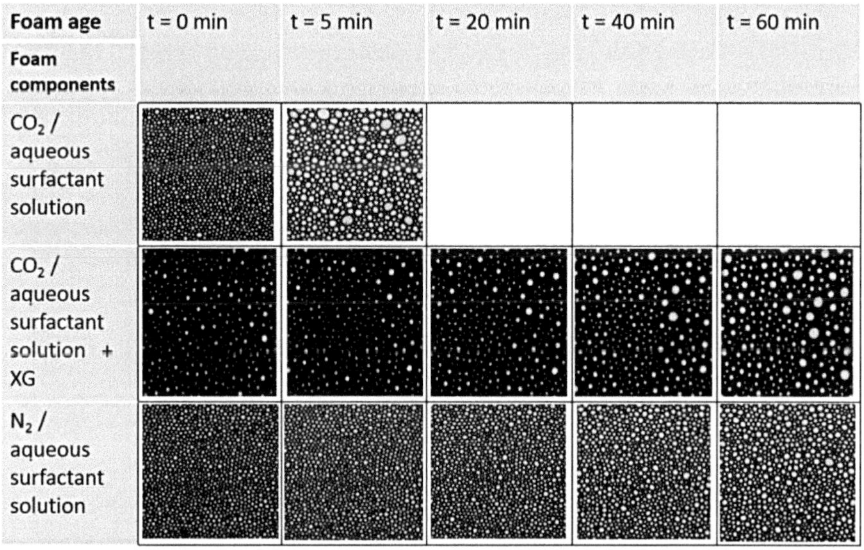

FIGURE 7.4 Foam stability in the presence of surfactant and polymer on droplet size.

Source: Reprinted from Sakthivel et al. (2019). Copyright (2019), with permission from ACS.

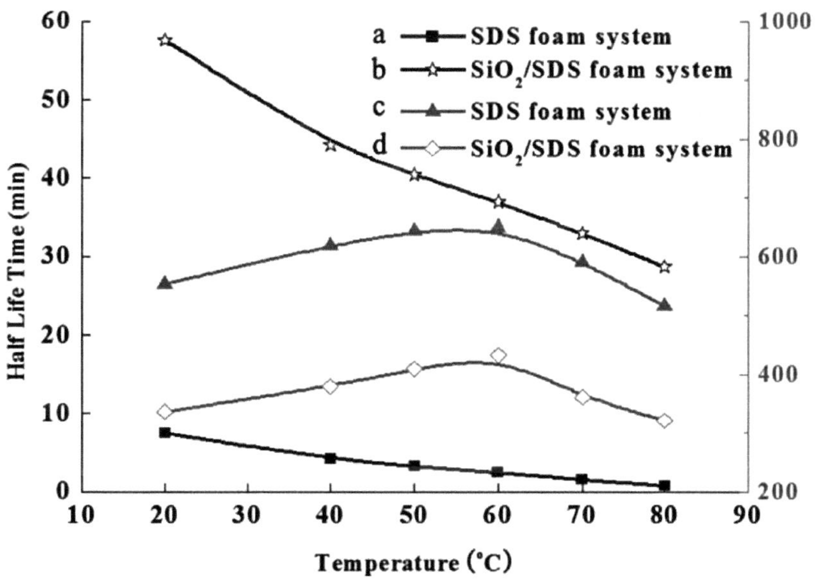

FIGURE 7.5 Foam half-life with foam volume at elevated temperature.

Source: Reprinted from Sun et al. (2014). Copyright (2014), with permission from ACS.

sample containing NPs is more stable than conventional foam. Ibrahim and Nasr-El-Din used a dual-core flood to determine the divergence ability of VES-stabilized foam in heterogeneous formations (permeability contrast of 700/100). The VES improved sweep efficiency while maintaining stable foam. After the water flooding stage, oil recovery from low-permeability and high-permeability cores was 19% and 55%, respectively. Residual oil saturation was 36% in low-permeability cores and 31% in high-permeability cores, indicating a low sweep efficiency. The addition of VES to the foam system increased sweep efficiency and reduced residual oil saturation to 23 and 27 vol % in the low- and high-permeability cores, respectively. According to the foam prepared using the setup provided in Figure 7.6, the overall oil recovery rate following foam injection was between 50% and 62% of the original oil in situ (OOIP). While water flooding generated the majority of the oil in the core with high permeability, foam flooding boosted the recovery factor by just 10%. The majority of the decrease in red color distribution occurred following foam flooding in the event of a low-permeability core. Wei et al. found that adding xanthan gum to a sulfobetaine-based surfactant foam system enhanced its foam durability at 90°C at 1450 pressure. Wei, et al. (2018) As the amount of polymer in the liquid phase increases, its viscosity increases as well. As a result, lamella drainage is reduced and the

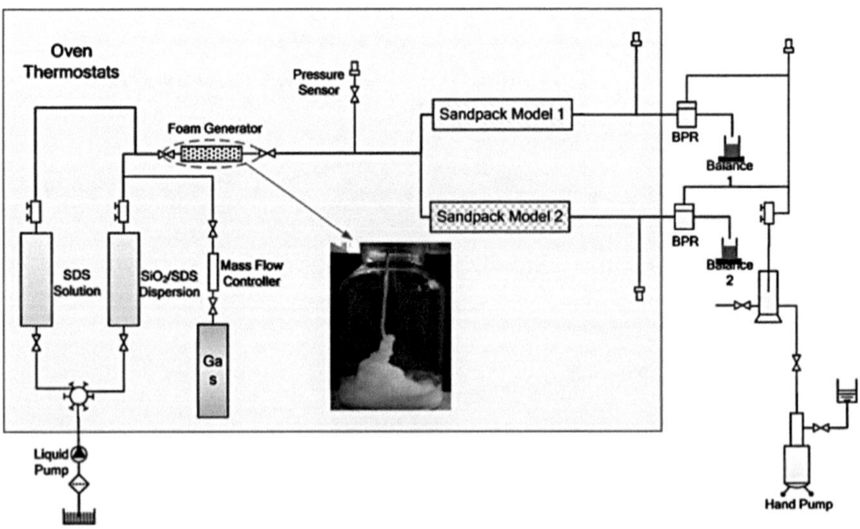

FIGURE 7.6 Schematic showing foam generation and flooding experimental set up for oil recovery.

Source: Reprinted from Sun et al. (2014). Copyright (2014), with permission from ACS.

half-life of foam stability is enhanced. However, when polymer concentration increases, surface-tension stresses increase proportionately, limiting the solution's foamability. Stable foam has a longer half-life than unstable foam, which raises the apparent viscosity of the displacing fluid and so boosts sweep efficiency. The polymer-foam system recovered 43.2% of the oil, but only 21.8% using the polymer-free foam method (Fig. 7.7).

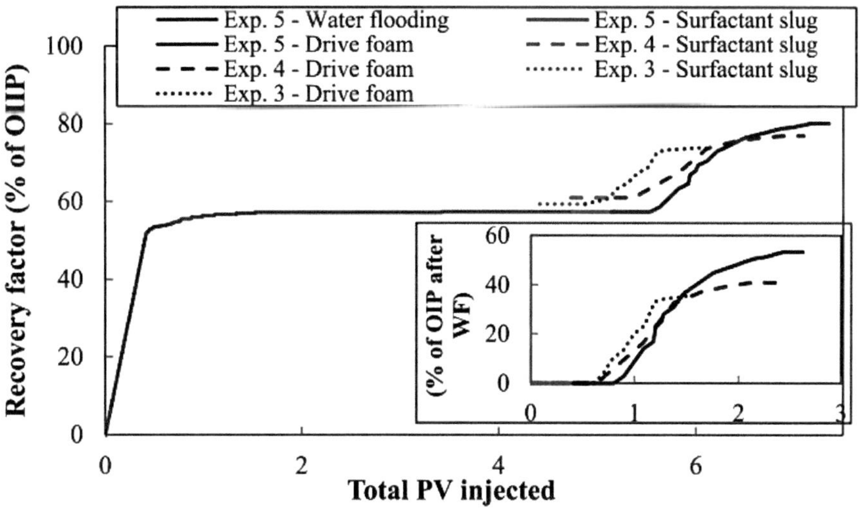

FIGURE 7.7 Oil recovery results with water flooding, slug injection, and foam flooding.
Source: Reprinted from Sun et al. (2019). Copyright (2019), with permission from ACS.

7.6 CONCLUSION

This chapter discusses how foam may be used to aid in oil recovery, with an emphasis on the following points: To create foam lamellae, a variety of procedures can be used, including snap-off, lamella division, and leave-behind operations. Because the lamellae lie parallel to the flow direction, the foam formed by the leave-behind technique is feeble. The effect of crude oil on foam stability may be calculated by plotting to enter, spreading, and bridging coefficients against the oil–gas–water IFT. NPs can be utilized to enhance foam's endurance in difficult settings, such as those with a high salt content or a high temperature. SiO_2 with a modified surface is the most effective and widely used NP for foam stability applications. VES can be used as a thickening agent in high-temperature applications to avoid lamella drainage and foam disintegration. CO_2 foam can be utilized to enhance oil recovery during the EOR process. On the contrary, surfactant-based foam is unstable and has a low sweep efficiency. As a consequence, NPs are added to the foam to increase its viscosity, resulting in increased stability.

KEYWORDS

- **foam**
- **EOR**
- **surfactant**
- **interfacial tension**
- **mobility control**
- **reservoir sweep efficiency**

REFERENCES

Aveyard, R.; Binks, B. P.; Fletcher, P. D. I.; Peck, T. G.; Rutherford, C. E. Aspects of Aqueous Foam Stability in the Presence of Hydrocarbon Oils and Solid Particles. *Adv. Colloid Interface Sci.* **1994,** *48*, 93–120. https://doi.org/10.1016/0001-8686(94)80005-7

Bayat, A. E.; Rajaei, K.; Junin, R. Assessing the Effects of Nanoparticle Type and Concentration on the Stability of CO_2 Foams and the Performance in Enhanced Oil Recovery. *Colloids Surfaces A Physicochem. Eng. Asp.* **2016,** *511*, 222–231. https://doi.org/10.1016/j.colsurfa.2016.09.083

Chaturvedi, K. R.; Trivedi, J.; Sharma, T. Single-Step Silica Nanofluid for Improved Carbon Dioxide Flow and Reduced Formation Damage in Porous Media for Carbon Utilization. *Energy* **2020,** *197*, 117276. https://doi.org/10.1016/j.energy.2020.117276

Chen, S.; Liu, H.; Yang, J.; Zhou, Y.; Zhang, J. Bulk Foam Stability and Rheological Behavior of Aqueous Foams Prepared by Clay Particles and Alpha Olefin Sulfonate. *J. Mol. Liq.* **2019,** *291*, 111250. https://doi.org/10.1016/j.molliq.2019.111250

Emrani, A. S.; Nasr-El-Din, H. A. An Experimental Study of Nanoparticle-Polymer-Stabilized CO_2 Foam. *Colloids Surfaces A Physicochem. Eng. Asp.* **2017,** *524*, 17–27. https://doi.org/10.1016/j.colsurfa.2017.04.023

Hirasaki, G. J.; Miller, C. A.; Puerto, M. Recent Advances in Surfactant EOR. *SPE J.* **2011,** *16* (04), 889–907. https://doi.org/10.2118/115386-PA

Ibrahim, A.F.; Nasr-El-Din, H. In *An Experimental Study for the Using of Nanoparticle/VES Stabilized CO_2 Foam to Improve the Sweep Efficiency in EOR Applications*, Proceedings of the SPE Annual Technical Conference Exhibition, Sept 2018. https://doi.org/10.2118/191534-MS

Janssen, M. T. G.; Mutawa, A. S.; Pilus, R. M.; Zitha, P. L. J. Foam-Assisted Chemical Flooding for Enhanced Oil Recovery: Effects of Slug Salinity and Drive Foam Strength. *Energy Fuels* **2019,** *33*, 4951–4963. https://doi.org/10.1021/acs.energyfuels.9b00645

Jensen, J. A.; Friedmann, F. In *Physical and Chemical Effects of an Oil Phase on the Propagation of Foam in Porous Media*, SPE California Regional Meeting, CRM 1987 473–486, 1987. https://doi.org/10.2118/16375-MS

Jones, S. A.; van der Bent, V.; Farajzadeh, R.; Rossen, W. R.; Vincent-Bonnicu, S. Surfactant Screening for Foam EOR: Correlation Between Bulk and Core-Flood Experiments. *Colloids Surfaces A Physicochem. Eng. Asp.* **2016,** *500*, 166–176. https://doi.org/10.1016/j.colsurfa.2016.03.072

Kumar, R. S.; Sinha, A. S. K.; Sharma, H.; Sharma, T. High Performance Carbon Dioxide Foams of Nanocomposites of Binary Colloids for Effective Carbon Utilization in Enhanced Oil Recovery Applications. *J. Mol. Liq.* **2021,** *343*, 117659. https://doi.org/10.1016/J.MOLLIQ.2021.117659

Liu, J.; Xie, L.; Yao, Y.; Gan, Q.; Zhao, P.; Du, L. Preliminary Study of Influence Factors and Estimation Model of the Enhanced Gas Recovery Stimulated by Carbon Dioxide Utilization in Shale. *ACS Sustain. Chem. Eng.* **2019,** *7*, 20114–20125. https://doi.org/10.1021/acssuschemeng.9b06005

Moradi, B.; Pourafshary, P.; Jalali Farahani, F.; Mohammadi, M.; Emadi, M. A. In *Application of SiO2 Nano Particles to Improve the Performance of Water Alternating Gas EOR Process*, SPE Oil and Gas India Conference and Exhibition; Society of Petroleum Engineers, 2015. https://doi.org/10.2118/178040-ms

Osei-Bonsu, K.; Shokri, N.; Grassia, P. Fundamental Investigation of Foam Flow in a Liquid-Filled Hele-Shaw Cell. *J. Colloid Interface Sci.* **2016,** *462*, 288–296. https://doi.org/10.1016/j.jcis.2015.10.017

Sakthivel, S.; Adebayo, A.; Kanj, M. Y. Experimental Evaluation of Carbon Dots Stabilized Foam for Enhanced Oil Recovery. *Energy Fuels* **2019,** *33*, 9629–9643. https://doi.org/10.1021/ACS.ENERGYFUELS.9B02235/SUPPL_FILE/EF9B02235_SI_001.PDF

Shojaei, M. J.; Osei-Bonsu, K.; Richman, S.; Grassia, P.; Shokri, N. Foam Stability Influenced by Displaced Fluids and by Pore Size of Porous Media. *Ind. Eng. Chem. Res.* **2018,** *58*, 1068–1074. https://doi.org/10.1021/ACS.IECR.8B05265

Sun, Q.; Li, Z.; Li, S.; Jiang, L.; Wang, J.; Wang, P. Utilization of Surfactant-Stabilized Foam for Enhanced Oil Recovery by Adding Nanoparticles. *Energy Fuels* **2014,** *28*, 2384–2394. https://doi.org/10.1021/ef402453b

Sydansk, R. D.; Romero-Zerón, L. n.d. Reservoir Conformance Improvement 138.

Tang, J.; Vincent-Bonnieu, S.; Rossen, W. R. Experimental Investigation of the Effect of Oil on Steady-State Foam Flow in Porous Media. *SPE J.* **2019,** *24*, 140–157. https://doi.org/10.2118/194015-PA

Wei, P.; Pu, W.; Sun, L.; Pu, Y.; Li, D.; Chen, Y. Role of Water-Soluble Polymer on Foam-Injection Process for Enhancing Oil Recovery. *J. Ind. Eng. Chem.* **2018,** *65*, 280–289. https://doi.org/10.1016/j.jiec.2018.04.039

Xue, Z.; Panthi, K.; Fei, Y.; Johnston, K. P.; Mohanty, K. K. CO_2-Soluble Ionic Surfactants and CO_2 Foams for High-Temperature and High-Salinity Sandstone Reservoirs. *Energy Fuels* **2015,** *29*, 5750–5760. https://doi.org/10.1021/acs.energyfuels.5b01568

CHAPTER 8

Recent Trends in EOR: Nanofluid Flooding

RAVI SHANKAR KUMAR[1,2], MUHAMMAD ARIF[1], KUSHAL DAS[2], and TUSHAR SHARMA[2]

[1]*Department of Petroleum Engineering, Khalifa University, Abu Dhabi, United Arab Emirates*

[2]*Enhanced Oil Recovery Laboratory, Department of Petroleum Engineering, Rajiv Gandhi Institute of Petroleum Technology, Jais, Amethi, , India*

ABSTRACT

In this chapter, the role of a modern technique, nanofluid flooding has been elucidated for chemical enhanced oil recovery (EOR). Nanofluids are the suspensions of nanoparticles in a base fluid, predominantly water, are used to mobilize oil by altering the wettability of the formation, lowering the interfacial tension and application of disjoining pressure on the isolated/stranded globules of oil. This chapter discusses the preparation, characterization, stability, and factors affecting the stability of nanofluids used in oilfield applications, particularly for EOR. Nanofluid preparation involves two methods: step-by-step and two-steps. The latter disperses pre-prepared powdered nanoparticles (NPs) in a suitable base fluid, often polyacrylamide (PAM). Maintaining nanofluid stability is challenging due to NP agglomeration, leading to sedimentation, rheological, and thermal instability, potentially lowering oil recovery. This chapter also discusses the methods to assess nanofluid stability, including visual appearance, dynamic light scattering (DLS), and ultraviolet–visible (UV–vis) spectroscopy. Visual appearance observes changes in color, while DLS provides quantitative data on NPs' size

Advancements in Chemical Enhanced Oil Recovery.
Tushar Sharma, Krishna Raghav Chaturvedi, Tarek Ganat, and Imtiaz Ali (Eds.)

distribution and zeta potential. UV–vis spectroscopy measures absorbance, indicating the weight percentage of NPs. Stability is crucial for successful EOR applications. Subsequently, in this chapter factors influencing nanofluid stability are discussed. Agglomeration and degradation are influenced by reservoir conditions such as temperature, salinity, and pH. Studies using different metal NPs (Al_2O_3, TiO_2, SiO_2) highlight their impact on oil recovery rates. However, the efficiency depends on reservoir conditions, influencing factors like wettability alteration, disjoining pressure, and adsorption on rock surfaces. Moreover, this chapter also discusses wettability alteration, IFT reduction, and disjoining pressure effects due to silica nanoparticles. Oil recovery experiments using nanofluids in sand pack cores demonstrate the impact of temperature and saline conditions on efficiency. Challenges in nanofluid application include agglomeration, economic viability, and the need for further research on mixed NPs. The chapter emphasizes the importance of stable and well-dispersed nanofluids, economic considerations, and fundamental understanding of processes. Finally, this chapter concludes the use of nanotechnology in EOR shows great potential, improving dispersion stability, rheological properties, wettability alteration, and IFT reduction. Ongoing challenges include developing stable nanofluids economically and addressing fundamental process understanding. Future research prospects were also discussed that includes modeling flow behavior through porous media and further fieldwork on nanoparticle assisted EOR methods.

8.1 INTRODUCTION

Nowadays along with the growing population mitigating the energy demand is a big challenge for any country as the storage of nonrenewable sources is depleting continuously. Though renewable energy has come up with some new solutions but still, globally, hydrocarbon is the most prominent source of energy. So, along with exploring new resources, focusing on maximizing the oil production from present reservoirs is essential to fulfill the upcoming energy demand worldwide (Subramanian et al., 2019). Conventionally, oil is recovered by three different stages, namely, primary, secondary, and tertiary or enhanced oil recovery (EOR) (Sheng, 2013). It is found that by primary and secondary recovery methods, only one–third of the original oil in place (OOIP) can be recovered and the rest of the oil remains in the formation due to the combined action of capillary forces, reservoir heterogeneity, and unfavorable mobility contrast. So, to extract the remaining unrecovered oil and maximize oil recovery, EOR comes into play under the tertiary recovery methods. In general, EOR methods are divided into two sub-parts namely thermal and nonthermal. Thermal EOR is generally employed for shallow-depth reservoirs or in the high-pay zone and its main principle is to lower the viscosity of the crude oil by using heat energy. However, for the reservoirs at greater depth and in narrow pay zones, thermal recovery does not work whereas nonthermal EOR can do the job. Nonthermal EOR is now becoming a great potential recovery method to extract the residual oil. In nonthermal EOR processes, chemical EOR has turned out to be the most effective one (Speight, 2013). Wettability alteration and interfacial tension (IFT) reduction which are one of the two most important things to be done for a good oil gate can be done more efficiently and economically by this Chemical EOR methods (Nwidee et al., 2016; Stoll et al., 2008). The wettability of a reservoir rock is indicated by the fluid-rock interaction which defines the spreading tendency of the injected fluid whereas the IFT can be observed by the fluid–crude oil interaction (Kumar et al., 2021a). So, to increase the oil recovery, the oil–wet porous reservoir has to be changed to water–wet condition by altering the wettability using chemical slugs and also to lower the IFT between the injected fluid and crude oil. In this way, water–wet conditions and lowered IFT between fluid–crude oil favor a reservoir to produce the unrecovered oil. There are many chemicals such as polymers, alkali, surfactants have been used to date for chemical EOR to enhance the productivity of oil (Yousefvand and Jafari, 2015). Polymers are the most used chemicals for EOR purposes as it has the ability to enhance the displacement efficiency by increasing the viscosity of the injecting fluid and making a difference in viscosity measure between the crude oil and the injecting fluid which is a good criterion for the increment of oil recovery. The alkali and surfactants can change the wettability of the rock surfaces and reduce the IFT between injected fluid and crude oil resulting in an increase of oil recovery. Moreover, the application of these three [alkali (A), surfactant (S), polymer (P)] simultaneously can achieve a significant level of recovery and it has been practiced in the reservoirs nowadays. As an example, in the Viraj oil field (India), at real reservoir conditions, alkaline–surfactant–polymer (ASP) flooding resulted in 50% total oil recovery from the earlier 32% of OOIP during water flooding (Pratap and Gauma, 2004).

In spite of these successful EOR processes and increased oil production, energy demand is still there as the population is also rising. This can only be resolved by applying new technologies in the oil industry and nanotechnology is one of them. Nanoparticles (NPs), which is a branch of nanotechnology, have made a great leap due to their strikingly good properties that can be very much potential for oilfield applications such as exploration, drilling, cementing, gas hydrate, and oil recovery (Kapusta et al., 2012). As the recovery rate depends upon the change in wettability of the reservoir rock to water–wet and reduction in IFT, NPs have proved their efficiency in both of these. NPs are generally employed by making suitable nanofluids by adding certain base fluids (water, surfactant, or polymer) to them. These nanofluids possess enhanced rheological properties than the conventional injection fluids and that is why NPs are more advantageous in case of oil recovery from current reservoirs. Different NPs such as TiO_2 (Bayat et al., 2014), ZnO (Ilyas et al., 2019), Al_2O_3 (Gbadamosi et al., 2019), Fe_3O_4 (Wang et al., 2018), GO (Radnia et al., 2018), and SiO_2 (Kumar et al., 2020a; Kumar and Sharma, 2018; Sharma et al., 2016) have explored by different researchers for oil recovery applications and significant improvement in oil recovery was reported. NPs are found to have high surface energy which ease them to get adsorbed by the rock, removing the oil particles from the pore spaces. Subsequently, the reservoir condition changes to water–wet from oil–wet. Also due to the difference in surface energies of NPs and oil particles, the IFT reduces significantly. Lu et al. (2017) conducted a flooding experiment using silica NPs and found improvement in oil recovery. In addition to this, Lu et al. (2017) also suggested that cyclic recovery can be more effective for oil recovery as compared to continuous nanofluid injection. Moreover, the wettability alteration and IFT reduction can also be achieved using silica NPs (Li et al., 2017). NPs show a film-like structure at the time being transported through porous media. This assembly of wedge-shaped-film-like structures is responsible for the sweep efficiency of the crude oil from the rock surfaces via adsorption. These adsorbed NPs create a coating-like structure on the rock surface which makes the surfaces more water–wet than intermediate or oil–wet. This chapter thus insights into the nanofluid flooding aspect of the formulation methodology of stable nanofluid for the oil and gas industry by keeping wettability alteration, IFT reduction, and improved oil recovery as a primary objective.

8.1.1 NANOFLUID PREPARATION AND CHARACTERIZATION

Nanofluid preparation is the primary but important step toward the oilfield application. Generally, there are two procedures that are followed for nanofluid preparation—step-by-step method and two-steps method. In the first method, it is very complicated to generate the NPs and it is carried within the base fluids. The latter two step method is accepted by many because of its economic viability where preprepared powdered NPs are dispersed in suitable base fluid. In a previous work, the schematic of the preparation of nanofluid in a base fluid of Polyacrylamide (PAM) by using two-steps method (Kumar et al., 2020b, 2020a, 2021b; Kumar and Sharma, 2018, 2020a; Subramanian et al., 2019). After the preparation, the main challenge is to have a good stability of the nanofluid which is one of the main controlling factors for the EOR applications. The NPs, dispersed in the base fluid have a natural tendency to agglomerate due to the van der Waals force which results in forming a cluster of large NPs. Consequently, sedimentation, rheological, and thermal instability are seen to be encountered lowering the oil recovery.

8.1.2 STABILITY OF NANOFLUID

There are several methods available for looking into the dispersion stability of any nanofluid in which visual appearance, dynamic light scattering (DLS), and ultraviolet–visible (UV–vis) spectroscopy study are more focused. In many previous reports, it was shown that these methodologies stand for a good measure for the nanofluid stability analysis.

8.1.2.1 VISUAL APPEARANCE STUDY

Visual appearance study of NPs is the initial analysis of the distribution of NPs in the dispersed phase of the base fluid with its settling time. After the fresh preparation of the nanofluid, it is transferred to a transparent vessel, and changes in the nanofluid color are observed periodically. For observing the disparity in the color of the nanofluid, images are also captured showing the color of the nanofluid at the respective time. Saha et al. (2018) reported total dispersion stability of 10 days of Silica nanofluid in the presence of reservoir formation water on the basis of visual appearance analysis only. Similarly, for silica and silica–titania nanofluids, different sets of experiments were performed by Kumar and Sharma (2018) and it was stated that the agglomeration of NPs is a natural phenomenon. However, these agglomerated large-sized clustered NPs settle down due to the gravitational forces and show sedimentation as shown in Figure 8.1. The silica NPs showed higher sedimentation in the nanofluid; however, the homoagglomeration of silica NPs can be controlled using TiO_2 NPs as it was proposed in our previous study (Kumar and Sharma, 2018).

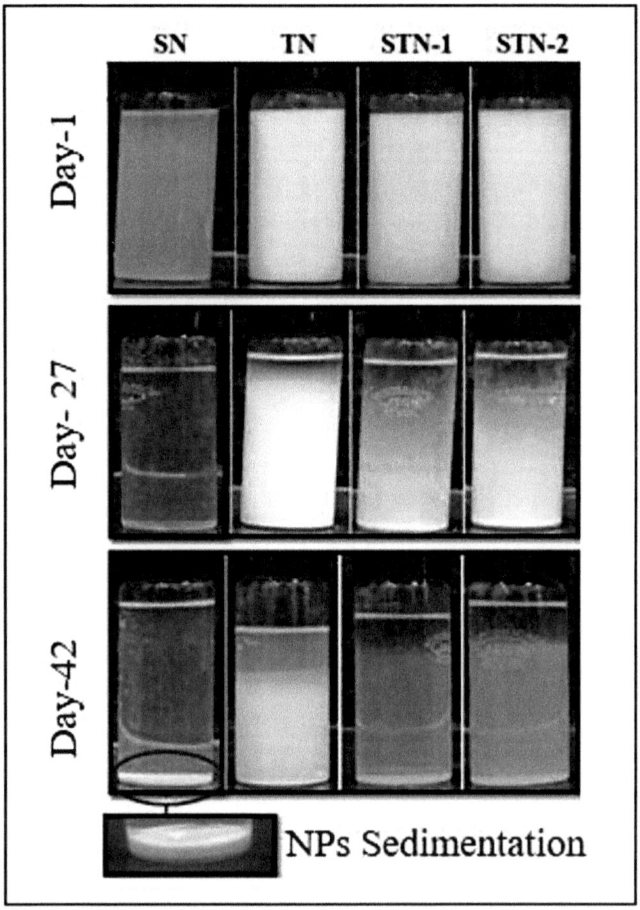

FIGURE 8.1 Visual appearance images of nanofluid with time. SN = silica nanofluid (0.5 wt% SiO_2), TN = titania nanofluid (0.5 wt% TiO_2), STN-1 (0.5 wt% SiO_2 +0.05 wt% TiO_2), and STN-2 (0.5 wt% SiO_2 + 0.1 wt% TiO_2).

Source: Reprinted from Kumar et al. (2018). Copyright (2018), with permission from Elsevier.

8.1.2.2 DLS STUDY

As macroscopic analysis (visual appearance study) of NPs does not give accurate, promising data as it is just a qualitative measure, and also large variations are expected in results. So, to minimize these constraints and get quantitative and reliable data about the nanofluid stability DLS study has been done. By virtue, NPs possess surface potential energy, measured as zeta (ζ) potential, which helps NPs to remain in dispersed form within the nanofluid through repulsive forces. As soon as these repulsive forces tend to be weaker between the NPs, they start to agglomerate in large-size clusters. The zeta (ζ) potential and size distribution of NPs can be determined using a particle size analyzer, which works on the principle of DLS technique. The large clusters of NPs can create some hindrance in the porous reservoir and it is one of the concerns for limited oil production. Thus, zeta (ζ) potential and NP size are important governing factors for the EOR. Sharma et al. (2016) conducted a DLS study for silica nanofluid and found a significant contribution of size over oil recovery.

8.1.2.3 ULTRAVIOLET–VISIBLE SPECTROSCOPY

UV–vis spectroscopy is another quantitative measure that determines the nanofluid stability. This methodology relies upon the absorbance of light to the proportion of nanofluid. Usually, the freshly formulated nanofluid sample is exposed to the incident light and according to the NPs higher or lower concentration in the base fluid, higher or lower absorbance is recorded. Absorbance is a relative measure of the weight percentage of the NPs within the nanofluids. So, as much as dense the NP solution is, the absorbance is also high, whereas lower populated NPs in solution indicate a lower absorbance value. Kumar et al. (2020a) also found that wt% of suspended NPs is solely responsible for the absorbency of light. It is inferred that; a higher absorbance value indicates more suspended NPs in the solution and a lower absorbance value indicates a lower wt% of NPs in suspension. Moreover, the agglomeration of the NPs can cause a decrease in absorbance value.

8.2 FACTORS AFFECTING NANOFLUID STABILITY

8.2.1 AGGLOMERATION AND DEGRADATION

In EOR processes, the efficiency of the injected fluids plays an important role which is also controlled by some reservoir conditions such as temperature, salinity, and pH. Because of these conditions, injected nanofluid may get contaminated, and at higher temperatures, salinity, and lower pH, it loses its dispersion stability. At elevated temperatures, the intermolecular repulsive forces between the NPs get weakened and this causes agglomeration resulting in low efficient EOR. In the case of a saline environment, the charged NPs in the nanofluid solution are preferentially screened in the reservoir brine. This restricts the NPs' agglomeration and sedimentation to some extent (shown in Fig. 8.1). By this process, the viscosity of the nanofluid can be reduced. Generally, the higher dispersion stability of NPs made the nanofluid more stable and prevented agglomeration of NPs. Kumar et al. (2020b) observed the effect of temperature on nanofluid stability when two different NPs are mixed together. They reported that silica–titania NPs based nanofluid can minimize the temperature effect responsible for the instability of nanofluids at higher temperatures.

8.2.2 ADSORPTION AND RETENTION OF NP IN POROUS MEDIA

Adsorption and retention of NPs indirectly affect the EOR efficiency depending upon the type of nanofluid injected. During flow through the porous media, NPs interact with the rock surfaces through electrostatic interaction, steric interaction, and van der Waal forces, due to which the concentration of injected nanofluid solution gets decreased. In the case of nanofluids like surfactants, adsorption is being encountered whereas the bigger polymer macromolecules tend to retain themselves due to the mechanical entrapment. When the NPs present in the bulk phase of the nanofluid get transferred to the solid–liquid interface, adsorption takes place at that site. The NPs, due to their high surface energy are favorable for the adsorption phenomenon at the interface than the bulk phase. In the process of NP retention, the NPs get removed or reduced from the aqueous phase in the flow medium, which is hazardous to the environment. Also, this retention phenomenon and inaccessible pore spaces give an idea of the flow of the nanofluid through the porous media. There are several factors such as electrolyte concentration (salinity), temperature, pH, composition of reservoir fluids, and the presence of clay mineral content affect the adsorption process. So, it can be said that, adsorption and retention NPs, these two are definitely one of the controlling factors of economic EOR efficiency.

8.2.3 EFFECT OF NPS AGGLOMERATION/SEDIMENTATION ON RHEOLOGICAL PROPERTIES

The viscosity of the nanofluid is affected by several factors, in which the presence of different ions in brine is the most. The variation in rheological properties is due to the different forces of interaction combined which depends on the saturation of the EDL by the ions of NPs and electrolytes and also the Brownian motion of the layered NPs present in the solution. Olayiwola and Dejam, (2020) studied the effect of different ions on the viscosity of nanofluid. According to them, cations and anions both can cause changes in viscosity results. For the same cation part, it was found that the viscosity of the nanofluid with SO_4^{2-} ion is higher than that composed of Cl^-. It follows the order of increasing viscosity- Na>Mg>K>Ca with the same anion.

8.3 SELECTION OF NANOMATERIALS

There are several NPs used for many oilfield applications but choosing the most efficient NPs for EOR activities in conventional reservoir conditions is very important. Bayat et al. (2014) in their work with three metal NPs (Al_2O_3, TiO_2, and SiO_2), examined the displacement efficiency as well as the quantity and quality of the recovered oil from an intermediate limestone core at variable temperature conditions. Compared with the conventional water flooding with brine (NaCl 0.3 wt%) which yielded an average of 47.3% oil at 26?C, the use of Al_2O_3, TiO_2, and SiO_2 NPs at the same conditions increased the recovery rate to 52.6%, 50.9%, and 48.7%, respectively in the displacement test. It can be observed that the oil recovery rate using Al_2O_3 nanofluid is better than the rest two, as it shows 1.5 and 2.2 times greater recovery than that of TiO_2 and SiO_2, respectively. The reason behind this higher percentage of recovery rate is the ability of Al_2O_3 NPs to lower capillary forces of oil at the time of displacement test. Alomair et al. (2014) observed the recovery rate by doing several flooding experiments using four different nanofluids (Al_2O_3, TiO_2, SiO_2, and NiO) at different concentrations. An increase of recovery was noted for the SiO_2 and Al_2O_3 NPs than the conventional water flooding method. SiO_2 increased the recovery by 0.958% while Al_2O_3 increased the recovery by 4.895%. After this potential result, they observed the result of the flooding experiment by mixing both the NPs where it showed a spike in oil recovery amounting to 23.724% greater than water flooding (Alomair et al., 2014). In the study of SiO_2, NiO, and Fe_3O_4 NPs on the recovery of heavy oil, it was found that flooding using SiO_2, NiO, and Fe_3O_4 NPs increased the oil recovery by 22.6%, 14.6%, and 8.1%, respectively with respect to water injection. In this experiment, a glass micromodel has been used in the core flooding setup for the stimulation of oil production. The higher amount of oil recovery by Silica NPs can be due to their ability to wettability alteration and IFT reduction. Nazari Moghaddam et al. (2015) studied the wettability phenomenon for eight different NPs and took $CaCO_3$ and SiO_2 NPs for EOR purposes. The performance study of these two NPs revealed that oil recovery was increased by 8–9% when nanofluid was injected which was also supported by the spontaneous imbibition results. When $CaCO_3$ and SiO_2 are present in the base fluid the recovery was to have increased by 4% and 6%, respectively. Roustaei and Bagherzadeh (2015) also studied the EOR performance for modified SiO_2 NPs by using two different Iranian light and intermediate oil reservoir samples. The two core samples were saturated, one with light oil and another one with intermediate oil, and the application of modified Silica NPs increased the oil recovery significantly. The oil recovery by the first plug of light oil saturation, increased by a further 25% than the water flooding which only recovered 54%. For the intermediate oil-saturated second plug, the recovery of 41% by water flooding was further increased by 14% by the nanofluid injection. It was inferred that the oil recovery rate also depends on the type of oil present in the reservoirs. According to them, this study's results surely showed the efficiency of NPs toward the recovery of oil after the primary and secondary recovery methods.

The proper selection of NPs for EOR purposes turns out to be a determining factor for its retention (by adsorption) on rock surfaces. The surface charge of reservoir rock or type (sandstone or carbonate) and the surface potential of the NPs reduce this retention phenomenon to some extent. If there are opposite charges on the rock surface and NPs a minimum repulsion does not exist by which NPs could remain in the bulk phase. Preliminary laboratory tests are always to be done before the application of EOR to check the efficiency of the NPs on wettability alteration, disjoining pressure, and adsorption trend on respective rock surfaces. Reservoirs encountering high temperature and high salinity formation fluids result in reduction surface

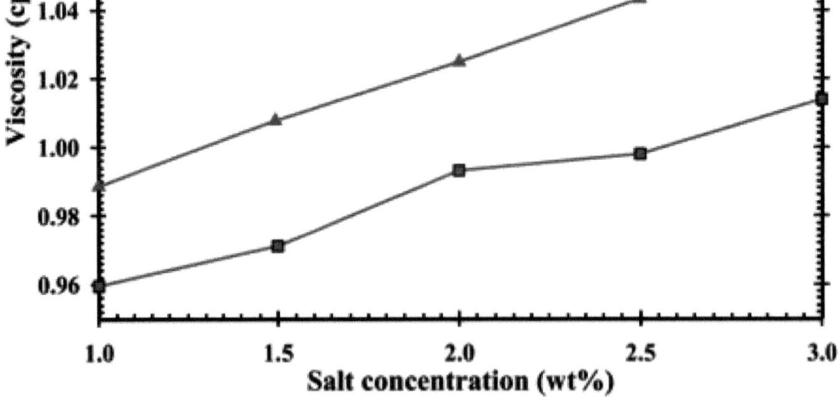

FIGURE 8.2 Variation of viscosity of 0.05 wt% SiO_2 nanofluid over salt concentration.

Source: Reprinted from Olayiwola et al. (2020). Copyright (2020), with permission from ACS.

charge of the injected NPs, which led to agglomeration of these NPs. For this, NPs or nanocomposites that can be resistive against high temperature and salinity can be designed. In heterogeneous reservoirs and fractured reservoirs, to obtain maximum recovery foam, emulsion, and gels made of NPs are recommended to obstruct the high permeable zone and make the injected fluids interact with the unswept oil. In summary, the selection of NPs for efficient oil recovery totally depends on the reservoir conditions and the suitability of different NPs in that reservoir conditions.

8.4 NANOFLUIDS IN EOR

Commonly the nanofluids used in enhancing the oil recovery are produced by simply using NPs suspension in a base fluid on its own without any chemicals or heat application. The various NPs available for nanofluid flooding follow many complex mechanisms for oil recovery processes that were discussed earlier like wettability alteration, IFT reduction, and disjoining pressure. Silica NPs are examined for wettability alteration of solid surfaces and many of the researchers reported that wettability alteration can be a slow process that is affected by the size of NPs, pH, and ionic strength. In some other work, it was shown that the hydrophilic SiO2 NPs can change the wettability of rock surfaces from oil–wet to water–wet. This eases for a thin oil film to be trapped near the pore walls as well as in the pore throats. Moreover, the effectivity toward wettability alteration of carbonate rocks has also been tested. This was done by contact angle measurement which proved that by using a low concentration of silica NPs the contact angle can be reduced from 156° (oil–wet) to 41° (water–wet). Wettability which is a measure of solid–liquid interaction, is measured in the form of contact angle made by the wetting phase to the solid surface (here rock surface) and it is always measured from denser to lighter phase. The wetting conditions for the solid surfaces are decided by the contact angle measurement which is governed by Young's law. It states three conditions—first, if $\theta < 90°$ it implies a condition of water–wet, secondly if $\theta > 90°$ it implies oil–wet and lastly, $\theta = 90°$ is the intermediate wet condition. When the nanofluid is injected, the presence of NPs accelerates the surface adsorption rate during the disjoining pressure phenomenon and makes the rock surfaces water–wet from the previous condition. Thus, a rich water–wet condition can be established through complete removal of the oil phase from the rock surface by using the nanofluids. So, wettability is one of the major controlling factors for the rate of oil recovery by affecting the capillary pressure, relative permeability, and connate water saturation. Another important parameter for the EOR process is the oil viscosity. So, for understanding the correlation between wettability and oil viscosity a study has been done by Maghzi et al. (2011) to find the wettability alteration result by silica NPs during water flooding to heavy oils. It was found that SiO_2 NP-based nanofluid effectively altered the wettability to water–wet from oil–wet. Cao et al. conducted tests for zirconium oxide and silica by dispersing in surfactant, alkali, and brine solution which showed a potential wettability alteration result under high pressure and high temperature (HPHT) reservoirs. Although, in practice, surfactant flooding is used for the IFT reduction, NPs-based nanofluid flooding has also found its potential for this application. Due to the high surface hydrophilicity of silica NPs, it can be used here as a replacement of conventional surfactant agents. In general, these hydrophilic and hydrophobic or any two different functional groups are connected on the two sides of the NPs to lower IFT. Bayat et al. (2014) experimented with the specimens on Al_2O_3, TiO_2, and SiO_2 nanofluid–crude oil systems to see the IFT results. Similarly, according to the study of Sharma et al. (2016), the silica NPs get adsorbed on the surface of crude oil and lower the IFT between crude oil and injected fluid. Saha et al. (2018) prepared a silica nanofluid in the base solution of 5000 ppm XG which resulted in a good oil recovery rate. In every case, the silica NPs are found to have lesser IFT at the oil–nanofluid interface with respect to the oil–water interface. The quality of silica nanofluid also determines its performance toward surface activity. The work done in this field of literature proved that functionalized Silica NPs are better surface reactive agents. In this literature, the various steps such as self-assembly, masking step, or phase separation regarding the synthesis of silica nanofluid, as well as its unique features of it, have also been described.

Structural disjoining pressure is very much responsible for the oil recovery from the porous media as many mechanisms are dependent on it. The wedge-shaped NP structure increases the disjoining pressure in the wedge film; the higher pressure being near the tip of the wedge. The presence of NPs during the disjoining pressure phenomenon accelerates the wettability alteration of rock surfaces.

8.4.1 OIL RECOVERY EXPERIMENTS

An oil recovery experiment is to be performed to understand the efficiency of the desired nanofluids at reservoir condition. The schematic for the flooding experiment setup has been given in Figure 8.3. A conventional EOR

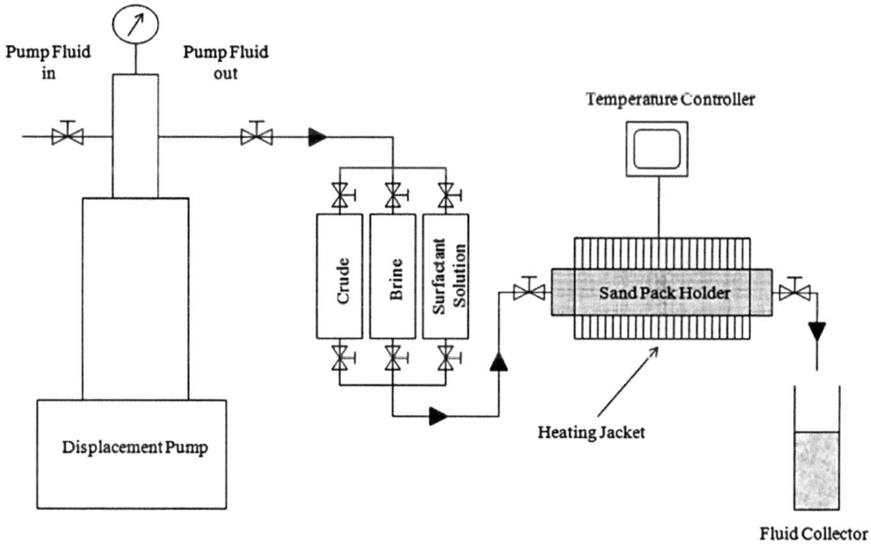

FIGURE 8.3 Schematic of core flooding apparatus used for flooding experiments.

Source: Reprinted from Goswami et al. (2018). Copyright (2018), with permission from Elsevier.

apparatus consists of a syringe pump, accumulators, and a core sample holder with a heating jacket. Several studies reported the use of sand pack, which is made up of fine clean sand as a core plug for oil recovery observation. The porosity and permeability measurements are done by using Darcy's law. The nanofluid is injected at a constant rate and at a low value to avoid early fracturing of cores or encountering a fingering effect. The result obtained by Kumar et al. (2020b) of the nanofluid flooding in sand pack cores is shown in Figure 8.4. This resulted in 68% maximum oil recovery at 40°C temperature and 0 wt% of NaCl, whereas the recovery reduces to 60% at the same temperature when 5 wt% NaCl solution is used as shown in Figure 8.4. In this experiment, an injection strategy of 2 pore volume (PV) of water, 0.5 PV of chemical slug (nanofluid), and 2.5 PV of chase water are employed. The difference in recovery results of using water, hydrolyzed polyacrylamide (HPAM) (1000 ppm), HS1, and HTS1 can be found graphically in Figure 8.4 (Kumar et al., 2020b). So, the temperature and saline conditions of the reservoirs do affect the efficiency of oil recovery by conventional fluids including silica nanofluid. The stability of nanofluids is also one of the major factors responsible for nanofluid recovery where reservoir conditions are also taken into account.

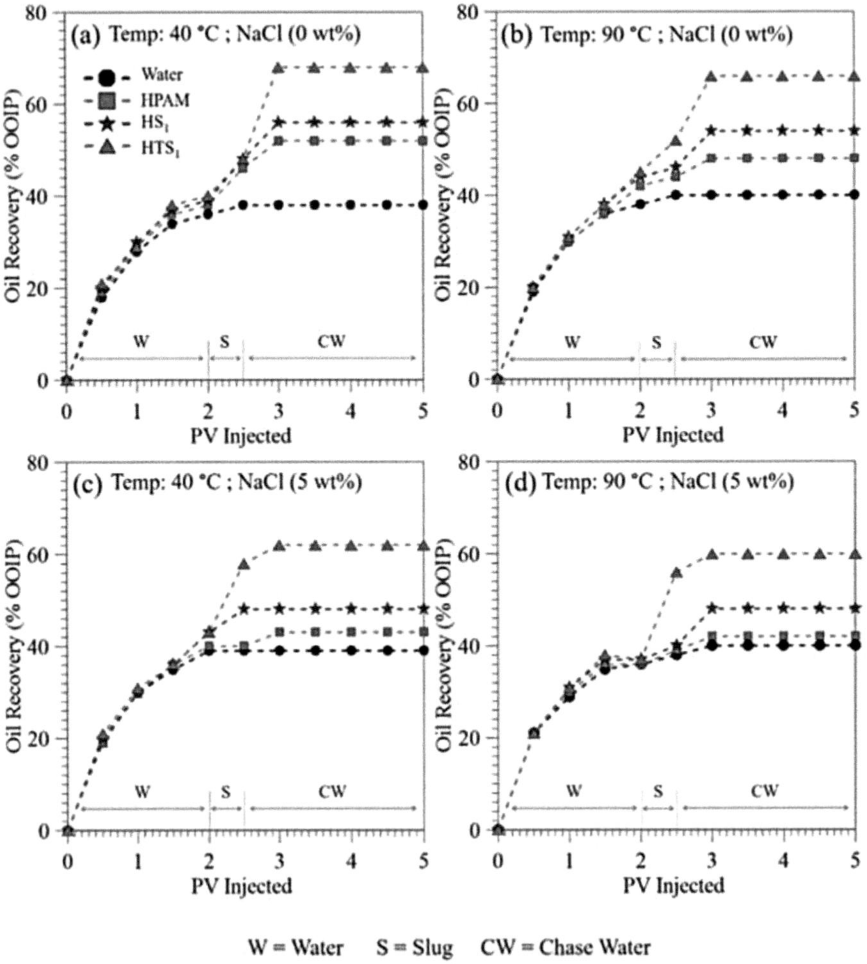

FIGURE 8.4 Oil recovery experiments at different temperatures and salinity for water, HPAM (1000 ppm), HS1 (1.0 wt % SiO_2), HTS1 (1.0 wt% SiO_2 + 0.1 wt% TiO_2); obtained recovery is in % via water flooding = 2 PV, chemical slug = 0.5 PV and 2.5 PV chase water flooding.

Source: Reprinted from Kumar et al. (2020b). Copyright (2020), with permission from ACS.

8.4.1.1 CHALLENGES AND FUTURE SCOPE

Nanotechnology itself has emerged as a potential field for various applications including the petroleum industry. The surface-active NPs have been successfully employed for the EOR methods. However, only very few of the

NPs have been recommended for field tests, the rest are still under research on a laboratory scale. Various challenges have been faced in the application of nanofluid flooding in EOR methods:

- NPs possess the tendency to form agglomeration and settle down due to gravity thus destabilizing the nanofluid. It is very challenging to make an improved, stable, and well-dispersed nanofluid (homogeneous).
- Though the study of several NPs for EOR purposes showed a lot of potential but still it requires a lot of investigations and modelling analysis for the case of a mixture of two NPs. Also, the economic viability should be considered as mixing of two NPs in nano-assisted EOR processes can increase the cost.
- Some areas like the mechanism of the NPs-based EOR method, interaction with reservoir formation, and initial reservoir fluid properties need more focus of study to get clearer inference from these.

From this point of view, the following areas of research need to be emphasized in the upcoming time:

- Formulation of more functional and stable nanofluid having greater dispersion stability toward the better recovery of oil through EOR methods.
- Better dispersion stability and rheological properties can be achieved by mixing the NPs but in an economical way to increase the production from depleted reservoirs and increase the cost effectivity.
- The fundamentals of the processes should be made clear and for this more feasible and vigorous authorization is needed.

8.5 CONCLUSIONS

From the discussion of this chapter, it is concluded that surely the advancement of EOR methods by the application of NPs is a great achievement. The use of nanotechnology in chemical EOR processes has come up with great potential results in oil recovery. In this chapter, the mechanism of oil recovery by nano-assisted EOR is also discussed. The application of NP improved the dispersion stability and enhanced the rheological properties along with the efficient wettability alteration and IFT reduction. This chapter also focuses on the ongoing and upcoming challenges of the nano-assisted EOR method and also the possible solution of it. The chapter also discussed the future research prospects in various areas like modeling the flow behavior of nanofluids through porous media, which can give a better understanding of the process of designing and fieldwork on the nano-assisted EOR method.

KEYWORDS

- **foam**
- **EOR**
- **Surfactant**
- **interfacial tension**
- **mobility control**
- **reservoir**

REFERENCES

Alomair, O. A.; Matar, K. M.; Alsaeed, Y. H. In *Nanofluids Application for Heavy Oil Recovery*, SPE Asia Pacific Oil & Gas Conference and Exhibition; Society of Petroleum Engineers, 2014. https://doi.org/10.2118/171539-MS

Bayat, A. E.; Junin, R.; Samsuri, A.; Piroozian, A.; Hokmabadi, M. Impact of Metal Oxide Nanoparticles on Enhanced Oil Recovery from Limestone Media at Several Temperatures. *Energy Fuels* **2014,** *28*, 6255–6266. https://doi.org/10.1021/ef5013616

Gbadamosi, A. O.; Junin, R.; Manan, M. A.; Augustine, A.; Oseh, J. O.; Usman, J. Effect of Aluminium Oxide Nanoparticles on Oilfield Polyacrylamide: Rheology, Interfacial Tension, Wettability and Oil Displacement Studies. *J. Mol. Liq.* **2019,** 111863. https://doi.org/10.1016/j.molliq.2019.111863

Goswami, R.; Chaturvedi, K. R.; Kumar, R. S.; Chon, B. H.; Sharma, T. Effect of Ionic Strength on Crude Emulsification and EOR Potential of Micellar Flood for Oil Recovery Applications in High Saline Environment. *J. Pet. Sci. Eng.* **2018,** *170*, 49–61. https://doi.org/10.1016/j.petrol.2018.06.040

Ilyas, S. U.; Narahari, M.; Theng, J. T. Y.; Pendyala, R. Experimental Evaluation of Dispersion Behavior, Rheology and Thermal Analysis of Functionalized Zinc Oxide-Paraffin Oil Nanofluids. *J. Mol. Liq.* **2019,** *294*, 111613. https://doi.org/10.1016/j.molliq.2019.111613

Kapusta, S.; Balzano, L.; Te Riele, P. In *Nanotechnology Applications in Oil and Gas Exploration and Production*, Society of Petroleum Engineers - International Petroleum Technology Conference, IPTC 2012; OnePetro, 2012; pp 2770–2774. https://doi.org/10.2523/iptc-15152-ms

Kumar, R. S.; Al-Arbi Ganat, T.; Sharma, T. Performance Evaluation of Silica Nanofluid for Sand Column Transport with Simultaneous Wettability Alteration: An Approach to Environmental Issue. *J. Clean. Prod.* **2021a,** *303*, 127047. https://doi.org/10.1016/j.jclepro.2021.127047

Kumar, R. S.; Chaturvedi, K. R.; Iglauer, S.; Trivedi, J.; Sharma, T. Impact of Anionic Surfactant on Stability, Viscoelastic Moduli, and Oil Recovery of Silica Nanofluid in Saline Environment. *J. Pet. Sci. Eng.* **2020a,** *195*, 107634. https://doi.org/10.1016/j.petrol.2020.107634

Kumar, R. S.; Goswami, R.; Chaturvedi, K. R.; Sharma, T. Effect of Nanoparticle on Rheological Properties of Surfactant-Based Nanofluid for Effective Carbon Utilization: Capturing and Storage Prospects. *Environ. Sci. Pollut. Res.* **2021b,** *2021*, 1–16. https://doi.org/10.1007/S11356-021-14570-6

Kumar, R. S.; Narukulla, R.; Sharma, T. Comparative Effectiveness of Thermal Stability and Rheological Properties of Nanofluid of SiO_2-TiO_2 Nanocomposites for Oil Field Applications. *Ind. Eng. Chem. Res.* **2020b,** *59*, 15768–15783. https://doi.org/10.1021/acs.iecr.0c01944

Kumar, R. S.; Sharma, T. Stability and Rheological Properties of Nanofluids Stabilized by SiO_2 Nanoparticles and SiO_2-TiO_2 Nanocomposites for Oilfield Applications. *Colloids Surfaces A Physicochem. Eng. Asp.* **2018,** *539*, 171–183. https://doi.org/10.1016/j.colsurfa.2017.12.028

Kumar, R. S.; Sharma, T. Stable SiO_2–TiO_2 Composite-Based Nanofluid of Improved Rheological Behaviour for High-Temperature Oilfield Applications. *Geosystem Eng.* **2020b,** *23*, 51–61. https://doi.org/10.1080/12269328.2020.1713909

Li, Y.; Li, N.; Pan, W., Yu, Z.; Yang, L.; Tang, B. Hollow Mesoporous Silica Nanoparticles with Tunable Structures for Controlled Drug Delivery. *ACS Appl. Mater. Interfaces* **2017,** *9*, 2123–2129. https://doi.org/10.1021/acsami.6b13876

Lu, T.; Li, Z.; Zhou, Y.; Zhang, C. Enhanced Oil Recovery of Low-Permeability Cores by SiO_2 Nanofluid. *Energy Fuels* **2017,** *31*, 5612–5621. https://doi.org/10.1021/acs.energyfuels.7b00144

Maghzi, A.; Mohebbi, A.; Kharrat, R.; Ghazanfari, M. H. Pore-Scale Monitoring of Wettability Alteration by Silica Nanoparticles During Polymer Flooding to Heavy Oil in a Five-Spot Glass Micromodel. Transp. *Porous Media* **2011,** *87*, 653–664. https://doi.org/10.1007/s11242-010-9696-3

Nazari Moghaddam, R.; Bahramian, A.; Fakhroueian, Z.; Karimi, A.; Arya, S. Comparative Study of Using Nanoparticles for Enhanced Oil Recovery: Wettability Alteration of Carbonate Rocks. *Energy Fuels* **2015,** *29*, 2111–2119. https://doi.org/10.1021/EF5024719

Nwidee, L. N.; Theophilus, S.; Barifcani, A.; Sarmadivaleh, M.; Iglauer, S. EOR Processes, Opportunities and Technological Advancements. In *Chemical Enhanced Oil Recovery (CEOR) - A Practical Overview*; InTech, 2016. https://doi.org/10.5772/64828

Olayiwola, S. O.; Dejam, M. Experimental Study on the Viscosity Behavior of Silica Nanofluids with Different Ions of Electrolytes. *Ind. Eng. Chem. Res.* **2020,** *59*, 3575–3583. https://doi.org/10.1021/ACS.IECR.9B06275/SUPPL_FILE/IE9B06275_SI_001.PDF

Pratap, M.; Gauma, M. S. In *Field Implementation of Alkaline-Surfactant-Polymer (ASP) Flooding : A Maiden Effort in India*, SPE Asia Pacific Oil Gas Conference Exhibition APOGCE 97–101, 2004. https://doi.org/10.2118/88455-MS

Radnia, H.; Rashidi, A.; Solaimany Nazar, A. R.; Eskandari, M. M.; Jalilian, M. A Novel Nanofluid Based on Sulfonated Graphene for Enhanced Oil Recovery. *J. Mol. Liq.* **2018,** *271*, 795–806. https://doi.org/10.1016/j.molliq.2018.09.070

Roustaei, A.; Bagherzadeh, H. Experimental Investigation of SiO_2 Nanoparticles on Enhanced Oil Recovery of Carbonate Reservoirs. *J. Pet. Explor. Prod. Technol.* **2015,** *5*, 27–33. https://doi.org/10.1007/s13202-014-0120-3

Saha, R.; Uppaluri, R. V. S.; Tiwari, P. Silica Nanoparticle Assisted Polymer Flooding of Heavy Crude Oil: Emulsification, Rheology, and Wettability Alteration Characteristics. *Ind. Eng. Chem. Res.* **2018,** *57*, 6364–6376. https://doi.org/10.1021/acs.iecr.8b00540

Sharma, T.; Iglauer, S.; Sangwai, J. S. Silica Nanofluids in an Oilfield Polymer Polyacrylamide: Interfacial Properties, Wettability Alteration, and Applications for Chemical Enhanced Oil Recovery. *Ind. Eng. Chem. Res.* **2016,** *55*, 12387–12397. https://doi.org/10.1021/acs.iecr.6b03299

Sheng, J. In *Enhanced Oil Recovery Field Case Studies*; Elsevier Inc., 2013. https://doi.org/10.1016/C2010-0-67974-0

Speight, J. G. Nonthermal Recovery of Heavy Oil. In *Heavy Oil Production Processes*; Elsevier, 2013; pp 63–92. https://doi.org/10.1016/b978-0-12-401720-7.00004-6

Stoll, W. M.; Hofman, J. P.; Ligthelm, D. J.; Faber, M. J.; van den Hoek, P. J. Toward Field-Scale Wettability Modification - The Limitations of Diffusive Transport. *SPE Reserv. Eval. Eng.* **2008,** *11*, 633–640. https://doi.org/10.2118/107095-pa

Subramanian, K. R. V.; Rao, T. N.; Balakrishnan, A.; Kumar, R. S.; Sharma, T. Nanofluids: Applications and Its Future for Oil and Gas Industry. In *Nanofluids and Their Engineering Applications*; CRC Press, 2019; pp 465–478. https://doi.org/10.1201/9780429468223-22

Wang, Z.; Wu, Y.; Luo, P.; Tian, Y.; Lin, Y.; Guo, Q. Poly (Sodium p-styrene Sulfonate) Modified Fe_3O_4 Nanoparticles as Effective Additives in Water-Based Drilling Fluids. *J. Pet. Sci. Eng.* **2018,** *165*, 786–797. https://doi.org/10.1016/J.PETROL.2018.03.001

Yousefvand, H.; Jafari, A. Enhanced Oil Recovery Using Polymer/nanosilica. *Procedia Mater. Sci.* **2015,** *11*, 565–570. https://doi.org/10.1016/j.mspro.2015.11.068

CHAPTER 9

Nanomaterials for Polymer and Surfactant Flooding

RAVI SHANKAR KUMAR[1,2], MUHAMMAD ARIF[1], KUSHAL DAS[2], and TUSHAR SHARMA[2]

[1]*Department of Petroleum Engineering, Khalifa University, Abu Dhabi, United Arab Emirates*

[2]*Enhanced Oil Recovery Laboratory, Department of Petroleum Engineering, Rajiv Gandhi Institute of Petroleum Technology, Jais, Amethi, India*

ABSTRACT

Nanotechnology has emerged as a revolutionary approach in the oil and gas industry, particularly in enhanced oil recovery (EOR). This chapter explores the application of nanomaterials, specifically nanoparticles (NPs), with dimensions ranging from 1 to 100 nm, in various oilfield processes. The focus is on the preparation of nanofluids tailored for different reservoir conditions and their potential to improve rheological properties at real field conditions. The critical factor for successful nanofluid application in EOR is the rheological stability, which depends on the surface activity of the nanoparticles. This chapter presents experimental studies on nanofluid flow behavior through porous media, addressing factors such as nanoparticle size, stability, and their effects on oil recovery efficiency. SiO_2 nanoparticles are highlighted as particularly promising for EOR applications. The discussion extends to the challenges faced in nanofluid stability, emphasizing issues such as agglomeration over time and the impact of salinity on stability. Moreover, strategies were discussed to enhance stability, such as modifying surface properties, incorporating surfactants, and utilizing viscous base fluids. The

Advancements in Chemical Enhanced Oil Recovery.
Tushar Sharma, Krishna Raghav Chaturvedi, Tarek Ganat, and Imtiaz Ali (Eds.)

stability of nanofluids is crucial for maintaining their effectiveness in EOR. This chapter delves into the promising combination of nanoparticles with polymers in polymer-nanoparticle (PN) flooding techniques. The chapter also identifies the main challenges associated with the use of nanomaterials (i.e., their stability) when used in combination with the other chemicals and explores the synergy between the binary and ternary combinations of nanomaterials, polymers and surfactants. The interactions between nanoparticles and polymers are studied for their impact on dispersion stability, wettability alteration, and interfacial tension reduction, all contributing to improved oil recovery efficiency. Despite the potential benefits of nanotechnology in EOR, challenges persist, including the lack of a comprehensive understanding of the oil displacement mechanism via nanoparticles. Finally, this chapter concludes by highlighting the ongoing research and the need for further exploration, offering insights into the prospects of nanofluids in advancing EOR methods. In summary, the chapter provides a comprehensive overview of the application of nanotechnology in EOR, addressing mechanisms, applications, and challenges. It serves as a valuable resource for researchers, engineers, and professionals in the petroleum industry seeking to harness the potential of nanomaterials for enhanced oil recovery.

9.1 INTRODUCTION

Nanomaterials are very small particles with dimensions in the order of 1–100 nm. These nanoparticles can enhance several processes due to their smaller sizes but greater surface area-to-volume ratio. For this, nanotechnology, through the nanoparticles, is employed in various oilfield applications including polymer and surfactant flooding. It has shown potential results of improving the rheological properties of fluids at real field conditions. Nanoparticles are used for enhanced oil recovery (EOR) purposes by adding with different base fluids and designing suitable nanofluids for different reservoir condition. One of the most important factors for the nanofluid is their rheological stability or the dispersion of nanoparticles in the base fluid. This depends on the functionality (surface activity) of the nanoparticles. As nanoparticles are generally treated or functionalized, a layer or shield is formed surrounding the particle, which prevents particle–particle interaction and further agglomeration of the nanoparticles. Various experiments are performed to understand the flow behavior of nanofluid through porous medium. Kanj et al. performed the nano-fluid core flooding experiments to investigate the usable size of nanoparticles in reservoir rocks (Kanj et al., 2009). Besides these two, many other research works also proved that nanotechnology has emerged as a promising alternative for the oilfield applications in the recent years. The nano–EOR applications also can be a great technological advancement in this field, as compared to conventional flooding techniques.

Nanotechnology finds numerous uses in several industries including increasing focus on the petroleum industry. Studies proved that SiO_2, Al_2O_3, MgO, and Fe_2O_3 nanoparticles have shown good results for EOR applications but among them SiO_2 nanoparticle is the most promising one (Kumar and Sharma, 2018). It has been also observed that some combination of nanoparticles is showing good efficacy in oil recovery than SiO_2 only (Kumar et al., 2020b). According to the present study, it can be predicted that in future this nanotechnology and nano-emulsion can make greater efficiency in chemical EOR and micellar flooding. In the past decade, along with other nanoparticles (NPs), silica-based nanoparticles have been recognized as an efficient nano–EOR agent which surely offers promising results in near future in EOR processes. Adding NPs to a particular base fluid like surfactant or polymer enhances the fluid behavior in reservoir conditions by reducing the interfacial tension and wettability alteration and increasing the mobility of the capillary-trapped oil (Kumar et al., 2021a). Nanotechnology still remains as an alternative method to extract the unrecovered oil trapped in the pore spaced of reservoir rocks and has been applied as a novel EOR method in the past decade. There is still much lack of understanding of the oil displacement mechanism via NPs due to which spontaneous application of this nano–EOR methods can't be accomplished. Recent pilot and full-field tests showed that by the application of polymeric micro-spheres and nano-spheres reduction of water cut, increase of sweep efficiency and improvement of oil recovery can be achieved to a great extent. They reported that these micro- and nano-spheres have the swelling ability at the time of interaction with water resulting in reduction of water permeability. This is due to its ability to reduce the capillary forces and change the water flow paths. Thus, water can get passage to flow into the bypassed area and this enhances displacement efficiency. It is observed that, in the past decade, application of nanoparticle in EOR has been in focus many times in the petroleum industry. Laboratory-scale research works on the different rock types like sandstone and carbonate, which help to understand the complex mechanism associated with the nanoparticles, are still in progress. Researchers have found several mechanisms responsible for fluid behavior, such as, interfacial tension reduction, wettability alteration, oil viscosity, and reduction of mobility ratio. Moreover, the oil/water emulsion and foam can be stabilized by the use of NPs which improves the mobility control in EOR methodologies. Also, due to the ability of interfacial tension and adsorption reduction, NPs have also been used with surfactants to enhance the functions of it. According to a study by Tian et al., the polymeric microspheres and nanospheres are more advantageous because of the non-degradation of the particles at elevated temperature and salinity. Besides this, combinations of nanoparticles and surfactant solution are found to more promising by many researchers. Le at al. studied about the synergetic blends of SiO_2 NPs and surfactants at high-temperature conditions for EOR purposes (Le et al., 2011). The IFT reduction has been found to be 90% more when the NP combination and surfactant is used simultaneously compared to the single application of surfactant as an injection fluid (Kumar et al., 2021a). It has been observed that sometimes the IFT value becomes as low as zero when surfactant-based NP combination is employed. Also, on the basis of previous application of NPs in hydrocarbon upstream sector, many researchers have contributed their efforts for oil recovery applications. Therefore, the basic formulation methodology of nanofluid and their implication in EOR techniques has been summarized in this chapter. The objectives of this chapter are to emphasize the novelty used in oil recovery and to discuss the mechanism underlying nanofluid stability. Further, the challenges of nanofluid stability is well illustrated and their effect on oil recovery performance was studied.

9.1.1 NANOFLUIDS ENHANCED POLYMER FLOODING

In the recent years, nanotechnology through the use of nanoparticles has emerged as a successful advantageous alternative to many industries like medical, electronic, material science, etc., including recent focus in the oil and gas industries. Nanoparticles has found uses starting from drilling and completion, cementing, reservoir exploration and sensing, stimulation to flow assurance, and in many other sections too. Recently, the EOR processes have also observed potential results of application of NPs. The various mechanical, electrical, optical, chemical, physiochemical, and thermal properties have shown many advantages compared to the conventional chemicals (surfactants and polymer) usually used as the EOR agents (Sharma and Sangwai, 2017). The NPs by virtue of their greater functionality can increase the recovery efficiency by targeting either the trapped oil in capillaries of reservoir rocks or the bypassed oil. One of the most important facts is the sustainability of the nano-sized particles in the fluid solution at elevated temperature, pressure, and salinity conditions over the surfactants and polymer fluids (Sadeghalvaad and Sabbaghi, 2015).

Polymer-nanoparticle (PN) flooding techniques involves the injection of a PN slug first (Maghzi et al., 2011). After that, a polymer buffer and chase water flooding has been done subsequently. Generally, polymers increase the sweep efficiency by lowering the mobility ratio. For achieving a good mobilization and recovery of tertiary oil, capillary number should be increased in the porous media, for which proper designing of nanoparticles is of prime importance. The PN interaction both in the solution and at the slid/liquid interface is of great importance as the first one controls the flow behavior in the porous media and displacement efficiency; whereas the latter one is responsible for the loss of chemicals by the rock surfaces. This loss of NP by adsorption can account for non-feasibility of economic viability of nano-assisted EOR field application. Many studies have been performed both theoretically and experimentally to understand this PN interaction in solution and several papers has also been published. It is observed that ultra-low IFT value can be achieved by the application of NP with higher concentration value though this can cause pore blockage to some extent. Kumar et al. has studied the behavior of hydrophilic NP mixed with polymers. However, the studies on PN flooding are only focused on the screening and evaluation of the polymer and NP and their interaction (Kumar et al., 2020b). The reduction in mobility ratio and IFT is greatly influenced by several reservoir conditions like brine salinity, temperature, concentration of chemicals present in oil components, and many other (Kumar et al., 2020a).

Nanoparticles have been considered as an excellent flooding agent in EOR processes supported by many researchers; due to certain features they possess, such as, cheap and environment friendly, they can be designed or modified to get the enhanced desirable properties like changed wettability, increased mobility, and reduced surface tension. They can have easy passage through reservoir rock without blocking the pores. Nanoparticles due to their small sizes but high surface-to-volume ratio possess enhanced thermal properties.

The behavior of NPs has been closely studied when it is applied in the formation and their effect on oil recovery has also been noted. Li et al. stated that NPs have the ability of enhancing the recovery factor altering various factors in the formation (Li et al., 2013). The hydrocarbon industry might be facing the challenges to recover the tertiary oil that is difficult to extract with available technology. After a certain period, most of the oil reservoirs will be achieving a stage when the recovery rate will be declining. So, it is of prime importance to find a suitable and economical way of extracting the unrecovered tertiary oil before the abandonment of a well. For this, nanotechnology comes into play. Nanoparticles size typically ranges in 1–100 nm, which is smaller than the rock pore throat in micron size (Hu, 2017). A nanofluid is a NP-fluid suspension which is generated by dispersing nanoparticles in suitable base fluids. Nanofluid flooding have shown potential to improve oil recovery from the existing reservoir as enhanced oil recovery (IOR/EOR) technique in the petroleum industry (Huang et al., 2015). Several researchers have come up for performing experimental study to demonstrate the oil recovery mechanism in porous medium using nanofluid. In this chapter, several mechanisms associated with EOR processes like wettability alteration and interfacial tension reduction viscosity control has also been demonstrated. Hendraningrat et al. reported that nanoparticles can be used as an EOR chemical agents (Hendraningrat and Torsæter, 2015). They further suggested that interfacial tension (IFT) reduction and wettability alteration are two of the major reasons behind the improved oil recovery.

Nanoparticles can be employed for EOR purposes after synthesizing a suitable nanofluid out of it. Das et al. described nanofluids can be used as heat transfer fluid or displacing medium like water or oil (Das et al., 2003). The common nanoparticles used to formulate nanofluid is SiO_2, TiO_2, ZnO, Al_2O_3, CuO, and MgO. Ahmadi et al. reported that MgO can reduce the fine migration by attaching the fines to surface layers (Ahmadi et al., 2016). Hendraningrat and Torsaeter studied the stability of Al_2O_3 and MgO nanoparticles (Hendraningrat and Torsæter, 2015); and found that the addition of PVP dispersants can improves the stability of nanofluid. This dispersant alters the properties of the metal oxides nanoparticles which results into improved nanofluid stability. It was also observed that the combination of dispersant such as polymers with nanoparticles may enhance the recovery efficiency via wettability alteration of the rock from oil wet to water wet. Nwidee et al. conducted an experiment with ZrO_2 NPs for wettability alteration of oil-wet rock (limestone) (Nwidee et al., 2016). According to their study, ZrO_2 (0.05 wt%) nanofluid was able to do the wettability alteration of the strongly oil-wet limestone rock to water-wet and the wettability alteration was confirmed using contact angle study where the contact angle reduced from 152° to 44°. This result considers ZrO_2 to be a good EOR agent. Ogolo et al. studied the nanoparticles' ability to trap fine migrations for sandstone and noted the metal oxide NPs hold potential properties for EOR (Ogolo,

2013). This was performed using distilled water, brine, and ethanol as the base fluid. Moreover, they found that trapping of fines depends upon several factors such as pH and adsorption properties. Hu et al. studied the effect of NPs on oil recovery with combination of nanoparticle and polymer to record the recovery potentials, using aluminum oxide and polyacrylamide (PAM) (Hu, 2017). From the study it was inferred that at first only by polymer flooding 2% increase in the recovery factor (RF) can be achieved. Next, by the application of the combination of nanoparticles and polymer, 66.29% of OOIP can be recovered with a 40% reduction in cost, whereas the polymer flooding can yield 67.16% of OOIP, which has 2% more recovery factor. Lastly, this combination is also an aid for the alteration of wettability to more water-wet.

For the last two decades, the use of the combination of chemicals in a single EOR method is widely accepted. Polymers increase the viscosity of the aqueous phase, i.e., nanofluid, resulting in maintaining a superior mobility ration and viscosity contrast (Lake, 2010). It is observed that nanoparticles have the tendency to increase the viscosity of the wetting phase. But it also possesses the properties of wettability alteration of rock surfaces and interfacial tension reduction, which favors the recovery efficiency of oil. The objective of this chapter is to understand this new EOR method; how the interactions of different nanoparticles among each other and in the solution both, i.e., their synergy. In the case of nanoparticle–polymer solution, something similar to the surfactant–polymer interaction takes place. The interaction of NP and polymer enhances the dispersion stability of nanofluid, which has vital role on wettability alteration and IFT reduction in oil recovery applications. Saha et al. formulated silica nanofluid in 5000 ppm xanthan gum (XG) polymer and found significant change in IFT (Fig. 9.1) and wettability (Fig. 9.2). Furthermore, he conducted oil recovery experiments using those formulated nanofluid and found 54.13% of cumulative oil recovery (Fig. 9.3) (Saha et al., 2018).

9.1.2 CHALLENGES FOR NANOFLUID STABILITY

AGGLOMERATION WITH TIME

As we know that the NPs are very promising for EOR applications which requires prerequisite, i.e., stable nanofluid. However, the NPs present in the nanofluid generally show agglomeration with time (Fig. 9.5). The agglomeration of NPs is generally caused by the reduction of surface potential of

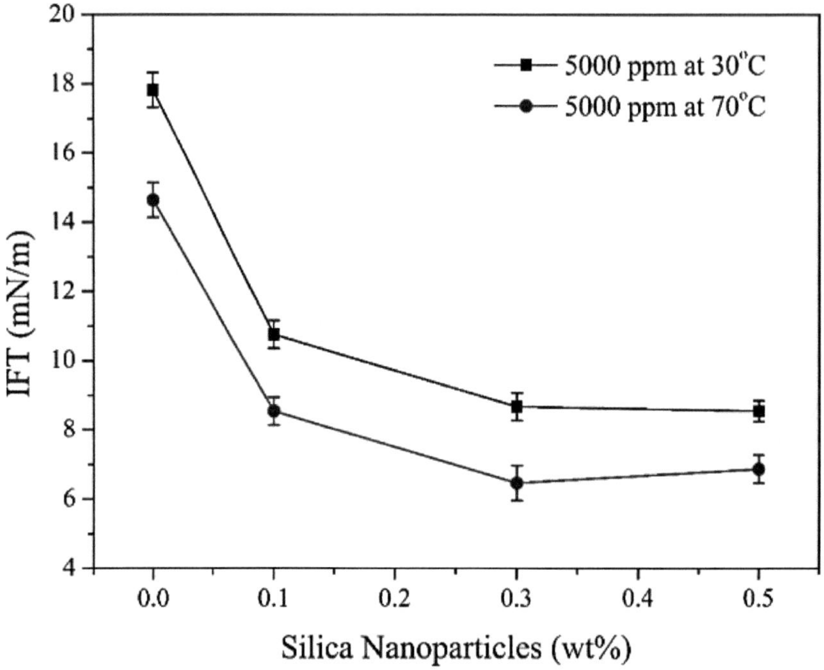

FIGURE 9.1 Interfacial tension reduction with nanoparticle concentration.

Source: Reprinted from Saha et al., 2018. Copyright (2018), with permission from ACS).

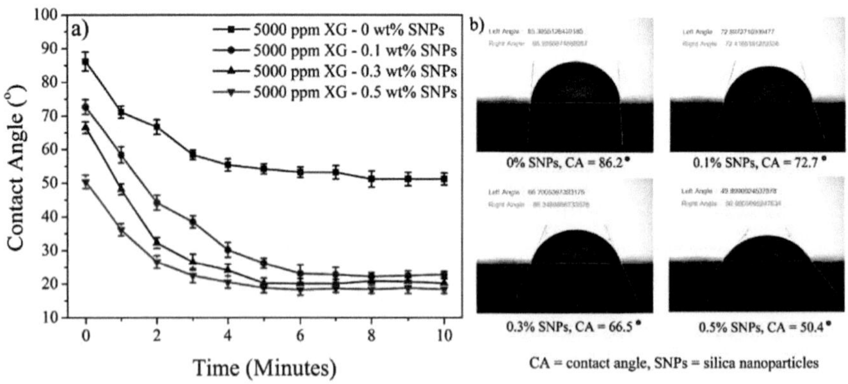

FIGURE 9.2 Contact angle results showed change in wettability of reservoir rock with different nanofluid samples.

Source: Reprinted from Saha et al., 2018. Copyright (2018), with permission from ACS.

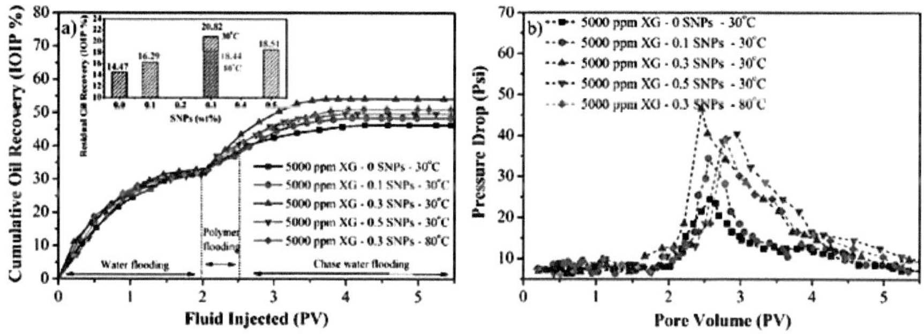

FIGURE 9.3 Oil recovery and pressure drop result showed enhancement in oil production with surfactant polymer and nanoparticle.

Source: Reprinted from Saha et al., 2018. Copyright (2018), with permission from ACS).

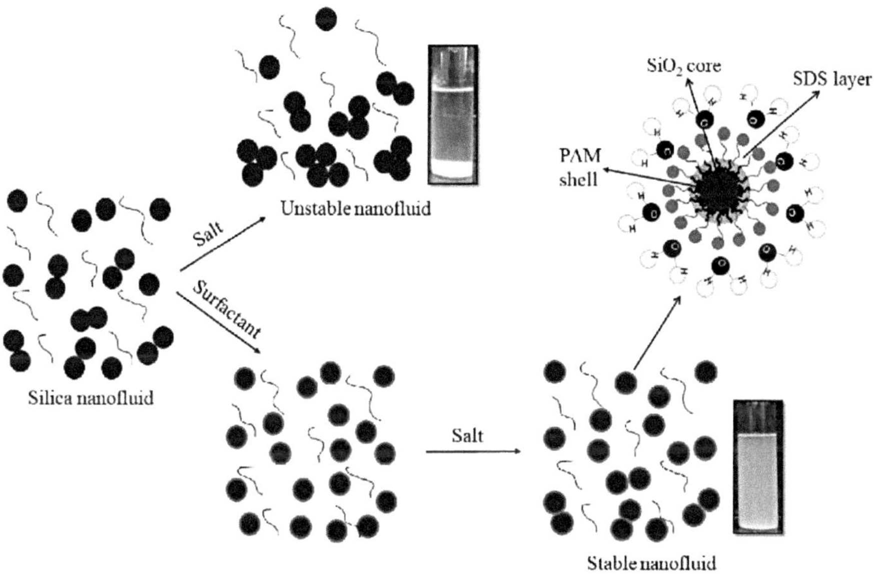

FIGURE 9.4 Schematic showing improvement in stability of silica nanofluid prepared in base fluid of polyacrylamide using anionic surfactant SDS.

Source: Reprinted from Kumar et al., 2020. Copyright (2020), with permission from Elsevier).

NPs within the nanofluid. Thus, it is very clear that the agglomerated NPs may lose its original property and show sedimentation of NPs. The sedimentation of these NPs further can reduce the stability of nanofluid and limit their application in EOR projects.

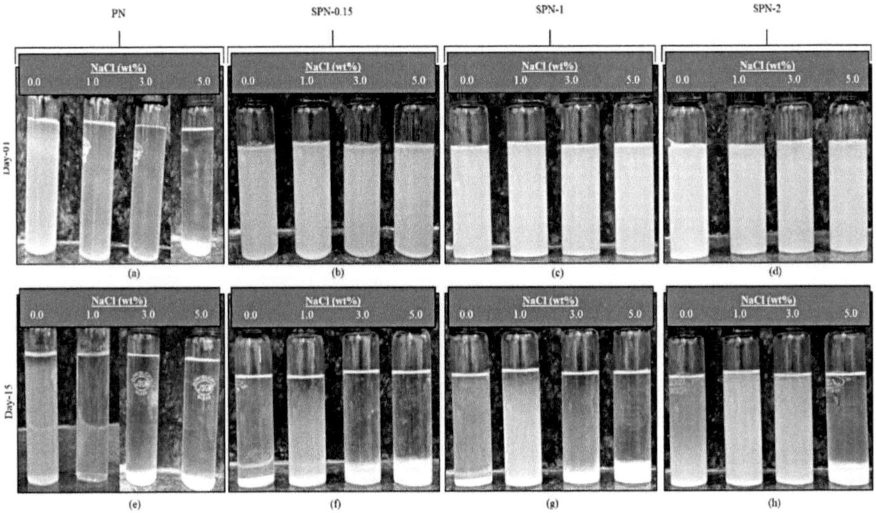

FIGURE 9.5 NPs agglomeration followed by sedimentation of NPs clusters and their improvement using anionic surfactant.

Source: Reprinted from Kumar et al., 2020. Copyright (2020), with permission from Elsevier.

The long-term stability of the nanofluid is recommended for oil recovery and can be achieved by modifying the surface properties of NPs in the nanofluid. In general, the NPs possess high surface potential within the nanofluid and help NPs to remain in suspension due to repulsive forces. The base fluid viscosity is also considerable for enhancing the stability of nanofluid. The higher viscosity of base fluid used to formulate the nanofluid may reduce agglomeration followed by sedimentation and maintain viscosity contrast during flow through porous media. Also, inclusion of different NPs of similar charge may also reduce the agglomeration of NPs (Kumar and Sharma, 2018).

AGGLOMERATION CAUSED BY SALINITY

A stable nanofluid can be affected by the saline environment of the reservoir. The reservoir salinity is very detrimental to nanofluid stability. The majority of saline reservoir contains numerous components, in which monovalent and divalent ions are major factors. The solute monovalent and divalent of saline reservoirs generally screen out the surface charge of NPs and make them neutralized. Therefore, the existing repulsive force of NPs within the nanofluid may reduce, which promotes severe agglomeration (Al-Anssari et al., 2017). The heavy agglomeration of NPs caused by salinity is very detrimental to oil recovery as this may block the pore throat of oil reservoir and reduce the oil recovery. The reduction in oil recovery in saline reservoirs using NPs can be enhanced by improving the stability of the nanofluid. In this regard, a suitable nanofluid which possess higher stability may curtail the effect of salinity and that can be performed by enhancing the surface charge of NPs in the nanofluid. Al-anssari et al. suggested that the surface charge of NPs can be improved using the incorporation of surfactant (Al-Anssari et al., 2017). In one of our study, silica nanofluid was stabilized by this technique in saline environment using anionic surfactant sodium dodecyl sulfate. The surfactant generally produces micelles and that micelles improves the surface potential

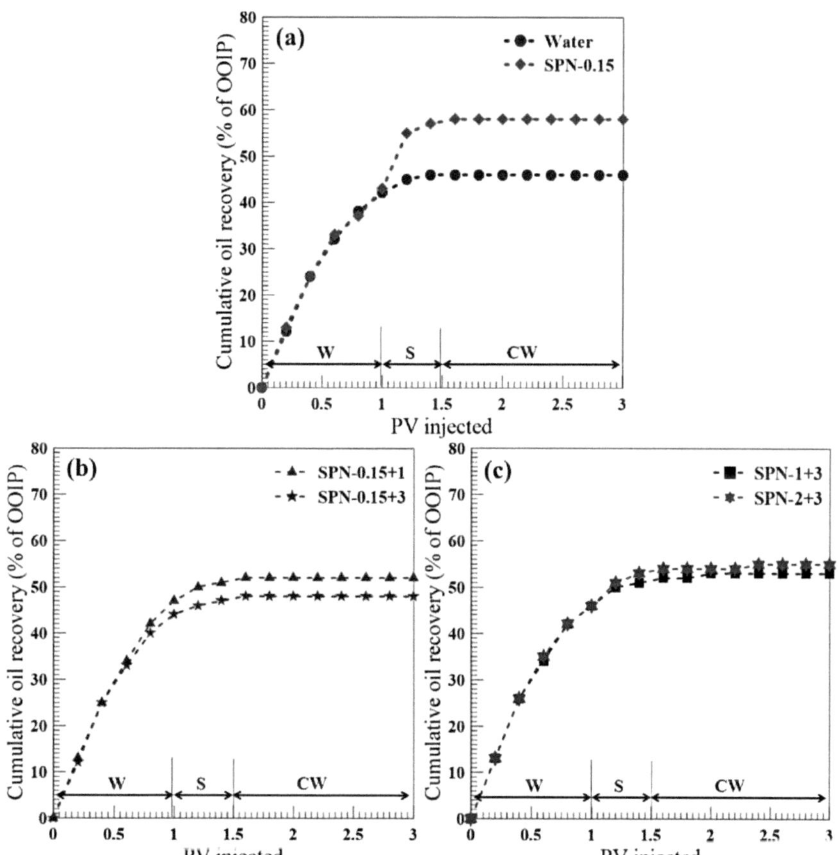

FIGURE 9.6 Oil recovery result in saline media using silica nanofluid, reduction in oil recovery due to agglomeration and improvement in oil recovery using surfactant.

Source: Reprinted from Kumar et al., 2020. Copyright (2020), with permission from Elsevier).

of NPs in the nanofluid. The addition surface charge of NPs resulting high repulsive forces between the NPs and thus enhance the stability of nanofluid. The enhanced stability of nanofluid significantly improves the oil recovery and curtail the effect of agglomeration in saline environment (Kumar et al., 2020a).

Thus, it is very clear that a nanofluid can be stable for longer duration if

- Nanofluid is prepared in viscous base fluid such as polymers (Fig. 9.5).
- Inclusion of binary NPs system for nanofluid may also improves the stability (Fig. 9.7).
- Nanofluid stability may be improved using surfactant (Fig. 9.4).

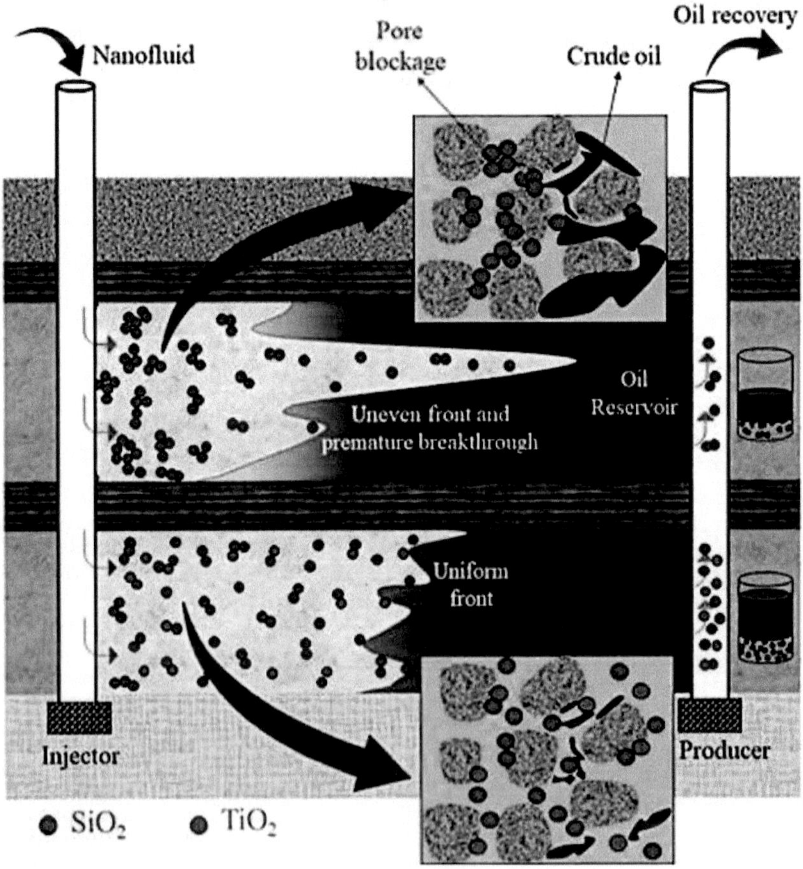

FIGURE 9.7 Schematic showing stability mechanism of nanofluid prepared by SiO2-TiO2 nanocomposite and oil recovery.

Source: Reprinted from Kumar et al., 2021. Copyright (2021), with permission from Elsevier).

A nanofluid prepared using the combination of binary NPs of similar charges in the base fluid of polymers, and inclusion of surfactant may enhance the stability of nanofluid also can improves the oil recovery from the reservoir (Kumar et al., 2021b, 2021a, 2020a, 2020b; Kumar and Sharma, 2020, 2018). A detailed discussion on the stability of a silica nanofluid is provided in previous works (Chaturvedi and Sharma, 2021).

9.2 CONCLUSION

This chapter has discussed that surely the advancement of EOR methods by the application of nanoparticles is a great achievement. The use of nanotechnology in chemical EOR processes has come up with great potential results in oil recovery. In this chapter, the mechanisms of oil recovery by nanoparticle, polymer, and surfactant are discussed. The application of NP is efficient for wettability alteration and interfacial tension reduction. This chapter focuses on the ongoing results and discusses the challenges for nanofluid applicability in EOR.

KEYWORDS

- **nanomaterials**
- **polymer flooding**
- **surfactant flooding**
- **high temperature**
- **high salinity**
- **enhanced properties**

REFERENCES

Ahmadi, M. A.; Ahmad, Z.; Phung, L. T. K.; Kashiwao, T.; Bahadori, A. Evaluation of the Ability of the Hydrophobic Nanoparticles of SiO_2 in the EOR Process Through Carbonate Rock Samples. *Pet. Sci. Technol.* **2016,** *34,* 1048–1054.

Al-Anssari, S.; Arif, M.; Wang, S.; Barifcani, A.; Iglauer, S. Stabilising Nanofluids in Saline Environments. *J. Colloid Interface Sci.* **2017,** *508,* 222–229.

Chaturvedi, K. R.; Sharma, T. Rheological Analysis and EOR Potential of Surfactant Treated Single-Step Silica Nanofluid at High Temperature and Salinity. *J. Pet. Sci. Eng.* **2021,** *196,* 107704.

Das, S. K.; Putra, N.; Thiesen, P.; Roetzel, W. Temperature Dependence of Thermal Conductivity Enhancement for Nanofluids. *J. Heat Transfer* **2003,** *125,* 567–574.

Hendraningrat, L.; Torsæter, O. A Stabilizer that Enhances the Oil Recovery Process Using Silica-based Nanofluids. *Transp. Porous Media* **2015,** *108,* 679–696.

Hu, Z. Nanoparticle-assisted Enhanced Oil Recovery. **2017**.

Huang, T.; Han, J.; Agrawal, G.; Sookprasong, P. A. In *Coupling Nanoparticles with Waterflooding to Increase Water Sweep Efficiency for High Fines-containing Reservoir - Lab and Reservoir Simulation Results, Proceedings - SPE Annual Technical Conference and Exhibition.* Society of Petroleum Engineers (SPE), 2015; pp 1101–1114. DOI: 10.2118/174802-ms

Kanj, M. Y.; Funk, J. J.; Al-Yousif, Z. Nanofluid Coreflood Experiments in the ARAB-D. *Soc. Pet. Eng. - SPE Saudi Arab. Sect. Tech. Symp.* **2009**. DOI: 10.2118/126161-MS

Kumar, R. S.; Al-Arbi Ganat, T.; Sharma, T. Performance Evaluation of Silica Nanofluid for Sand Column Transport with Simultaneous Wettability Alteration: An Approach to Environmental Issue. *J. Clean. Prod.* **2021a,** *303,* 127047.

Kumar, R. S.; Chaturvedi, K. R.; Iglauer, S.; Trivedi, J.; Sharma, T. Impact of Anionic Surfactant on Stability, Viscoelastic Moduli, and Oil Recovery of Silica Nanofluid in Saline Environment. *J. Pet. Sci. Eng.* **2020a,** *195,* 107634.

Kumar, R. S.; Narukulla, R.; Sharma, T. Comparative Effectiveness of Thermal Stability and Rheological Properties of Nanofluid of SiO_2-TiO_2 Nanocomposites for Oil Field Applications. *Ind. Eng. Chem. Res.* **2020b,** *59,* 15768–15783.

Kumar, R. S.; Sharma, T. Stability and Rheological Properties of Nanofluids Stabilized by SiO_2 Nanoparticles and SiO_2-TiO_2 Nanocomposites for Oilfield Applications. *Colloids Surf. A Physicochem. Eng. Asp.* **2018,** *539,* 171–183.

Kumar, R. S.; Sharma, T. Stable SiO_2–TiO_2 Composite-based Nanofluid of Improved Rheological Behaviour for High-temperature Oilfield Applications. *Geosystem Eng.* **2020,** *23,* 51–61.

Kumar, R. S.; Sinha, A. S. K.; Sharma, H.; Sharma, T. High Performance Carbon Dioxide Foams of Nanocomposites of Binary Colloids for Effective Carbon Utilization in Enhanced Oil Recovery Applications. *J. Mol. Liq.* **2021b,** *343,* 117659.

Lake, L. W. Enhanced Oil Recovery. **2010**.

Le, N. Y. T.; Pham, D. K.; Le, K. H.; Nguyen, P. T. Design and Screening of Synergistic Blends of SiO_2 Nanoparticles and Surfactants for Enhanced Oil Recovery in High-Temperature Reservoirs. *Adv. Nat. Sci. Nanosci. Nanotechnol.* **2011,** *2.* DOI: 10.1088/2043-6262/2/3/035013

Li, S.; Hendraningrat, L.; Torsaeter, O. Improved Oil Recovery by Hydrophilic Silica Nanoparticles Suspension: 2-Phase Flow Experimental Studies. *SPE J.* **2013**. DOI: 10.2523/iptc-16707-ms

Maghzi, A.; Mohebbi, A.; Kharrat, R.; Ghazanfari, M. H. Pore-Scale Monitoring of Wettability Alteration by Silica Nanoparticles During Polymer Flooding to Heavy Oil in a Five-Spot Glass Micromodel. *Transp. Porous Media* **2011,** *87,* 653–664.

Nwidee, L. N.; Al-Anssari, S.; Barifcani, A.; Sarmadivaleh, M.; Iglauer, S. Nanofluids for Enhanced Oil Recovery Processes: Wettability Alteration Using Zirconium Oxide. *Offshore Technol. Conf. Asia* **2016,** 2808–2814. DOI: 10.4043/26573-MS

Ogolo, N. A. The Trapping Capacity of Nanofluids on Migrating Fines in Sand. *Proc. - SPE Annu. Tech. Conf. Exhib.* **2013,** *7,* 5528–5542.

Sadeghalvaad, M.; Sabbaghi, S. The Effect of the TiO_2/Polyacrylamide Nanocomposite on Water-based Drilling Fluid Properties. *Powder Technol.* **2015,** *272,* 113–119. https://doi.org/10.1016/J.POWTEC.2014.11.032

Saha, R.; Uppaluri, R. V. S.; Tiwari, P. Silica Nanoparticle Assisted Polymer Flooding of Heavy Crude Oil: Emulsification, Rheology, and Wettability Alteration Characteristics. *Ind. Eng. Chem. Res.* **2018,** *57,* 6364–6376.

Sharma, T.; Sangwai, J. S. Silica Nanofluids in Polyacrylamide with and Without Surfactant: Viscosity, Surface Tension, and Interfacial Tension with Liquid Paraffin. *J. Pet. Sci. Eng.* **2017,** *152,* 575–585.

CHAPTER 10

Foam Stabilization Using Engineered Nanomaterials

RAVI SHANKAR KUMAR[1], MUHAMMAD ARIF[2], and TUSHAR SHARMA[1]

[1]*Enhanced Oil Recovery Laboratory, Department of Petroleum Engineering, Rajiv Gandhi Institute of Petroleum Technology, Jais, Rae Bareli, India*

[2]*Department of Petroleum Engineering, Khalifa University, Abu Dhabi, United Arab Emirates*

ABSTRACT

High energy demands leads to additional production of crude oil. Among water flooding, chemical flooding, microbial flooding and foam flooding, foam flooding has the greatest potential in enhanced oil recovery (EOR) projects. This is due to its potential to penetrate in highly permeable layer and thereby driving the fluid to lower permeable layer. But application of foam flooding possesses certain limitations in heterogeneous reservoir and to overcome this issue, the role of nanoparticles (NPs) was explored. Therefore, this chapter delves into the utilization of nanomaterials for foam stabilization, exploring their role in enhancing the stability of nanofluids. The application of NPs promotes the improved flow behavior of foam within porous media and thereby increases oil recovery. The study investigates the potential applications of these stabilized nanofluids in the oilfield industry, focusing on their effectiveness in various processes such as enhanced oil recovery and well stimulation. The role of nanomaterials, particularly silica, has been identified to strengthen the foam lamellae and improve the displacement potential of the foams in porous media. NPs also help in stabilization of CO_2 foams and silica NPs has the ability to form tight, coherent particle

Advancements in Chemical Enhanced Oil Recovery.
Tushar Sharma, Krishna Raghav Chaturvedi, Tarek Ganat, and Imtiaz Ali (Eds.)

shell at fluid-fluid interface. The chapter explores the possible measure to control CO_2 mobility during EOR studies. As the synergy of silica NPs and CO_2 foams offers exceptionally stable CO_2 foams, but with reduced foam mobility. Thus, inclusion of NPs tends to improve sweep efficiency during CO_2 injection. Using experimental studies, this chapter further emphasizes the significance of stable foams in oilfield applications, the research sheds light on the promising outcomes achieved through the integration of nanomaterials in adverse field conditions. The chapter also highlights the application of surface modified silica NPs for improved mobility control as the foams are much stronger, highly stable and improved foaming tendency with a finer bubble structure. The chapter also covers application of binary NPs in enhancing foam stability even at elevated temperature.

10.1 INTRODUCTION

Increased use of energy resources necessitates the production of additional crude oil. However, water flooding, the most commonly utilized method of oil recovery, can only recover around 40% of crude oil from reservoirs. Because of the great mobility of water relative to oil and the poor sweep efficiency of water floods, oil recovery is typically insufficient. Meanwhile, due to the high interface tension between oil and water, emulsifying oil into the water phase is problematic, resulting in a low displacement efficiency. Several oil recovery systems, including chemical flooding, microbiological flooding, and foam flooding, have been researched. Foam flooding has the most potential of these because of its high apparent viscosity, which allows it to penetrate high-permeability layers, selectively block water channels, increase wave coefficient, and finally drive fluid to low-permeability layers. As a consequence, water replaces oil residue in depleted reservoirs, improving oil recovery. Despite the numerous advantages of foam flooding, its inability to be used in enhanced oil recovery (EOR) has remained a fundamental hurdle. Foam is described as gas bubbles dispersed in liquid with a bubble diameter less than the container's diameter. When using heterogeneous porous media such as actual oil field rock, the bubble size often approaches that of the pores, resulting in the development of lamellar liquid films. For four reasons, lamellas are notorious for their failure to retain stability over time: Liquid drainage and film rupture is caused by an imbalance of gravity and capillary force; gas diffusion is caused by non-uniform bubble size distribution; surfactant loss occurs due to adsorption on rocks;

and surfactant degradation occurs at high temperatures. To minimize or mitigate these effects, other techniques have been attempted, including the use of modified polymers, surfactants, nanoparticles, or a combination of these reagents. Recently, the use of nanoparticles was explored, and it was shown that they had significant potential for foam stabilization under extreme temperature and salinity circumstances. To stabilize CO_2-in-water foams, Andrew et al. employed silica nanoparticles coated with polyethene glycol. Surfactants were frequently used to make macroscopic foams by adsorbing on particle surfaces. The successive production of surfactant-coated particles at the air–water interface causes the foam to be stable over time. Chen et al. show that interacting with nonionic surfactants boosts foam stability (Chen et al., 2015). Carn et al. used a range of methodologies on both the macroscopic and microscopic scales to explore the drainage of SiO2 nanoparticle cationic surfactant mixtures via calibrated aqueous foams (Carn et al., 2009). Sun et al. conducted N2 foam flooding tests using sodium dodecyl sulfate (SDS) and hydrophobic SiO2 nanoparticles. According to the findings, the inclusion of hydrophobic nanoparticles can improve the flow behavior of multiphase foam in porous media, hence increasing oil recovery (Sun et al., 2014). The majority of articles, however, concentrated on traditional surfactants such as sodium dodecyl sulfate (SDS), cetrimonium bromide (CTAB), and others, which were well-known for their high toxicity and inefficiency in severe settings with high salinity and temperature (Kumar et al., 2020).

Furthermore, CO_2 flooding is a common enhanced oil recovery (EOR) technique due to its easy miscibility with crude oil, which results in greater oil volume and decreased oil viscosity (Chaturvedi et al., 2021). An early CO_2 breakthrough, on the other hand, results in a low sweep efficiency and a high operating cost due to gravity override, viscous fingering, and formation heterogeneity (Seyyedsar et al., 2017). CO_2 foams or chemical gels are the most often used agents for improving the effectiveness of the macroscopic sweep by limiting CO_2 transport in high-permeability zones. While polymer gels are frequently employed to plug thief zones by directing CO_2 to other zones, they tend to stay close to the wellbore and have minimal influence on deep formation management (Li et al., 2019). CO_2 foam, which is a dispersion of discontinuous CO_2 in the liquid phase, may solve the aforementioned issues and is widely regarded as an effective agent for CO_2 flood mobility management. CO_2 foam is commonly used to increase the sweep efficiency of heterogeneous reservoirs by reducing the relative permeability and increasing the viscosity of the gas. When the temperature and pressure in the reservoir are above the critical CO_2 levels (31.1°C and 7.37 MPa, respectively), gaseous CO_2 is usually converted to a thick, liquid-like fluid (supercritical CO_2) ((Kang et al., 2021). Supercritical CO_2 has a much lower viscosity and a much greater density (0.5–0.9 g/cm^3) than gaseous CO_2. As a result, the physical characteristics of supercritical CO_2 and gaseous CO_2 at ambient circumstances are fundamentally different. Furthermore, when supercritical CO_2 is disseminated in a surfactant solution, CO_2 foam exhibits significantly higher solubility, pH, texture, and motility than gaseous CO_2 foam.

However, in unfavorable reservoir conditions (high temperature and salinity), the stability of CO_2 foam frequently deteriorates (Chang, 2016; Chen et al., 2015). The tensile strength of a bubble film is characterized by foam stability, which is influenced by a variety of factors such as bulk viscosity, surface viscoelasticity, the Marangoni effect, and disjoining pressure. Furthermore, the temperature, salinity, pressure, and pore structure of a porous medium influence film stability. Various foaming agents, including surfactants, have been used to increase the foam's stability. The apparent viscosity of SC-CO_2 foam stabilized with surfactants and polymers is much greater than that of pure gaseous foam (by several orders of magnitude). Surfactants and polymers, on the other hand, lose strength and solubility when temperatures and salinity rise, resulting in poor foam stability. Because of their great chemical stability at high temperatures and salinity, nanoparticles are the most efficient agents for CO_2 foam stabilization. Because of their resistance to film deformation, silica nanoparticles can form tight, coherent particle shells at the gas–liquid interface, resulting in better foam stability. Solid particles' efficacy is determined by their size, shape, concentration, and wettability. When solid particles are wetted (= 0°) or dewetted (= 180°), they frequently stay distributed in the liquid or gas phase, with no persistent liquid film-forming. The contact angle (contact angle) of solid particles begins to condense into liquid films between 0 and 180 degrees. Aqueous and gaseous particles prefer to congregate in their respective aqueous and gaseous phases (Aveyard and Clint, 1995). Li et al. discovered that at high salinities (25 104 mg/L) and temperatures (up to 90°C), hexadecyl hydroxypropyl sulfobetaine (HHSB; a zwitterionic surfactant) can form SC-CO_2 foam. However, after 90°C, the stability of SC-CO_2 foam degraded drastically. Hydrophilic silica nanoparticles were added to a dilute HHSB solution to improve the stability of SC-CO_2 foam at higher temperatures. Prior research has concentrated on the capability of NPs-stabilized CO_2 foam at atmospheric pressure and room temperature, as well as a thorough examination of NPs' influence on SC-CO_2

foam. The stability, rheological properties, and method of stabilization of SC-CO_2 foam were investigated when the NP concentration was increased.

The amount of residual oil beneath the surface that can be collected by traditional methods is enormous, as is commonly known (Hirasaki et al., 2011). In general, by employing primary and secondary oil production techniques fields may generate 45–50% of the original oil in situ (OOIP) (Stoll et al., 2008). As a result, the amount of oil produced is inadequate to fulfil the world's increasing energy demand. To collect this significant volume of residual oil, enhanced oil recovery (EOR) technologies are required. CO_2 is routinely used to help in the recovery of oil. It has the technological ability to increase swelling, decrease oil viscosity, evaporate, and extract the crude oil. Furthermore, due to its low oil solubility, CO_2 is a suitable gas for EOR applications. Despite its stated efficacy, CO_2 injection has significant limitations due to its insufficient volumetric sweep. The low density and viscosity of CO_2 in comparison to reservoir fluids, as well as reservoir variability such as high permeability and extensively fractured zones, all, contribute to this problem. Certain scholars advocated combining the two strategies to address the issue of low sweep efficiency. They advised beginning with conformity control, such as gel, and progressing to more extensive mobility control, such as CO_2 foam. The purpose of this chapter is to explore how to control CO_2 mobility during EOR procedures.

10.2 NANOPARTICLE-STABILIZED FOAMS

Surfactants can be employed to build foam and restrict gas mobility, resulting in higher sweep efficiency and oil recovery in different gas flooding field experiments. However, the severe environmental conditions found in reservoirs, such as high temperature, salinity, and adsorption on the rock surface, are the main barriers to surfactants' being widely used in gas injection systems. Research is being conducted for use in gas EOR as a surfactant-free alternative to surfactants as foam stabilizers due to their unique properties, which include exceptional chemical stability even under severe reservoir conditions, low retention on mineral surfaces, and the ability to form extremely stable foams. For this reason, commercially available silica nanoparticles have been extensively investigated. Espinosa et al., for example, demonstrated that silica nanoparticles may be used to form CO_2 foams in water without the need for surfactants (Espinoza et al., 2010). According to Andrew et al., 50% SiOH NPs adsorb substantially at the CO_2–water interface, resulting in the development of exceptionally stable CO_2 foams (Worthen et al., 2012). They also demonstrated the capacity to create stable CO_2 foam using silica nanoparticles at a concentration of 5000 parts per million. According to the findings of the core flooding research, the ensuing CO_2 foam may significantly hinder CO_2 mobility. The foam mobility was reduced as the nano-silica concentration and flow velocity rose, yet the foam resistance factor increased. Furthermore, Aminzadeh et al. demonstrated that, due to the CO_2–water foam generated, NPs might be employed to enhance sweep efficiency during liquid CO_2 injection into a brine-filled sandstone core for CO_2 sequestration. Following the initial CO_2 injection, the CO_2 trapping capability of NP-stabilized foam was increased (Aminzadeh et al., 2013). Le et al. investigated the use of SiO2 nanoparticles and surfactants in high-temperature EOR reservoirs (Le et al., 2011). They carried out tests in which they added several surfactant solutions to SiO2 NPs. Because of their resistance to adsorption on the rock surface and thermostability at 91°C, some of the combinations have demonstrated significant potential for EOR applications. Using micromodels, Nguyen et al. studied the stability and efficacy of NP-stabilized CO_2 foam in the EOR of a range of various oils (light, medium, and heavy oils) (Nguyen et al., 2014). Because NP-stabilized CO_2 foam flooding created much smaller oil-in-water emulsions, all three oils suggested considerable extra oil recovery and the possibility of significant reservoir homogeneity. The results established the pore-scale dynamics, efficacy, and potential for improved oil recovery employing NP-stabilized CO_2 foam flooding. The combination of NPs with surfactants may provide a unique method for foam reinforcement to control gas mobility. However, the application of nanoparticles to improve surfactant foam stability is still in its early stages. The synergistic impact of nanoparticles and surfactants on foam stabilization is unknown. More research is needed to have a better understanding of the processes behind foam production and stability in the presence of NPs. The goal of this work is to create a more stable foam by carefully adjusting the amounts of NPs and surfactants.

The Pickering emulsion approach has long been used to stabilize CO_2 foams containing nanoparticles (Fig. 10.1). It has recently been proven that nanoparticle-stabilized CO_2 foams have the potential to be used in EOR applications. These investigations revealed that nanoparticle foams are substantially more stable than surfactant foams due to the nanoparticles' high adsorption energy at the gas–liquid interface. Furthermore, carbon dioxide (CO_2) injection is a popular approach in the field of enhanced oil recovery (EOR). Depending on the reservoir conditions and crude oil makeup, CO_2

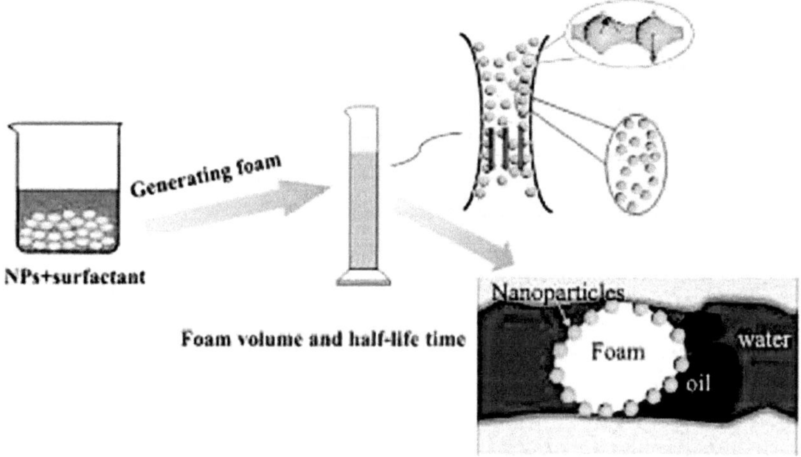

FIGURE 10.1 Schematic showing stabilization mechanism of foam using nanoparticles and its flow in porous media.

Source: Reprinted with permission from Zhang et al. (2021). Copyright 2020. Elsevier.

can be utilized as a miscible or immiscible dispersant. CO_2 injection can assist in boosting output by freeing more oil held in reservoir rock pores. Because of its lower viscosity, CO_2 has a less favorable mobility ratio than crude oil. Furthermore, because of the variability of the reservoir formation, early breakthrough occurs, resulting in a low oil-sweep efficiency. Surfactant use is one method for mitigating the negative effects of CO_2 floods. By preventing CO_2 from entering through high-permeability zones, it helps to lower the viscosity difference between the displacing and displaced fluids. The main surfactant EOR challenges nowadays are high surfactant retention in porous media and unstable foam properties under high-temperature reservoir settings. Researchers recently investigated a new type of CO_2 foam that uses nanotechnology to build long-lasting foam. This new technique has piqued the curiosity of many because of the foams' long-term stability under tough reservoir conditions. Dickson et al. were the first to show that micro-silica particles may be used to generate CO_2 foam. Supercritical CO_2 was used to shear nano-silica particles that had been distributed in the water. They created a CO_2 foam and examined it using a view cell and they discovered that it increases particle concentration, CO_2 density, and shear rate-enhanced foam stability. Later that year, Espinoza et al. (2010) proposed making nanoparticle-stabilized supercritical CO_2 foams (CO_2 in water foam) with commercially available surface-modified silica particles for possible mobility control applications (Espinoza et al., 2010). The foams were created by simultaneously injecting CO_2 and distributing nanoparticles in water with a glass-bead pack. They demonstrated that nanoparticles containing as little as 0.05 wt % CO_2 can produce supercritical CO_2 foams. Andrew et al. conducted foam production studies on nanoparticles with various surface coatings and observed that utilizing bigger 50% SiOH nanoparticles resulted in stronger, more stable CO_2 foams than using PEG-coated silica particles. Singh and Mohanty investigated the mechanism of foam stabilization in bulk and porous media using in situ surface-activated nanoparticles. Alumina-coated silica nanoparticles were combined with different amounts of propyl gallate (PG), and the hydrophilic surface of the particle was transformed to a partly hydrophobic surface by binding the PG molecule to it (Singh and Mohanty, 2017). When surface-modified nanoparticles (SNPs) were utilized, they noticed a considerable foaming tendency, with foam stabilized with SMNPs having a finer bubble structure than foam stabilized with surfactants. Several studies investigated the method of surfactant-mediated surface modification of nanoparticles at the same time. Synergistic interaction between surfactant and nanoparticles was discovered to produce stable foam. Worthen et al. (2012), for example, demonstrated the capacity of bare silica nanoparticles and a zwitterionic surfactant to create viscous and stable CO_2 foams with a fine texture, even though none of these species is capable of foam stabilization on their own (Worthen et al., 2012). Prigiobbe et al. investigated the flow dynamics of nanoparticle-supported CO_2 foam (Prigiobbe et al., 2016). To evaluate CO_2 foam transfer in porous media, we used a two-phase flow mechanistic model that includes the mass conservation law and the population balance equation. Their findings showed that the two-phase flow model closely matched the tests performed to develop a robust, high-quality foam.

Despite the numerous advantages of foam flooding, its inability to be used in improved oil recovery has remained a fundamental hurdle (EOR). Foam is described as gas bubbles dispersed in liquid with a bubble diameter less than the container's diameter. When using heterogeneous porous media such as actual oil field rock, the bubble size often approaches that of the pores, resulting in the development of lamellar liquid films. Lamellas are prone to long-term instability for four reasons: liquid drainage and film rupture due to an imbalance between gravity and capillary force; gas diffusion due to an uneven bubble size distribution; surfactant loss due to adsorption on rocks; and surfactant degradation at high temperatures. To minimize or mitigate these effects, other techniques have been attempted, including the use of modified polymers, surfactants, nanoparticles, or a combination of these reagents. Recently, the use of nanoparticles was explored, and it was shown that they had significant potential for foam stabilization under extreme temperature and salinity circumstances. To stabilize CO_2-in-water foams, Andrew et al. and Al-Yousef et al. (2017) employed silica nanoparticles coated with polyethene glycol. Surfactants were frequently used to make macroscopic foams by adsorbing on particle surfaces, which demonstrated higher stability (see Fig.10.2). The successive production of surfactant-coated particles at the air–water interface causes the foam to be stable over time.

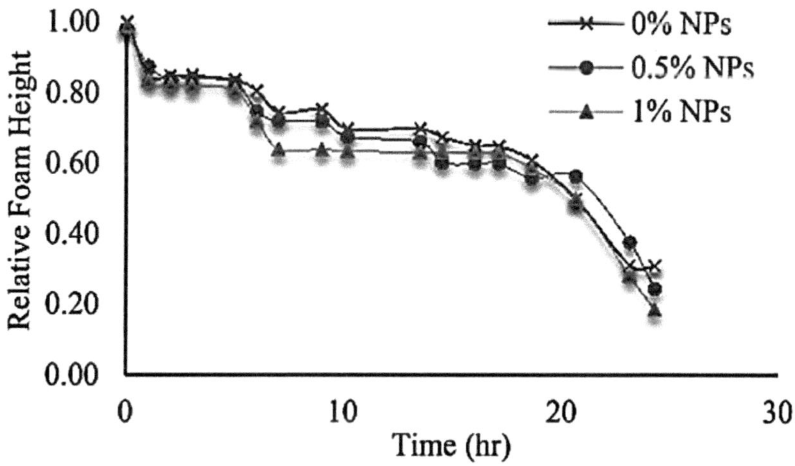

FIGURE 10.2 Relative foam height for foam in presence of surfactant and varying concentration of NPs with NaCl.

Source: Reprinted with permission from Al-Yousef et al. (2017). Copyright 2017. American Chemical Society

10.3 FOAM'S SALINE RESISTANCE AND THE IMPORTANCE OF NANOMATERIALS

Foams have been produced and are being used to stimulate unconventional reservoirs because they provide several advantages over water-based fracturing fluids, including decreased water consumption, faster cleanup, improved leak-off control, and proppant conveyance (Wang et al., 2016; Zhang et al., 2016). However, due to the high surface tension of the two phases, stable foams cannot be formed using water and CO_2. A foaming agent is utilized to minimize surface tension and stabilize the interfacial layers. Surfactants are amphiphilic molecules with a hydrophilic and hydrophobic tail that have been employed for decades to build and stabilize foams (Binks, 2002). When surfactant molecules are introduced to a two-phase solution, they aggregate with aligned tails at the interfaces, decreasing surface tension and thereby preventing foam bubbles from collapsing (Narasimman and Prabhakaran, 2012). See Figures 10.3 and 10.4 for various stages of foam destabilization. Water–CO_2 foams, on the other hand, are thermodynamically and kinetically unstable due to their high surface energy and proclivity to degenerate into distinct gas and liquid phases (Emrani and Nasr-El-Din, 2017). The main disadvantage of surfactant-stabilized foams is that the surfactants break down quickly in high-temperature, high-salinity reservoirs, remaining stable for just a few hours. Furthermore, because surfactants have a high degree of adsorption in porous media, excessive amounts of chemicals are used during the fracturing process. Though at a low volume of salts (0-1 wt % NaCl), the foam stability is expected to remain high (due to the stabilization by available Colloids) (see Fig. 10.5).

These surfactant limits necessitate the development of innovative foam stabilization techniques under severe reservoir conditions. Nanomaterials have been used to create extremely stable CO_2 foams in water. For example, nanoparticles, for example, have been shown to help sustain CO_2 foams that are exceptionally resistant to temperature and salinity fluctuations (Espinoza et al., 2010; Rognmo et al., 2018). Surfactants may dynamically adsorb and desorb from liquid–gas interfaces over time, while nanoparticle adsorption is typically irreversible. This irreversible adsorption of nanoparticles at

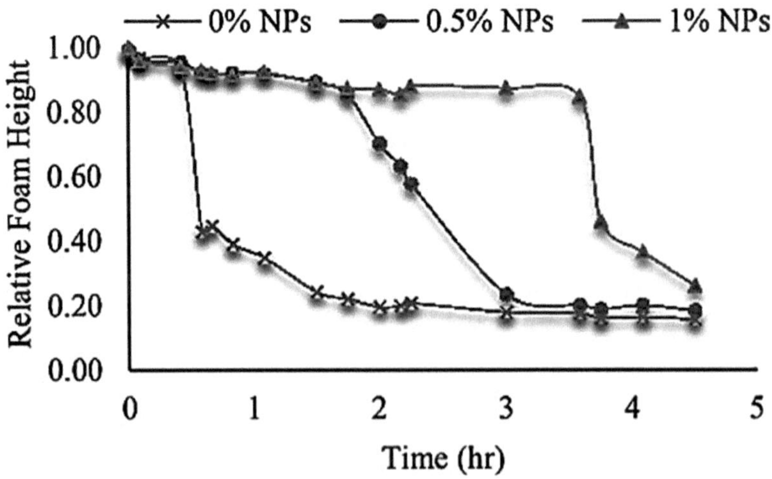

FIGURE 10.3 Foam life for prepared foam stabilized by silica NPs and surfactant in presence of crude oil.

Source: Reprinted with permission from Al-Yousef et al. (2017). Copyright 2017. American Chemical Society

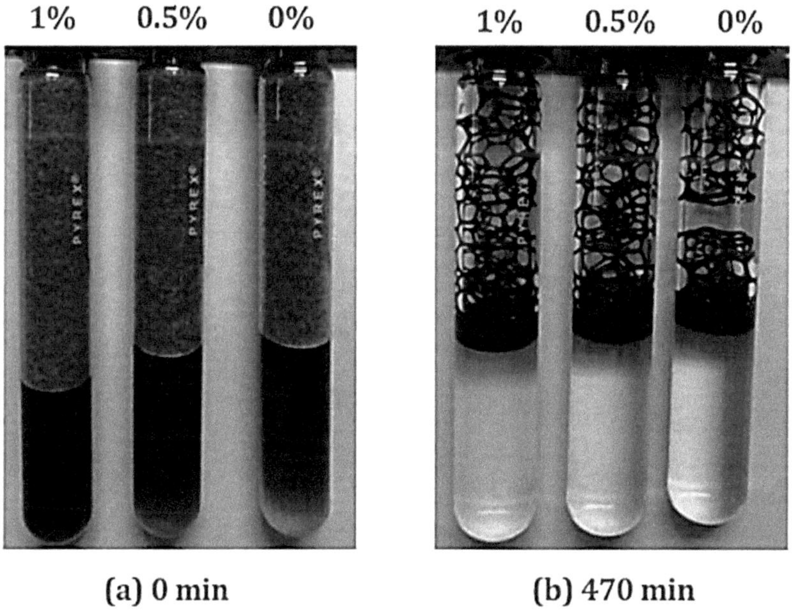

FIGURE 10.4 Foam decay with time for 0.50 wt % surfactant (0 wt % NPs) and mixture of surfactant and 0.50 and 1 wt % nanoparticles at 25°C using 1 wt % NaCl.

Source: Reprinted with permission from Al-Yousef et al. (2017). Copyright 2017. American Chemical Society

gas–liquid interfaces results in the formation of a strong network that serves as a physical barrier against bubble coalescence and considerably lowers film drainage under severe reservoir conditions (Emrani and Nasr-El-Din, 2017; Worthen et al., 2013). Another disadvantage of surfactant-stabilized foams is that their CO_2 solvation tails are insufficient. This is automatically resolved with nanoparticles, which are usually spherical with no tails and may be quickly adsorbed on surfaces with a high degree of affinity (Chen et al., 2015). A foam-assisted system generates a dynamic liquid film termed a lamella by injecting surfactant molecules deposited at the water–air (or gas) interface. The produced foams aid in the stable dispersion of trapped gas inside the thin liquid layer under suitable lamella structures, considerably restricting the gas's mobility; as a result, contact between the displaced oil and the displacing gas is enhanced, resulting in improved sweep efficiency (Kim and Kim, 2005). The equilibrium between bubble formation and liquid film drainage is largely responsible for foam stability. One of the most difficult challenges in using surfactant-assisted foams for EOR is ensuring that they grow and remain stable across a wide range of reservoir conditions. Once formed, the first foam height represents the foamability of the foam. The height of the foam is determined by the pace at which surfactant molecules diffuse from the bulk dispersion to the air–water interface. The surfactant's molecular structure and salt levels have a substantial impact on its ability to produce foam (Xu et al., 2018). Foam stability is a more difficult criterion to manage than foamability since it refers to the ability of created foams to remain stable in a reservoir for an extended length of time. Figure 10.5 represents the half-life of foam in presence of various stabilizing agents. Mechanical strength and viscosity increases in gas–water interfacial membranes are expected to improve foam stability in general. Foam instability in EOR is caused by the qualities of the oil, salt concentrations, salt ion types, temperature, and other environmental conditions, which results in foam drainage by gravity, liquid film rupture (Li et al., 2013), or gas diffusion.

Numerous studies have been undertaken to evaluate if the inclusion of various stabilizers, such as solid particles, increases foam stability in EOR applications (Kumar et al., 2021; Xue et al., 2015). Kumar and colleagues, 2021). (See Fig. 10.6, NP+ Surfactant systems demonstrated higher recovery than surfactant only solutions.) Over the last decade, not only have new

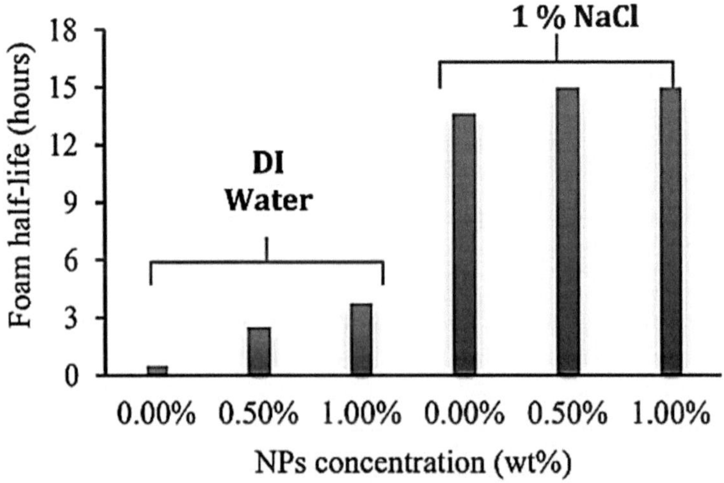

FIGURE 10.5 Foam half-lives of 0.50 wt % surfactant and mixture of surfactant and 0.50 and 1 wt % nanoparticles at 25°C and at DI water and 1% NaCl in the presence of crude oil.

Source: Reprinted from Al-Yousef et al. (2017). Copyright 2017. American Chemical Society

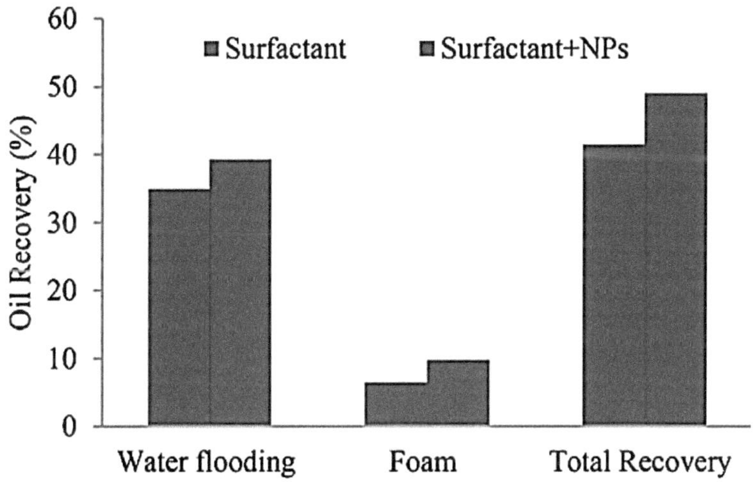

FIGURE 10.6 Oil recovery results using NPs stabilized foams.

Source: Reprinted with permission from Al-Yousef et al. (2017). Copyright 2017. American Chemical Society.

generation NP materials been successfully manufactured using low-cost materials such as silica and fly ash (Binks, 2002), but they have also demonstrated significant potential for foam stabilization under adverse conditions such as elevated temperatures and high saline concentrations (Xiao et al., 2017). Singh and colleagues (Singh and Mohanty, 2015) observed that raising the NP concentration significantly increased the half-life of their foam, which was made with NP and an AOS surfactant. Chen and colleagues (2016) discovered that optimal SiO2-NP/SDBS ratios in a carbondioxide foam system had a synergistic effect on foam stability. Non-adsorbed NP remains suspended in the bulk phase, increasing flow resistance and lengthening liquid drainage time (Lv et al., 2017). Despite this, the processes for foam stabilization via surfactants and NP are complicated because of the many interactions between surfactants and NP (i.e., surfactant–surfactant, NP–surfactant, and NP–NP interactions).

10.4 A BINARY NP SYSTEM IS USED TO IMPROVE FOAM STABILITY

Kumar et al. investigated the foam stabilization process at severe temperatures and discovered that a binary NPs system is suitable for high-temperature oil recovery reservoirs (Kumar et al., 2021). In that work, binary NPs

(SiO2 and TiO2) were demonstrated to be effective at stabilizing CO_2 foams at moderate pressure (70 bar) and temperature (90°C). Microscopic and rheological tests at 70 bar and 90°C validated the improved stability of the nanocomposites-stabilized CO_2 foam, which was then followed by UV–vis and contact angle studies to explore colloidal interaction with CO_2. By delaying the CO_2 coalescence process, TiO2 insertion significantly improved the thermal stability and rheological behavior of SF foam. Furthermore, the study shows that TiO2 NP lowers the size of foam bubbles by improving adsorption, as seen by lamella formation, and that foam is the least responsive to changes in shear, pressure, and temperature. Furthermore, the rheological features of these nanocomposites-stabilized CO_2 foams, such as viscosity and viscoelasticity, show potential for real-world applications. As a result, real-world circumstances should be used to evaluate CO_2 foam produced in a binary NP system (nanocomposites), which is useful for CO_2-based oil recovery applications.

10.5 CONCLUSION

The major goal of this chapter is to get a grasp of the fundamental process that underpins foam stability in oil recovery applications. It is now clearly evident that as the saline concentration of the solution increased, so did the size of the nanoparticle aggregates. Even though temperature increases did not influence the aggregation of the silica nanoparticles, they are highly sensitive to the saline reservoir. Furthermore, when the volumetric ratio of the gas phase increases, foam stability diminishes. As foam quality improved, the perceived viscosity rose about tenfold. Furthermore, the Ostwald ripening hypothesis and nanoparticle gravitational segregation may explain the increasing average size of foam bubbles.

KEYWORDS

- **nanomaterials**
- **foam stabilization**
- **nanofluids**
- **oilfield applications**
- **enhanced oil recovery**
- **well stimulation**

REFERENCES

Al-Yousef, Z.; Almobarky, M.; Schechter, D. Enhancing the Stability of Foam by the Use of Nanoparticles. *Energy Fuels* **2017,** *31,* 10620–10627.

Aminzadeh, B.; Chung, D. H.; Zhang, X.; Bryant, S. L.; Huh, C.; DiCarlo, D. A. Influence of Surface-Treated Nanoparticles on Displacement Patterns During CO_2 Injection. *Proc. - SPE Annu. Tech. Conf. Exhib.* **2013,** *4,* 2641–2660.

Binks, B. P. Particles as Surfactants—Similarities and Differences. *Curr. Opin. Colloid Interface Sci.* **2002,** *7,* 21–41.

Carn, F.; Colin, A.; Pitois, O.; Vignes-Adler, M.; Backov, R. Foam Drainage in the Presence of Nanoparticle-surfactant Mixtures. *Langmuir* **2009,** *25,* 7847–7856.

Chang, Q. Emulsion, Foam, and Gel. In *Colloid and Interface Chemistry for Water Quality Control*. Elsevier, 2016; pp 227–245.

Chaturvedi, K. R.; Ravilla, D.; Kaleem, W.; Jadhawar, P.; Sharma, T. Impact of Low Salinity Water Injection on CO_2 Storage and Oil Recovery for Improved CO_2 Utilization. *Chem. Eng. Sci.* **2021,** *229,* 116127.

Chen, Y.; Elhag, A. S.; Cui, L.; Worthen, A. J.; Reddy, P. P.; Noguera, J. A.; Ou, A. M.; Ma, K.; Puerto, M.; Hirasaki, G. J.; Nguyen, Q. P.; Biswal, S. L.; Johnston, K. P. CO_2-in-Water Foam at Elevated Temperature and Salinity Stabilized with a Nonionic Surfactant with a High Degree of Ethoxylation. *Ind. Eng. Chem. Res.* **2015,** *54,* 4252–4263.

Emrani, A. S.; Nasr-El-Din, H. A. An Experimental Study of Nanoparticle-Polymer-Stabilized CO_2 Foam. *Colloids Surf. A Physicochem. Eng. Asp.* **2017,** *524,* 17–27.

Espinoza, D. A.; Caldelas, F. M.; Johnston, K. P.; Bryant, S. L.; Huh, C. Nanoparticle-Stabilized Supercritical CO_2 Foams for Potential Mobility Control Applications. In *SPE Improved Oil Recovery Symposium*; Society of Petroleum Engineers: Tulsa, Oklahoma, USA, 2010.

Hirasaki, G. J.; Miller, C. A.; Puerto, M. Recent Advances in Surfactant EOR. *SPE J.* **2011,** *16* (4), 889–907.

Kang, X.; Kang, W.; Li, Z.; Yang, H.; Xie, A.; Li, M.; Zhu, T.; Yurievich, K.G.; Viktorovich, L. A. Stability Influence Factors and Mechanism of Produced Emulsion from CO_2 Flooding. *J. Mol. Liq.* **2021,** *333,* 115974.

Kim, K. D., Kim, H. T. Comparison of the Growth Mechanism of TiO_2-coated SiO_2 Particles Prepared by Sol-Gel Process and Water-in-Oil Type Microemulsion Method. *Colloids Surf. A Physicochem. Eng. Asp.* **2005,** *255,* 131–137.

Kumar, R. S.; Chaturvedi, K. R.; Iglauer, S.; Trivedi, J.; Sharma, T. Impact of Anionic Surfactant on Stability, Viscoelastic Moduli, and Oil Recovery of Silica Nanofluid in Saline Environment. *J. Pet. Sci. Eng.* **2020,** *195,* 107634.

Kumar, R. S.; Sinha, A. S. K.; Sharma, H.; Sharma, T. High Performance Carbon Dioxide Foams of Nanocomposites of Binary Colloids for Effective Carbon Utilization in Enhanced Oil Recovery Applications. *J. Mol. Liq.* **2021,** *343,* 117659.

Le, N. Y. T.; Pham, D. K.; Le, K. H.; Nguyen, P. T. Design and Screening of Synergistic Blends of SiO_2 Nanoparticles and Surfactants for Enhanced Oil Recovery in High-Temperature Reservoirs. *Adv. Nat. Sci. Nanosci. Nanotechnol.* **2011,** *2.*

Li, S.; Hendraningrat, L.; Torsaeter, O. Improved Oil Recovery by Hydrophilic Silica Nanoparticles Suspension: 2-Phase Flow Experimental Studies. *Soc. Petroleum Eng.* **2013.** DOI: 10.2523/iptc-16707-ms

Li, J.; Jiang, Z.; Wang, Y.; Zheng, J.; Huang, G. Stability, Seepage and Displacement Characteristics of Heterogeneous Branched-Preformed Particle Gels for Enhanced Oil Recovery. *RSC Adv.* **2018,** *8,* 4881–4889.

Li, W.; Wei, F.; Xiong, C.; Ouyang, J.; Shao, L.; Dai, M.; Liu, P.; Du, D. A Novel Supercritical CO_2 Foam System Stabilized With a Mixture of Zwitterionic Surfactant and Silica Nanoparticles for Enhanced Oil Recovery. *Front. Chem.* **2019,** *7,* 718.

Lv, Q.; Li, Z.; Li, B.; Shi, D.; Zhang, C.; Li, B. Silica Nanoparticles as a High-Performance Filtrate Reducer for Foam Fluid in Porous Media. *J. Ind. Eng. Chem.* **2017,** *45,* 171–181.

Narasimman, R.; Prabhakaran, K. Preparation of Carbon Foams by Thermo-Foaming of Activated Carbon Powder Dispersions in an Aqueous Sucrose Resin. *Carbon N. Y.* **2012,** *50,* 5583–5593.

Nguyen, P.; Fadaei, H.; Sinton, D. Pore-Scale Assessment of Nanoparticle-Stabilized CO_2 Foam for Enhanced Oil Recovery. *Energy Fuels* **2014,** *28,* 6221–6227.

Prigiobbe, V.; Worthen, A. J.; Johnston, K. P.; Huh, C.; Bryant, S. L. Transport of Nanoparticle-Stabilized CO_2 Foam in Porous Media. *Transp. Porous Media* **2016,** *111,* 265–285.

Rognmo, A. U.; Heldal, S.; Fernø, M. A. Silica Nanoparticles to Stabilize CO_2-Foam for Improved CO_2 Utilization: Enhanced CO_2 Storage and Oil Recovery from Mature Oil Reservoirs. *Fuel* **2018,** *216,* 621–626.

Seyyedsar, S. M.; Farzaneh, S. A.; Sohrabi, M. Investigation of Low-Density CO_2 Injection for Enhanced Oil Recovery. *Ind. Eng. Chem. Res.* **2017,** *56,* 5443–5454.

Singh, R.; Mohanty, K. K. Foam Flow in a Layered, Heterogeneous Porous Medium: A Visualization Study. *Fuel* **2017,** *197,* 58–69.

Singh, R.; Mohanty, K. K. Foams Stabilized by In-Situ Surface-Activated Nanoparticles in Bulk and Porous Media. *SPE J.* **2015,** *21*, 121–130.

Stoll, W. M.; Hofman, J. P.; Ligthelm, D. J.; Faber, M. J.; van den Hoek, P. J. Toward Field-Scale Wettability Modification—The Limitations of Diffusive Transport. *SPE Reserv. Eval. Eng.* **2008,** *11,* 633–640.

Sun, Q.; Li, Z.; Li, S.; Jiang, L.; Wang, J.; Wang, P. Utilization of Surfactant-Stabilized Foam for Enhanced Oil Recovery by Adding Nanoparticles. *Energy Fuels* **2014,** *28,* 2384–2394.

Wang, C.; Cao, X. L.; Guo, L. L.; Xu, Z. C.; Zhang, L.; Gong, Q. T.; Zhang, L.; Zhao, S. Effect of Adsorption of Catanionic Surfactant Mixtures on Wettability of Quartz Surface. *Colloids Surf. A Physicochem. Eng. Asp.* **2016,** *509,* 564–573.

Worthen, A.; Bagaria, H.; Chen, Y.; Bryant, S. L.; Huh, C.; Johnston, K. P. Nanoparticle Stabilized Carbon Dioxide in Water Foams for Enhanced Oil Recovery. In *SPE Improved Oil Recovery Symposium*; Society of Petroleum Engineers, 2012.

Worthen, A. J.; Bryant, S. L.; Huh, C.; Johnston, K. P. Carbon Dioxide-in-Water Foams Stabilized with Nanoparticles and Surfactant Acting in Synergy. *AIChE J.* **2013,** *59,* 3490–3501.

Xiao, C.; Balasubramanian, S. N.; Clapp, L. W. Rheology of Viscous CO_2 Foams Stabilized by Nanoparticles under High Pressure. *Ind. Eng. Chem. Res.* **2017,** *56,* 8340–8348.

Xu, C.; Zou, W.; Yang, Y.; Duan, Y.; Shen, Y.; Luo, B.; Ni, C.; Fu, X.; Zhang, J. Status and Prospects of Deep Oil and Gas Resources Exploration and Development Onshore China. *J. Nat. Gas Geosci.* **2018,** *3,* 11–24.

Xue, Z.; Panthi, K.; Fei, Y.; Johnston, K. P.; Mohanty, K. K. CO_2-Soluble Ionic Surfactants and CO_2 Foams for High-Temperature and High-Salinity Sandstone Reservoirs. *Energy Fuels* **2015,** *29,* 5750–5760.

Zhang, C.; Li, Z.; Sun, Q.; Wang, P.; Wang, S.; Liu, W. CO_2 Foam Properties and the Stabilizing Mechanism of Sodium bis(2-ethylhexyl)sulfosuccinate and Hydrophobic Nanoparticle Mixtures. *Soft Matter.* **2016,** *12,* 946–956.

Zhang, Y.; Liu, Q.; Ye, H.; Yang, L. L.; Luo, D.; Peng, B. Nanoparticles as Foam Stabilizer: Mechanism, Control Parameters and Application in Foam Flooding for Enhanced Oil Recovery. *J. Pet. Sci. Eng.* **2021,** 108561. DOI: 10.1016/J.PETROL.2021.108561

CHAPTER 11

Emulsions for EOR

RAMESH NARUKULLA and TUSHAR SHARMA

Enhanced Oil Recovery Laboratory, Rajiv Gandhi Institute of Petroleum Technology, Jais, India

ABSTRACT

An emulsion is a biphasic system comprising of two incompatible liquids in which one liquid is dispersed equally and consistently as globules (dispersed phase) over the second liquid (continuous phase), whereas the other liquid is not distributed evenly and consistently across the first liquid (continuous phase). In this chapter, the authors explore the application of emulsions in enhanced oil recovery (EOR) strategies, petroleum industries (upstream and downstream process) and their effects. As the recovery rates in emulsion-created era is much higher than no-emulsion era, investigating the unique properties of emulsions and their impact on oil displacement and mobility control. The emulsion forms an interfacial layer between the dispersed and continuous phase by the application of energy for generating emulsion. For generating emulsion, a wide range of surfactants are used that facilitates a strong barrier between dispersed and continuous phase and helps in stabilizing emulsion formation. The study aims to elucidate the effectiveness of emulsion based EOR techniques particularly for stranded oil and in constricted capillaries. Emphasizing the potential for improved reservoir sweep efficiency and enhanced displacement in challenging oil recovery scenarios, the research underscores the promising role of emulsions as a versatile tool in the pursuit of efficient and sustainable oil extraction methods. The chapter provides detailed application of emulsion and micro emulsions in EOR along with its stability. Formation of emulsion within reservoir is a natural process, affected by various parameters, that is, wettability of rock, composition of crude oil,

Advancements in Chemical Enhanced Oil Recovery.
Tushar Sharma, Krishna Raghav Chaturvedi, Tarek Ganat, and Imtiaz Ali (Eds.)

amount and distribution of residual oil. Stability of emulsion on microscopic, mesoscopic and macroscopical scales has been discussed. Mechanism of emulsion in improving sweep efficiency was also explored in microfluidic devices. The chapter outlines about the Pickering emulsion, that is, emulsion stabilized by the solid particles which offers increased stability. Formation of Pickering emulsion, factors affecting the emulsion stability along with its thermal and structural stabilities has been discussed.

11.1 INTRODUCTION

An emulsion is a biphasic system comprised of two incompatible liquids in which one liquid is scattered equally and consistently as globules (dispersed phase) over the second liquid (continuous phase), whereas the other liquid is not distributed evenly and consistently across the first liquid (continuous phase) (Adamson and Gast, 1997). In other words, an insoluble mixture consists of two liquids, one of which is dispersed and spreads as minute droplets in the other phase, while the other is continuous, resulting in an intractable combination (Jin et al., 2019). Emulsions are used in a number of ways in our daily lives; milk, creams, and lotions are all examples of frequently used emulsions. Numerous firms manufacture emulsions, particularly to meet the requirements of various applications. However, certain industries do not require emulsions, and the emulsions created in these industries may pose issues (Long et al., 2017). Emulsions can be prepared specifically for use in the petroleum industry's upstream processes, including emulsified acids, hydraulic fracturing, and better-enhanced oil recovery (EOR) (Long et al., 2017). On the contrary, the resulting emulsions have a detrimental effect on downstream operations. Once the oil–water stream is generated and reaches the surface in the form of stable emulsions, the separating procedure requires extra resources, time, and technical knowledge to prepare it for further processing. When crude oil is taken from the ground during drilling operations, a mixture of oil and water frequently forms at the wellhead. This affects how they are treated on the ground as well as in transit zones (Kokal, 2005).

Natural factors limit the quantity of oil that can be collected from petroleum reservoirs to between 20% and 60%; the major objective of various EOR technologies is to gather around 9.8 billion barrels of crude oil per day (Alvarado and Manrique, 2010). Chemical flooding agents modify the physical–chemical and interfacial properties of crude oil, as well as the physical–chemical properties of the displaced fluid. Additionally, they modify the way crude oil and the displacing fluid interact. These oilfield studies experienced a variety of difficulties while using chemical flooding processes such as alkali–surfactant–polymer (ASP) flooding and polymer flooding (Sheng, 2011). These strategies have been implemented in a variety of areas. It is conceivable for the polymer molecules to clog the medium during the polymer flooding process, thus limiting its permeability. This is because the crude oil layers and the polymer molecules have physically distinct physical properties. When alkali is immersed, it condenses as scale, contaminating the oil layers underneath. Polymer flooding is a frequently used approach in chemical EOR to control the fluidity of the displaced fluid. Due to the loss of viscosity caused by this polymer under high-salt and high-temperature reservoir conditions, it is not authorized to utilize it. In recent years, a range of surfactants, including ethoxylated amine and sulfonate surfactants, have been created to tolerate the reservoir's severe salinity and temperature conditions. By using these surfactants, it was feasible to develop solutions that remained stable at elevated temperatures and in salinized environments. Although the major limitation of surfactant use in floods is the inability to control their motion, this limitation is amplified by the fact that surfactant solutions can readily escape through porous oil layers. Importers have caused lasting formation damage and pollution, complicating the development of oil resources (Al-Anazi and Sharma, 2002; Sugar et al., 2020). To entirely overcome this problem, the flow of the displacing fluid must be regulated.

Following multiple laboratory and field-based ASP flooding studies, it was observed that emulsion generation is crucial for enhanced postflood recovery. When an emulsion was created in the generated fluid, recovery increased. However, when no emulsion was formed in the generated fluid, recovery dropped (Koczo et al., 1992; Zhou et al., 2019). It's worth mentioning that the recovery rate in the emulsion-created scenario was 5%, which was much greater than the recovery rate in the no-emulsion scenario. As a result, surfactant flooding's major purpose is to enhance emulsification performance. When utilized in high-salinity, high-temperature reservoirs, this is a potentially viable method of boosting oil recovery that is also environmentally friendly. Apart from that, Pickering emulsions, which are emulsions stabilized by solid particles, have been employed to attain increased stability in recent years. By incorporating solid particles into oil–water emulsions, it has been demonstrated that their viscosity and heat stability are greatly increased, making them more suitable for use in EOR applications (Jiang et al., 2020; Sharma et al., 2014).

Despite scientific advances, several essential features of the emulsion process and seepage in EOR continue to be a cause of consternation. Along with discussing the Pickering emulsion' application in EOR, this chapter discusses emulsion stability and how emulsions develop. Additionally, it also aims to examine seepage features, rheological characteristics, and EOR of emulsions at elevated depths. To conclude, this article discusses the use of emulsions in EOR, as well as the results of field and laboratory research.

11.2 APPLICATIONS OF EMULSIONS & MICROEMULSIONS IN EOR

Global energy demand is projected to be satisfied by the world's energy mix, which is already shifting away from fossil fuel dominance toward a more stable energy balance. Flooding with ordinary water alone will not recover all of the oil in a reservoir; the remaining oil is in the form of oil droplets, which are supported by capillary forces and account for around 30% of the reservoir's initial oil (OOIP), while water flooding generates approximately 40% of the oil. On the contrary, EOR technologies can be utilized to compensate for this loss of oil. Frequently, the terms "primary," "secondary," and "tertiary" oil recovery are used. These stages of recovery illustrate how oil is extracted in stages from a reservoir (Babadagli, 2020). The energy required to displace oil during the earliest stages of production comes from the reservoir's intrinsic energy reserves. Natural energy consumption decreases when oil is removed from a reservoir, complicating well extraction and limiting practical recovery (Rellegadla et al., 2021). To sustain production, the reservoir must be fed with external energy. This kicks off the second stage, which requires the use of supplementary recovery treatments. This is only a method of supplementing the reservoir's natural energy supply. Water flooding is the most commonly used approach for secondary recovery (Austad et al., 2012). Globally, it is estimated that between two and three billion barrels of oil will remain unrecovered following primary and secondary operations. The amount of oil that has not yet been recovered highlights the vital nature of third-generation recovery processes. The purpose of this book has been to argue in favor of the use of EOR or tertiary oil recovery techniques. This technology for tertiary or improved oil recovery uses heat energy, liquid chemicals, and/or gas injection (no reservoir). This chapter will explore how to successfully utilize emulsions and microemulsions to aid in recovery. Additionally, previous research has proven that emulsions or microemulsions are viable EOR techniques (Kumar and Mandal, 2016; Schumi et al., 2020).

The production of emulsions and microemulsions requires energy to combine the water, oil, and emulsifier. The emulsifier (i.e., amphiphilic proteins, and surfactants) generates an interfacial layer between the dispersed and continuous phases by using the energy necessary to generate the emulsion. The interfacial layer acts as a barrier between the dispersed and continuous phases of the emulsion, and stabilizing it throughout the emulsion's formation. Due to the thermodynamic and rheological characteristics of emulsion and/or microemulsion injection into reservoirs, it is a highly respected method in the oil recovery sector. The primary objective of the oil recovery process is to maximize the efficacy of macroscopic and microscopic displacement (Muggeridge et al., 2014). While macroscopic displacement attempts to remove oil on a large scale without trapping it, microscopic displacement works at the pore level. Emulsion formation in a reservoir is a natural chemical and physical process that is influenced by a range of factors, including the rock's wettability, the shape of the hole throat, the amount of residual oil, the crude oil properties, and the distribution of residual oil. The process of creating reservoir emulsions is more involved than mechanical stirring. There are a few publications on the emulsion generation process in porous media, and the majority of these studies establish the mechanism by eye observation or microfluidic testing. This process is illustrated in Figure 11.1.

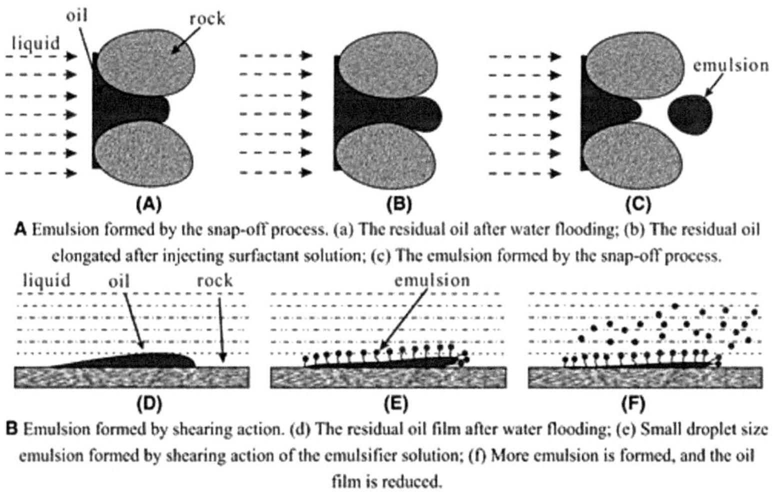

FIGURE 11.1 Illustration showing the various mechanisms through which emulsions are formed in porous media. (1) Emulsion formed as a result of the snap-off process. (A) The residual oil following water flooding; (B) The residual oil elongated following surfactant solution injection; and (C) The emulsion formed during the snap-off technique. (2) Emulsion formed as a result of a shearing action. (D) The residual oil film after water flooding; (E) A small droplet size emulsion generated by the emulsifier solution's shearing action; (F) More emulsion is formed, and the oil film is reduced.

Source: Reprinted with permission from Zhou et al. (2019). Copyright 2019. Elsevier.

Emulsion formation may also be simulated using microfluidic devices. However, other reservoir variables, such as high temperature and pressure, residual oil dispersion, and pore and throat characteristics, must be modeled. In the past, researchers have developed a novel approach for examining the characteristics of emulsions under extreme pressure and temperature conditions. They used a pressure–volume–temperature (PVT) cell to conduct their experiment. Experiments indicated that if the crude oil is capable of precipitating asphaltenes, an emulsion can form. Emulsions can form under reservoir conditions, especially when the crude oil is prone to precipitating asphaltenes (Abdallah and Taylor, 2007). Emulsion formation is heavily influenced by the chemical composition of crude oil, specifically the ratio of gelatine to asphaltene. Researchers have explored whether spontaneous emulsification occurs following injection of an ASP solution into the small visual center. The experiment indicated that as the solution moved, residual oil was progressively absorbed by the oil filament and that the oil filament grew and passed through the throat. The oil filament is divided into several oil-in-water (o/w) emulsions as it travels (Binks and Rodrigues, 2005; Lan et al., 2007; Pea et al., 2009). Bidhendi et al. (2018) and Pea et al. (2009) investigated the flow behavior of long oil droplets in two-phase flow in porous media using quartz capillary tubes and hypothesized that snap off may be a mechanism for emulsion formation (Bidhendi et al., 2018; Pea et al., 2009). The viscosity and capillary pressures exerted on the residual oil determine whether it is retained or migrates as a result of flooding. When the viscous force is raised sufficiently, the saturation of the nonwetting phase drops. The oil droplet or emulsion disperses and flows when the nonwetting phase becomes discontinuous. The capillary count, velocity, viscosity ratio, and pore throat ratio all affect the quantity of residual oil that snaps off (Amirian et al., 2017).

In previous work, researchers have also demonstrated drop snapping off in a restricted capillary with an elastic interface, demonstrating that the interface's rheology has a significant effect on drop snapping off and that a suitably viscoelastic interface may result in the best interfacial stability against snapping off (Li et al., 2021; Wu et al., 2021). Zhou et al. used a displacement experiment and a microscopic visual to explore the emulsion formation process (Zhou et al., 2015). The key formation processes, according to the findings, are residual oil snapping and emulsifier solution shearing. Further explanations include droplet formation in a cross-junction microchannel using the lattice Boltzmann approach and investigating the influence of surface tension on droplet length and the interval between droplets. The results indicated that as surface tension increased, so did the length and spacing of droplets. Furthermore, researchers modeled emulsion production using a ternary Lattice Boltzmann technique, and the simulation results were congruent with experimental evidence on emulsion formation (see Fig. 11.2). The results revealed that increasing the outer fluid flow rate imposed a viscous shear force against the interface tension, causing the droplet to pinch off at the exit orifice; however, as the fluid viscosity increased, the pinch-off occurred later (Li et al., 2021; Zhou et al., 2019).

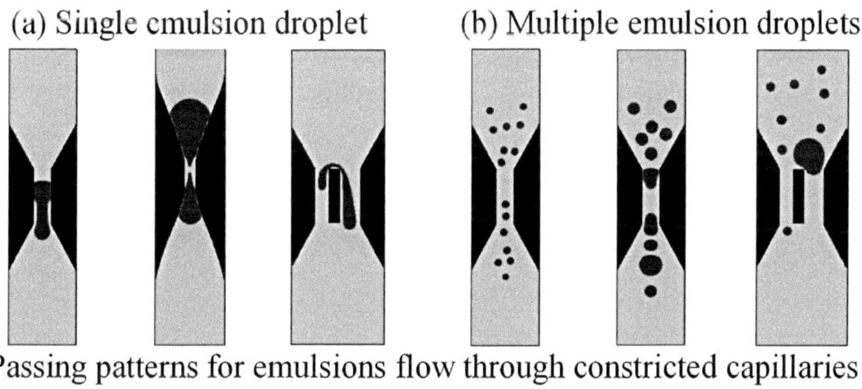

FIGURE 11.2 Emulsion generation during flow through constricted capillaries.

Source: Reprinted with permission from Wei et al. (2020). Copyright 2020. Elsevier.

In previous works, emulsion flooding has been found to expedite recovery by more than 15% when compared to conventional water flooding in displacement trials (Hou et al., 2021; Nguyen et al., 2014). Displacement in o/w emulsions is primarily characterized by a decrease in the mobility of the displacing liquid, which results in a drop in interfacial tension. The non-Newtonian fluid characteristics and interface potential of the emulsion are crucial for capillary network optimization and oil recovery. However, emulsion flooding holds several unique challenges of its own. For starters, the emulsion might choke the large throat, whereas the continuous phase would have removed any leftover oil. Second, because of the pushing and pulling action of the emulsion, any remaining oil near the edge or corner may be successfully displaced (Fredriksen et al., 2018; Mohebbifar et al., 2015).

Using a microfluidic device, researchers have also explored the mechanism of emulsion sweep improvement (Anna et al., 2003; Nie et al., 2008). When the droplet size exceeds the throat width of the high permeability channels, the results show that stable emulsion droplets can effectively block high permeability channels while facilitating flow via low permeability channels. However, spontaneous emulsification had a greater displacement effect on cores with varied permeability (Mohajeri et al., 2015). The residual oil was spontaneously emulsified into the emulsion, which had droplet sizes far smaller than the diameters of the pore and throat, hence lowering the residual oil's capillary force. Interfacial layer disruption, elastic deformation, and a decrease in interfacial tension all contribute to an improvement in displacement efficiency. A pore throat-sized emulsion when it comes into contact with residual oil, and it develops an elastic tension as a result of viscoelastic deformation. This elastic strain functions as a propellant, perhaps propelling the remaining oil. Injecting an emulsion slug following water flooding reduced oil saturation in two separate regions—near the injection well and along the displacement front. By reorienting the injected fluid, an emulsion injection can significantly reduce oil saturation. Taherpour et al. studied the wettability and viscosity variation of oil recovery emulsions (Taherpour and Hashemi, 2018). Sharma et al. produced novel Pickering emulsions that are thermally stable at elevated temperatures, which makes them ideal for application in EOR. Furthermore, EOR was investigated at 13.6 MPa reservoir pressure and temperatures ranging 39.85–89.85°C utilizing surfactant-polymer (SP) and o/w core flooding Pickering emulsions. At all test temperatures, Pickering emulsions retrieved more than 60% more oil cumulatively than SP flooding (Sharma et al., 2015b, 2015a). Kumar et al. used light mineral oil, carboxymethylcellulose (CMC), and silica nanoparticles to construct surfactant-polymer nanoparticle (SPN) Pickering emulsions in the presence of an anionic surfactant (Kumar and Mandal, 2016). The Pickering emulsion exhibits a consistent viscosity throughout a wide temperature range (30–100°C) and a constant loss modulus value, indicating the emulsion's enhanced flow capability. The sand pack flooding experiment demonstrated that the Pickering emulsion significantly speeds recovery by over 24% following conventional water flooding.

The impact of droplet sizes in the emulsion flooding displacement process varies greatly. While a little droplet can greatly increase effective driving force in both dry and wet reservoirs, a large droplet can temporarily shut the throat. Emulsions can increase oil recovery due to the synergistic effect of small and large droplet-size emulsions. Researchers have investigated the injection method for polymers and emulsions in reservoirs with bottom water and discovered that the o/w emulsion operated as a water-plugging agent, whereas alternating emulsion and polymer injection created a more efficient displacement effect. When compared to water inundation, this alternate injection method has been shown to boost recovery by more than 10%. Accordingly, research on the influence of spontaneous emulsification on heavy oil recovery discovered that by using emulsions instead of water flooding, recovery may be boosted by 10–35% (Austad et al., 2012; Kokal, 2005; Mohebbifar et al., 2015).

Microemulsions are also being studied for their possible use in some aspects of crude oil extraction, most notably EOR (Eastoe et al., 2013; Santra et al., 2001). Macro- and microemulsion floods are good EOR recovery methods because of the high extraction efficiency obtained by lowering the oil–water interfacial tension. A macroemulsion is a thermodynamically unstable heterogeneous mixture of oil and water in which either oil or water droplets are suspended in the oil (oil–water emulsion) or the water (water/oil emulsion). By absorbing at the oil–water interface, surfactants stabilize the droplets, effectively charging their surfaces and preventing them from colliding. On the contrary, microemulsions are transparent homogeneous mixtures of hydrocarbons and water that include a high concentration of surfactants. Microemulsions were recently introduced into EOR processes, which utilize chemicals, primarily surfactants, to extract oil from natural oil reservoirs. Since their discovery, microemulsions have gained relevance both in basic research and in the oil business (Liu et al., 2008). Despite substantial research, the mechanism by which microemulsions function (i.e., predictive power) remains unknown. Additionally, while attempting to construct an appropriate microemulsion layer, other challenges occur, such as the demand for significant temperature gradients, the precipitation of many ionic surfactants when exposed to brine, and the incompatibility of the majority of nonionic surfactants. Additionally, the adsorption of microemulsion components onto rocks, as well as the variable salinity and temperature of oil deposits, provide difficulties. However, tremendous strides have been achieved in residual oil recovery during the previous three decades, most notably chemically EOR by microemulsions. Babadagli just wrote an essay on ways to bolster the world's dwindling oil reserves (Babadagli, 2020). According to his study, surfactant solution injection is the most often utilized chemical-based EOR approach since it is less costly than micellar or microemulsion injection. The route of administration of the chemical is crucial to optimizing the chemical-based EOR process. Continuous injection of chemical solutions leads to a rise in operational costs and/or material consumption. Injection of chemical solution is critical in any efficient EOR strategy when considering the porous volume (PV) number.

The bulk of microemulsions is composed of hydrocarbons, surfactants/cosurfactants, and brine. Because surfactants are adsorbed at the interface rather than in the bulk phase, they are the primary constituents of microemulsions. Based on the charge of the head group, surfactants are classified as anionic, cationic, nonionic, or zwitterionic. While anionic surfactants such as sodium dodecyl sulfate (SDS) are negatively charged by nature, their cationic cousins include a tiny amount of sodium. The most often used surfactants in the oil recovery process are anionic. They adsorb in opposed ways in sandstone and carbonate. Their ability for adsorption in sandstone is considerably smaller than that of carbonate.

Microemulsion flooding entails filling the reservoir with water that contains a trace quantity of surfactant and other additives such as hydrocarbon, medium-chain alcohol, or brine. The surfactant is crucial in forming the right type of microemulsion that minimizes the interfacial tension of the target oil. This is critical for both oil mobilization and oil evaporation from the reservoir rock. Microemulsion injection is frequently used in places where water injection has failed due to insufficient mobility relationships; however, in many cases where water injection has failed due to insufficient mobility relationships, microemulsion flooding may still be beneficial due to the essential mobility control.

At increased temperatures (110°C) and salinity (220 g/L NaCl), novel CO_2-soluble surfactants have been created for stable CO_2 emulsion/foam production. The novel surfactants are exclusively made of alkyl chains and amine heads; they include no fluorine, silicon, or polyalkyl-oxide groups. They outperform previously documented surfactants in terms of solubility, thermal stability, and capacity to emulsify in aggressive environments. At elevated temperatures (110°C) and low pressures, 0.2 wt% of the novel surfactants are soluble in 220 g/L NaCl brine with a pH of 8 and temperatures ranging 20–120°C, as well as in CO_2 (91 bar). Surfactants are thermally stable in anaerobic environments at 110°C and pH = 4. At temperatures ranging from 25–110°C and pressures of 150 psi, a strong CO_2 emulsion may be easily generated in bulk and porous media. In comparison to traditional surfactants, the performance of the novel surfactants can be greatly enhanced by salt and temperature (refer to chapter on surfactants). Additionally, the method for creating these novel compounds and optimizing surfactant formulations for CO_2 EOR via emulsions is viable (Xing et al., 2012; Zhao et al., 2012).

Microemulsion flooding exemplifies the unique properties of microemulsion systems, such as their high viscosity and ability to create low interfacial tension, which improves oil extraction efficiency.

11.3 STABILITY OF EMULSIONS

The relative stability of an emulsion is primarily determined by the rates at which droplets aggregate and solidify. As the time interval between aggregation and coalescence increases, the rate of aggregation and coalescence decreases, resulting in a more stable emulsion. Currently, the stability of emulsions is studied on three distinct scales: macroscopical, mesoscopic, and microscopical. To assess the stability of emulsions, the macroscopic phase separation technique is utilized. It is a time-tested technique. Mesoscopic droplet size can more accurately depict the flocculation, coalescence, and settling processes that occur during the dispersed phase. Microscopic interfacial films have garnered considerable interest in recent years as a means of assessing the stability mechanism of emulsions. The principal approaches include light scattering, low-field nuclear magnetic resonance, the Langmuir trough, the microscale pipette, and scanning electron microscopy (Sharma et al., 2014; Wu et al., 2019).

The strength of an emulsion's interfacial layer is the most critical component in determining its stability. During the emulsion seepage process on porous surfaces, the dispersed emulsions regularly swarm and collide with one another. When the interfacial film strength is insufficient, emulsions rapidly coalesce and disperse, producing a small displacement effect. When the interfacial film strength is strong, emulsions rapidly coalesce and separate. Among other factors, the molecular structure, properties, concentration, crude oil composition, emulsion droplet size, and water cut of the emulsifier used to form the film all affect the interfacial film strength of emulsions. The larger the size of the emulsion droplet, the weaker the interfacial coating and the greater the likelihood of an emulsion splitting apart. Gelatine, asphaltene, and waxes are all components of crude oil having specific interfacial qualities that differentiate them. They not only enhance crude oil emulsification but also forming durable interfacial coatings and induce directed adsorption at the oil–water interface. In most cases, the molecular structure and properties of emulsifiers play a substantial role in influencing the strength of interfacial coatings. When the emulsifier concentration is insufficient, the interfacial film strength is diminished, and the resulting emulsion becomes unstable owing to a decreased number of emulsifier molecules absorbed at the interface, as is frequently the case. When the emulsifier concentration reaches a particular point, both the strength of the interfacial layer and the stability of the emulsion improve. The viscosity and elasticity of the interfacial layer of emulsions have a considerable effect on their aggregation and drainage rates, as well as their stability, and are discussed in further detail below. When the interfacial viscosity and elasticity of an emulsion are high, the aggregation and drainage rates are low, resulting in a more stable emulsion. The rotational speed plays a critical role in ensuring the stability of emulsions. As the rotation speed rises, the droplet size reduces and the emulsion becomes more stable, as observed in previous work by Narukulla and coworkers (Narukulla et al., 2019, 2017).

11.4 PICKERING EMULSIONS

Pickering emulsions are emulsions with colloidal particles that develop at oil–water interfaces (Chevalier and Bolzinger, 2013). Pickering emulsion droplets are extremely stable due to their solid phase interface. As a result, the Pickering emulsion system is widely used in several industrial applications, such as foods, cosmetics, paints, and insecticides. Colloidal particles such as polymer, inorganic, and carbon compounds are included in Pickering emulsions. In comparison to microparticles, when the Pickering emulsion is created using nanoparticles, a more customizable interfacial structure may be developed. The packing density, the particle size and shape, and the surface wettability all affect the emulsion's stability. Researchers have developed a simple and cost-effective approach for synthesizing Janus nanoplate surfactants with improved emulsifiability and stability (Agrawal and Agrawal, 2019; Zhang et al., 2021). Nanoparticles can be added to o/w emulsions to boost their stability and viscosity. The ability of these colloidal surfactants to stabilize Pickering emulsions in a variety of oil–water systems has been demonstrated. Sharma et al. created and patented a novel o/w emulsion that is stabilized by a nanoparticle surfactant-polymer system. They used lubricating oil, the polymer polyacrylamide (PAM), the surfactant sodium dodecyl sulfate (SDS), and nanoparticles such as silica, clay, and copper oxide (CuO) in their experiments. The thermal stability of this Pickering emulsion was investigated; it was found to be more stable than emulsions stabilized with a surfactant-polymer system, and both hydrophobic clay nanoparticles and sodium chloride (NaCl) can be used to partially increase the thermal stability of Pickering emulsions. To show structural stability, Figure 11.3 illustrates Pickering emulsions and emulsions stabilized by solid particles.

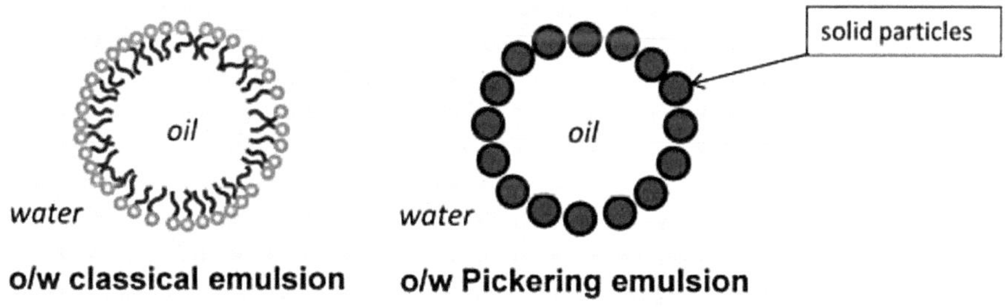

FIGURE 11.3 Schematic representing classic emulsions stabilized by solid particles.

Source: Reprinted with permission from Chevalier and Bolzinger (2013). Copyright 2013. Elsevier

When particles are hybridized with surfactants or have their surfaces changed with polymers, the Pickering emulsion enhances drop stability (Sharma et al., 2015b). Any surfactant adsorption on the particles' surfaces during this procedure may reduce their wettability, hence decreasing their emulsifying efficacy. For instance, electrostatic adsorption of cationic surfactants on silica nanoparticles renders the particle surface hydrophobic, hence decreasing the oil–water interface tension. Similarly, by coating particles with poly(acrylic acid), stable colloidal dispersions are formed by hydrogen bonding and ionic complexation. Poly(ethylene glycol) adsorption on the surface of silica particles may also help reduce interfacial tension at the oil–water interface.

The oil and gas sectors have recently expressed an interest in the usage of functional nanoparticles (Yekeen et al., 2018; Yu et al., 2012). There are several significant examples in the field of oil recovery. While flowing through reservoir rocks, well-designed nanoparticles can adsorb at oil–water interfaces within the rock pores, reducing interfacial tension and increasing oil recovery efficiency. Adsorption of nanoparticles at the oil–water interface in reservoirs can result in the formation of nanoparticle-sized emulsion droplets, enhancing fluidity in porous media. However, only a small amount of research has been conducted on the utilization of Pickering emulsions including a nanoparticle, surfactant, and polymer interface in oil reservoir applications.

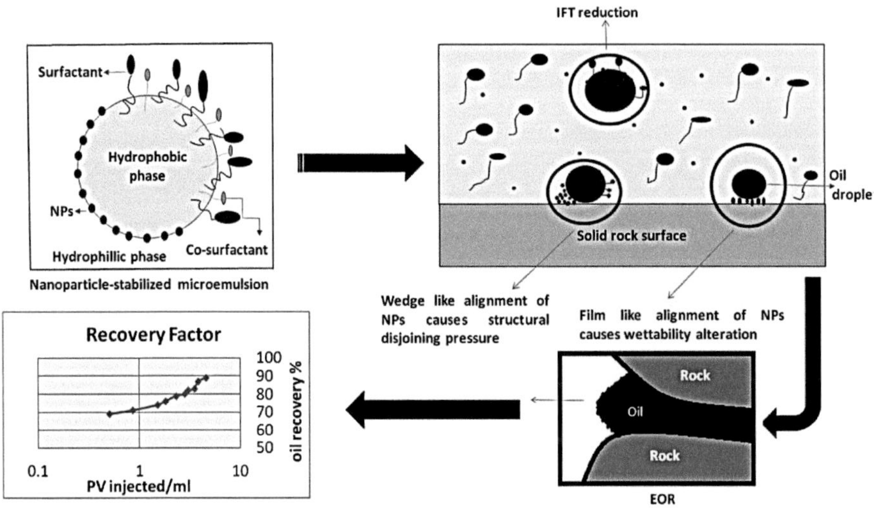

FIGURE 11.4 Visual representation of a nanoparticle-stabilized emulsion with applications in EOR for increasing oil production.

Source: Image reprinted with permission from Mariyate and Bera (2021). Copyright 2021. Elsevier

Because nanoparticles form a kinetically stable Pickering emulsion at the liquid-liquid interface, they are a viable alternative to conventional emulsifiers. Clay, silica, graphene, cellulose, and polymer latex colloidal nanoparticles all have their distinct methods of stability according to their size, shape, morphology, and adsorption (Chevalier and Bolzinger, 2013). When pH, temperature, and ionic strength are altered, nanoparticles can significantly increase emulsion stability. Additionally, the low concentration requirements and constant formulations of nanoparticles are beneficial. Pickering emulsions have grown in popularity over the last decade because of their simplicity of loading active chemicals and their capacity to efficiently stabilize and target molecules (Zhou et al., 2019). Along with the oilfield, Pickering emulsions have also found application in drug delivery and other modes of flow applications. A detailed study on the role of micro-emulsions has been reported by Mariyate and Bera who state that micro-emulsions (see Fig. 11.4), especially those stabilized by nanoparticles hold great promise for use in EOR due to their high stability and superior oil mobilization attributes (Mariyate and Bera, 2021).

KEYWORDS

- **emulsions**
- **enhanced oil recovery**
- **oil displacement**
- **mobility control**
- **reservoir sweep efficiency**

REFERENCES

Agrawal, G.; Agrawal, R. Janus Nanoparticles: Recent Advances in Their Interfacial and Biomedical Applications. *ACS Appl. Nano Mater.* **2019,** *2,* 1738–1757.

Anna, S. L.; Bontoux, N.; Stone, H. A. Formation of Dispersions Using "Flow-Focusing" in Microchannels. *Appl. Phys. Lett.* **2003,** *82,* 364–363.

Austad, T.; Shariatpanahi, S. F.; Strand, S.; Black, C. J. J.; Webb, K. J. Conditions for a Low-Salinity Enhanced Oil Recovery (EOR) Effect in Carbonate Oil Reservoirs. In *Energy and Fuels*. American Chemical Society, 2012; pp 569–575.

Babadagli, T. Philosophy of EOR. *J. Pet. Sci. Eng.* **2020,** *188,* 106930.

Chevalier, Y.; Bolzinger, M. A. Emulsions Stabilized with Solid Nanoparticles: Pickering Emulsions. *Colloids Surf. A Physicochem. Eng. Asp.* **2013,** *439,* 23–34.

Eastoe, J.; Hatzopoulos, M. H.; Tabor, R. Microemulsions. In *Encyclopedia of Colloid and Interface Science*. Springer Berlin Heidelberg: Berlin, Heidelberg, 2013; pp 688–729.

Fredriksen, S. B.; Rognmo, A. U.; Fernø, M. A. Pore-Scale Mechanisms During Low Salinity Waterflooding: Oil Mobilization by Diffusion and Osmosis. *J. Pet. Sci. Eng.* **2018,** *163,* 650–660.

Hou, J.; Han, M.; Wang, J. Manipulation of Surface Charges of Oil Droplets and Carbonate Rocks to Improve Oil Recovery. *Sci. Rep.* **2021,** *11,* 1–11.

Jankowski, A. Influence of Chosen Parameters of Water Fuel Microemulsion on Combustion Processes, Emission Level of Nitrogen Oxides and Fuel Consumption of CI Engine. *J. KONES* **2011,** *18,* 593–600.

Kokal, S. Crude-Oil Emulsions: A State-of-the-Art Review. *SPE Prod. Facil.* **2005,** *20,* 5–12.

Kumar, S.; Mandal, A. Studies on Interfacial Behavior and Wettability Change Phenomena by Ionic and Nonionic Surfactants in Presence of Alkalis and Salt for Enhanced Oil Recovery. *Appl. Surf. Sci.* **2016,** *372,* 42–51.

Liu, S.; Zhang, D. L.; Yan, W.; Puerto, M.; Hirasaki, G. J.; Miller, C. A. Favorable Attributes Alkaline-Surfactant-Polymer Flooding. *SPE J.* **2008,** *13,* 5–16.

Mariyate, J.; Bera, A. Recent Progresses of Microemulsions-Based Nanofluids as a Potential Tool for Enhanced Oil Recovery. *Fuel* **2021,** 306, *121640*.

McClements, D. J. Nanoemulsions Versus Microemulsions: Terminology, Differences, and Similarities. *Soft Matter* **2012,** *8,* 1719–1729.

Mohajeri, M.; Hemmati, M.; Shekarabi, A. S. An Experimental Study on Using a Nanosurfactant in an EOR Process of Heavy Oil in a Fractured Micromodel. *J. Pet. Sci. Eng.* **2015,** *126,* 162–173.

Mohebbifar, M.; Ghazanfari, M. H.; Vossoughi, M. Experimental Investigation of Nano-Biomaterial Applications for Heavy Oil Recovery in Shaly Porous Models: A Pore-Level Study. *J. Energy Resour. Technol.* **2015,** *137,* 014501.

Narukulla, R.; Ojha, U.; Sharma, T. Stable & Re-dispersible Polyacryloyl Hydrazide-Ag Nanocomposite Pickering Emulsions. *Soft Matter* **2017**. DOI: 10.1039/c7sm01364g

Narukulla, R.; Ojha, U.; Sharma, T. Enhancing the Stability and Redispersibility of o/w Pickering Emulsion Through Polyacryloyl Hydrazide-Tannic Acid Synergy. *Colloids Surf. A Physicochem. Eng. Asp.* **2019,** *568,* 204–215.

Nguyen, P.; Fadaei, H.; Sinton, D. Pore-Scale Assessment of Nanoparticle-Stabilized CO_2 Foam for Enhanced Oil Recovery. *Energy Fuels* **2014,** *28,* 6221–6227.

Nie, Z.; Seo, M. S.; Xu, S.; Lewis, P. C.; Mok, M.; Kumacheva, E.; Whitesides, G. M.; Garstecki, P.; Stone, H. A. Emulsification in a Microfluidic Flow-Focusing Device: Effect of the Viscosities of the Liquids. *Microfluid Nanofluid* **2008,** *5,* 585–594.

Santra, S.; Tapec, R.; Theodoropoulou, N.; Dobson, J.; Hebard, A.; Tan, W. Synthesis and Characterization of Silica-Coated Iron Oxide Nanoparticles in Microemulsion: The Effect of Nonionic Surfactants. *Langmuir* **2001,** *17,* 2900–2906.

Sharma, T.; Kumar, G. S.; Sangwai, J. S. Comparative Effectiveness of Production Performance of Pickering Emulsion Stabilized by Nanoparticle-Surfactant-Polymerover Surfactant-Polymer (SP) Flooding for Enhanced Oil Recovery for Brownfield Reservoir. *J. Pet. Sci. Eng.* **2015a,** *129,* 221–232.

Sharma, T.; Sangwai, J. S. Effects of Electrolytes on the Stability and Dynamic Rheological Properties of an Oil-in-Water Pickering Emulsion Stabilized by a Nanoparticle-Surfactant-Polymer System. *Ind. Eng. Chem. Res. Eng. Chem. Res.* **2015b,** *54,* 5842–5852.

Sharma, T.; Suresh Kumar, G.; Sangwai, J. S. Enhanced Oil Recovery Using Oil-in-Water (o/w) Emulsion Stabilized by Nanoparticle, Surfactant and Polymer in the Presence of NaCl. *Geosystem Eng.* **2014,** *17,* 195–205.

Taherpour, A.; Hashemi, A. A Novel Formulation of the Pickering Emulsion Stabilized with Silica Nanoparticles and Its Thermal Resistance at High Temperatures. *J. Dispers. Sci. Technol.* **2018,** *39,* 1710–1720.

Wei, B.; Hou, J.; Sukop, M. C.; Du, Q.; Wang, H. Flow Behaviors of Emulsions in Constricted Capillaries: A Lattice Boltzmann Simulation Study. *Chem. Eng. Sci.* **2020,** *227,* 115925.

Wu, D.; Mihali, V.; Honciuc, A. PH-Responsive Pickering Foams Generated by Surfactant-Free Soft Hydrogel Particles. *Langmuir* **2019,** *35,* 212–221.

Xing, D.; Wei, B.; McLendon, W.; Enick, R.; McNulty, S.; Trickett, K.; Mohamed, A.; Cummings, S.; Eastoe, J.; Rogers, S.; Crandall, D.; Tennant, B.; McLendon, T.; Romanov, V.; Soong, Y. CO_2-Soluble/Nonionic, Water-Soluble Surfactants that Stabilize CO_2-in-Brine Foams. *SPE J.* **2012,** *17* (4), 1172–1185.

Yekeen, N.; Manan, M. A.; Idris, A. K.; Padmanabhan, E.; Junin, R.; Samin, A. M.; Gbadamosi, A. O.; Oguamah, I. A Comprehensive Review of Experimental Studies of Nanoparticles-Stabilized Foam for Enhanced Oil Recovery. *J. Pet. Sci. Eng.* **2018**. DOI: 10.1016/j.petrol.2018.01.035

Yu, J.; An, C.; Mo, D.; Liu, N.; Lee, R. In Study of Adsorption and Transportation Behavior of Nanoparticles in Three Different Porous Media, *Proceedings - SPE Symposium on Improved Oil Recovery*. Society of Petroleum Engineers (SPE), 2012; pp. 311–323.

Zhang, X.; Fu, Q.; Duan, H.; Song, J.; Yang, H. Janus Nanoparticles: From Fabrication to (Bio)Applications. *ACS Nano.* **2021,** *15,* 6147–6191.

Zhao, J.; Xu, K.; Song, Y.; Liu, W.; Lam, W.; Liu, Y.; Xue, K.; Zhu, Y.; Yu, X.; Li, Q. A Review on Research on Replacement of CH4 in Natural Gas Hydrates by Use of CO_2. *Energies* **2012,** DOI: 10.3390/en5020399

Zhou, Y.; Yin, D.; Chen, W.; Liu, B.; Zhang, X. A Comprehensive Review of Emulsion and Its Field Application for Enhanced Oil Recovery. *Energy Sci. Eng.* **2019,** *7,* 1046–1058.

CHAPTER 12

Mobility Control Requirement in EOR Processes

TAREK GANAT[1] and IMTIAZ ALI[2]

[1]*Department of Petroleum Engineering, Universiti Teknologi PETRONAS, Malaysia*

[2]*Department of Petroleum and Gas Engineering, BUITEMS, Quetta, Balochistan, Pakistan*

ABSTRACT

This chapter explores an important role of mobility control in enhanced oil recovery (EOR) processes within petroleum engineering. Analyzing two-phase flow in porous structures, the study emphasizes Darcy's equation and the mobility concept. The mobility ratio, a key parameter, is introduced, influencing water flooding and miscible displacement processes. The chapter details mobility control methods such as polymer injection, foam, chemical injection, miscible gas, and steam flooding. It highlights the influence of mobility ratio on displacement efficiency, emphasizing the inverse relationship between mobility ratio and efficiency. Theoretical and numerical investigations validate a new criterion for mobility control, considering downstream pressure gradients and normalized oil saturation. Simulation results for homogeneous, layered, and heterogeneous formations highlight the advantage of this new criterion over traditional approaches. Addressing challenges and considerations for each method, the chapter provides a comprehensive overview of mobility control's significance in optimizing oil recovery processes.

Advancements in Chemical Enhanced Oil Recovery.
Tushar Sharma, Krishna Raghav Chaturvedi, Tarek Ganat, and Imtiaz Ali (Eds.)

12.1 MOBILITY CONTROL

In petroleum engineering, two-phase flow in porous structures has mainly been studied. Darcy law's equation illustrates fluid flow over any porous medium. Darcy developed a relationship between pressure differential (ΔP) and flow rate (q) with the assumption of a single-phase fluid flowing through a horizontal porous medium. The equation indicates a linear relationship between q and ΔP as follows:

$$q = -\frac{kA}{\mu}\frac{\Delta P}{L} \tag{12.1}$$

where k represents the permeability of the medium, A is the cross-sectional area, μ is the dynamic viscosity of the fluid, and L is the length of the sample. In eq 12.1, k/μ is known as the mobility (λ) of the fluid. In the case of multiphase flow, the mobility (λ) for every single phase can be stated as follows:

$$\text{Mobility} = \frac{\text{Effective permeability}}{\text{Fluid viscosity}} \tag{12.2}$$

$$\text{ë} = \frac{k}{\mu} \tag{12.3}$$

where k, μ, and λ represent the permeability, viscosity, and mobility, respectively.

Mobility ratio is the proportion of the injectant fluid mobility to the displaced fluid mobility. The mobility ratio in the presence of water and oil is the key fundamental consideration to influence the performance of a water flooding process. It is one of the most essential factors of a miscible displacement process. Methodically, mobility ratio can be expressed as

$$\text{Mobility ratio} = \frac{\text{Mobility of displacing fluid}}{\text{Mobility of displaced fluid}} \tag{12.4}$$

$$M = \frac{\text{ë}_d}{\text{ë}_o} \tag{12.5}$$

Using eq 12.5 for water and oil phase, we have

$$\text{M} = \frac{K_w / \mu_w}{K_o / \mu_o} \tag{12.6}$$

$$M = \frac{K_w \mu_o}{K_o \mu_w} \tag{12.7}$$

where M denotes mobility ratio, μ_o is the oil viscosity (cP); μ_w is water viscosity (cP); K_o is permeability to oil (mD); K_w is permeability to water (mD); λ_d and λ_o represent the mobility of displacing fluid and displaced fluid, respectively.

Any process that intends to minimize the mobility of injecting fluid or displacing fluid inside a reservoir is termed as mobility ratio control. The concept of mobility control is among the highly valuable concepts in enhanced oil recovery (EOR) processes. Volumetric displacement efficiency can be improved by controlling the mobility of displacing fluid. Generally, it is evaluated in terms of mobility ratio. Mobility ratio has an inverse relationship with displacement efficiency. Various chemicals are added to the displacing fluid to improve the mobility. These agents either enhance the viscosity of injectant fluid or reduce the permeability of rock to the injecting fluid.

There are two ways to enhance the displacing efficiency. The first one is the utilization of polymer and it is used when water is the base fluid. The other one is foam, which is applied when gas is the base fluid. Water alternating gas (WAG) can also be used in some cases to control mobility within the reservoir (Alamooti and Malekabadi, 2018).

Mobility ratio determines the consistency of the oil displacement mechanism. It may be accomplished via chemicals injection to alter the viscosity of displacing fluid and to reduce fluid relative permeability. Polymers are generally injected to control the mobility while materials such as surfactants, foams, and carbon dioxide are injected for wettability alteration.

According to eq. 12.7, the mobility ratio can be less than 1, equal to 1, or greater than 1. Mobility ratio greater than 1 indicates that water is extra mobile as compared to oil. This is an unfavorable state in EOR because water will bypass the oil and will cause a premature breakthrough. In this case, the oil displacement efficiency will be lower as shown in Figure 12.1a. $M \leq 1$ is often needed to ensure a high macroscopic sweep efficiency as shown in Figure 12.1b. The way polymer floods alter the recovery of oil by reducing the mobility ratio is shown in Figure 12.1. The addition of polymers induces an increase in the viscosity of the displacing fluid in the displacement process. Resultantly, this yields in such a balanced front of the injecting fluid which reduces the chances of viscous fingering and channeling within the reservoir, thus, causing an increased oil recovery.

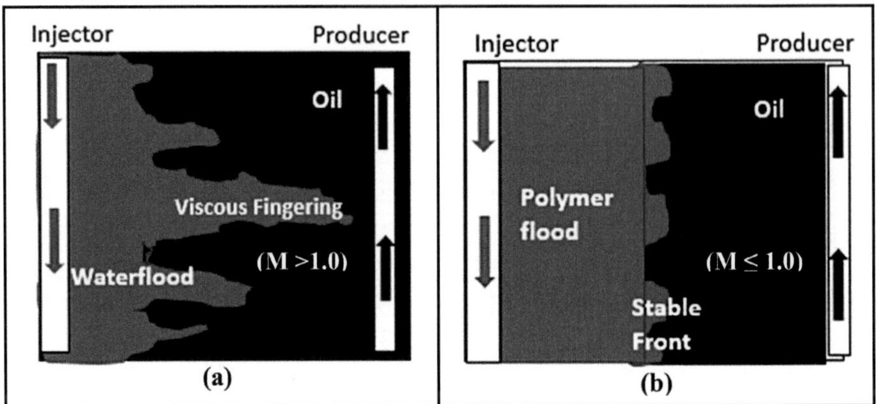

FIGURE 12.1 (a) Mobility ratio in case of typical water flooding, (b) polymer augmented water flooding

Source: Reprinted with permission from Sorbie, 2013; Gbadamosi et al. 2019. https://creativecommons.org/licenses/by/4.0/

Generally, for a water drive within a homogeneous reservoir, an unfavorable mobility ratio often exists because the injected water viscosity is lower than the oil viscosity. This result will induce the fraction of the water phase (water cut) during liquid production to rise rapidly. As a consequence, the sweep efficiency will be very low, due to viscous fingering. However, by injecting a viscous polymer solution, the mobility ratio can be improved. For waterflooding, the fraction of water phase (water cut) after water breakthrough can be expressed as (Wang, 2013).

$$f_w = \frac{\ddot{e}_d}{\ddot{e}_o + \ddot{e}_w} = \frac{kk_{rw} / \mu_o}{kk_{rw} / \mu_w + kk_{ro} / \mu_o} = \frac{1}{1 + \left(\mu_o / \mu_w . k_{rw} / k_{ro}\right)} \quad (12.8)$$

where f_w is the water fraction in liquid production or water cut, λ_o is the oil mobility, λ_w is the water mobility, k is the rock absolute permeability, k_{ro} and k_{rw} are the relative permeabilities in oil and water phase respectively, μ_o is the oil viscosity, and μ_w is the water viscosity.

Equation 12.8 indicates that the viscosity ratio between oil and water (μ_o/μ_w) strongly affects the rate of water cut (fraction of water phase). In other words, a higher viscosity ratio leads to increased water cut even at low water saturation. This condition may result in the oil field to terminate its production due to the ultimate water cut. The final oil recovery obtained may be far less than the ultimate displacement efficiency which could be achieved otherwise. In contrast, a lower viscosity ratio between oil and water will delay increases in the water cut. Thus, water cut achieves a given value in case of high-water saturation. Consequently, higher actual oil displacement efficiency can be achieved.

The relative permeabilities depend on two distinct zones in the reservoir at the time of water flooding. Craig proposed that the relative permeability to water would be acquired through the region swept by water, as the relative permeability to oil ought to be depending on an unswept zone, that is located in front of the displacement front. When the injecting fluid displaces the oil in the direction of the producers, the oil and water saturations vary with time as well as distance between the wells. Variations in fluid saturations are managed by relative permeability attributes as well as other rock and fluid characteristics (Bahadori, 2018).

12.2 MOBILITY CONTROL REQUIREMENT

Based on the conventional theory, when a polymer solution displaces other fluids (oil and water) forward, its believed that the injecting fluid mobility needs to be identical to or significantly lower when compared with the mobility of all fluids present. It indicates that the polymer mobility needs to be lower than or corresponding to the combined mobilities, that is, $\lambda_{polymer} \leq \lambda_{oil}$. In other words, when a single or multiple fluid displaces other fluids (water and oil) forward, the overall mobility of injecting fluids will be similar to or significantly lower compared to the overall mobility of the displaced fluids:

$$M = \frac{\sum \lambda_d}{\sum \lambda_o} \leq 1 \tag{12.9}$$

The problem in this concept is that it is dependent on the assumptions considering that the injecting and displaced fluids both move at a specific velocity and one fluid is displacing the second fluid. Conversely, when it comes to polymer flooding process, both the oil and water are displaced by polymer solution. Both oil and water move at distinct velocities. Hence, the typical theory could not be considered for the polymer flooding method or even to the common EOR techniques. When fluid saturations of various mobile fluids are unknown, the overall mobility of the fluids could not be determined since it is the function of saturations which are usually unknown. To overcome this issue, Gogarty (1969) and Gogarty et al. (1970) preferred to apply the minimum total mobility as given below:

$$M = \frac{\Sigma \lambda_{d(\min)}}{\Sigma \lambda_o} \leq 1 \quad (12.10)$$

When the injectant fluid such as polymer mixed in water, the viscosity of this mixture will change to a desired viscosity. This change in viscosity is a function of the polymer's quantity in the solution. This concept been demonstrated and validated by various authors through simulation approach at various conditions. It has also been validated through field experiments. This concept facilitates us to choose the appropriate polymer concentration for a workable EOR method (Sheng, 2010).

12.3 MOBILITY RATIO CONTROL PROCEDURES

12.3.1 POLYMERS INJECTION

A variety of polymers are available to increase the viscosity of water. Polymers are generally classified into two types including natural and synthetic polymers. When polymers are added to water, a high molecular solution is formed which can partially improve the viscosity of water (Fig. 12.2). In polymer flooding, two types of polymers are widely applied. These are hydrolyzed polyacrylamide (HPAM) and xanthan polymers. Polyacrylamide (PAM) perform two functions when added to water as an injecting fluid. Its primary function is to enhance the base fluid viscosity and additionally to decrease the rock permeability that responds to the chemical substances causes the decrease in mobility. On the other hand, xanthan polymers reduce the displacing fluid mobility while improving the viscosity. Usually, injection of polymers may enhance macroscopic volumetric displacement efficiency. In polymer injection method, polymers are constantly injected at their initial density for a particular period. Because, with the passage of time, the density of injected polymers decreases regularly during the injection. Water is then injected to chase the injected polymers solutions.

12.3.2 POLYMER FLOODING PROPERTIES

Polymer solution is well known to operate at intermediate shear rates as pseudoplastic fluids thus behaving like Newtonian fluids at lower or higher shear rates. All the particles are aligned along the flow path, thus in the higher shear rates, the viscosity is minimized. Xanthan gum (XG), HPAM, and seleroglucan are the polymers used in polymer flooding.

XG is soluble in cold and hot water upon stirring which produce a high viscous slurry even at very low concentration of XG. The most important property of XG is to control the rheological properties of fluids. PAM is a polymer that is water soluble and develops reactions characteristic of the amide group. Due to its comparatively high glass transition temperature, the polymer in the solid state is fragile and not a strong film former (Yen et al., 1989). PAM is often used for mobility control and also to control permeability (Yen et al., 1989). The stabilization of PAM in liquids is influenced by pH. Many of the amide groups are converted through hydrolysis to carboxylates at very low or high pH which further expands in water, and increase the viscosity of a solution. Additionally, without this hydrolysis, some viscosity variations have also been reported in freshly formulated PAM solutions in water and consequently aged PAM is the dominant polymer utilized on a commercial basis in large quantities. Anionic types of PAM are commonly used instead of the nonionic homopolymer itself. HPAMs are the copolymers obtained by partial polyacrylamide hydrolysis. HPAMs are the most common synthetic polymers in injection systems. Since these materials are commonly preferred due to their acceptable performance, viscosity behavior at elevated temperatures, salinity tolerance, and hardness in recovery processes (Sveistrup et al., 2016; Castro et al., 2020). Scleroglucan is a neutral biopolymer that is produced by fermentation of a plant pathogen fungus of the genus Sclerotium (Ferreira and Moreno, 2020). It has attracted scientists' attention primarily in evaluating its possible application for EOR. The viscosity of solutions containing this biopolymer is stable at elevated temperatures up to 130°C due to its nonionic nature. In terms of thermal, chemical, mechanical stability and oil recovery, the performance of scleroglucan offers a scenario of tremendous potential relative to other biopolymers and polyacrylamides (Viñarta et al., 2013; Liang et al., 2019). Figure 12.3 shows the chemical structures of the mentioned polymers.

In general, polymer flooding is used in reservoirs by applying synthetic or biopolymers where oil viscosity is below 150 cP, API gravity higher than 15, matrix permeability more than 10 mD, temperature range of 80–99°C. For higher oil recovery factor in polymer flooding, low salinity is preferred because of polymer degradation due to saline environment.

Temperature of reservoir also affects the viscosity of polymers. With increase in temperature, the viscosity is reduced. For some common polymers the viscosity trend is given in Figure 12.4.

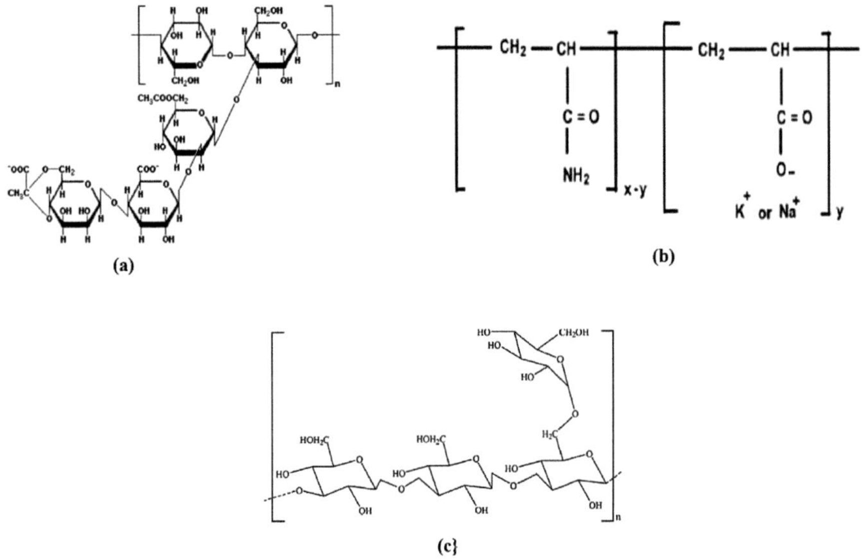

FIGURE 12.2 Structures of (a) Xanthan Gum, (b) HPAM, and (c) Seleroglucan
Source: Reprinted from Pu et al., 2018; Sun et al., 2012. Copyright 2018. Elsevier.

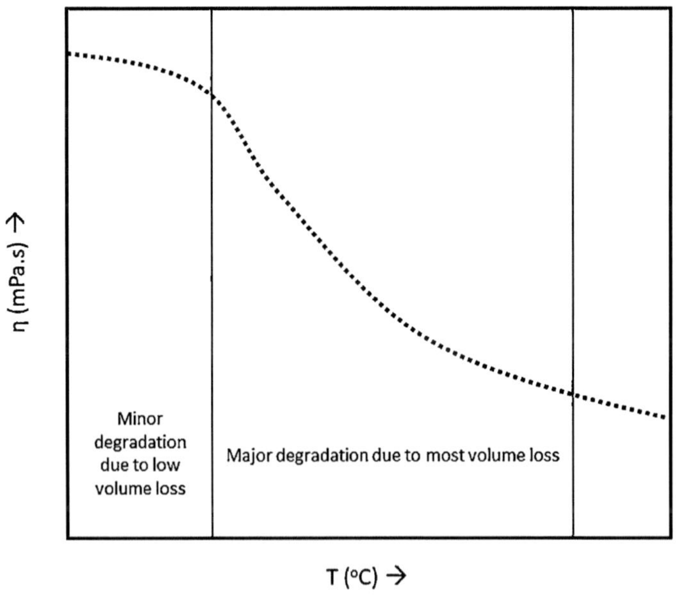

FIGURE 12.3 Polymers viscosity behavior as a function of temperature.
Source: Reprinted from Liang et al., 2019.

12.3.3 CHALLENGES OF POLYMER FLOODING

Polymer retained in porous media during a polymer flooding process either by precipitation, adsorption, or mechanical entrapment. The mentioned mechanisms cause the formation damage in the flooded region. The permeability of the rocks is reduced which is irreversible sometimes that leads to reduced oil recovery. Before applying the polymers in the field scale, the damage mechanism is studied through core flooding tests in the labs. The stability due because of polymers degradation is another issue linked to the polymer solutions. The degradation may be either chemical, mechanical, or biological. Oxidation or ionic ions effect are considered the chemical polymer's degradation. Mechanical degradation of polymers occurs when macromolecules are sheared at higher flowrates in the vicinity of wall. Biopolymers are degraded due to biological degradation. Moreover, several polymers also alter their volume ratios or viscosities under various conditions of reservoir which increase rock retention. This higher retention can cause severe formation damage by restricting the pore region (Xiao and Qiao, 2017).

12.3.4 FOAM INJECTION

Foam injection is a useful approach to control mobility ratio in case of gas injection as well as miscible process. The porous media is occupied when foam is injected into it. It has higher viscosity thus compresses the gas firmly into water and keep both separate by forming a light film. Hence the gas mobility declines in this region as the result of drop of permeability in the foamed zone. To further expand the foam, nitrogen or methane gas can also be injected in conjunction with steam.

Mobility control is acquired across the foamed area due to the fact that permeability of gas in this area is minimized substantially. The foam properties are governed by several factors such as the porous medium, surfactant features, and the foam performance. Such attributes must certainly be examined from laboratory evaluation. Practically, the surfactant solution can be introduced as a slug between gas-injection cycles or simultaneously. A gas such as nitrogen or methane, needs to be injected together with steam to produce foams in steam flooding procedures.

According to several simulation studies, it has been noticed that the injecting fluid mobility must be linked only to the mobility of displaced oil, instead of the overall mobility of the displaced fluids. Based on the consistency approach, a novel criterion concerning the mobility control requirement comes whenever a single fluid is displacing two mobile fluids.

12.3.5 CHEMICAL INJECTION

Due to the higher cost of polymers only a small slug of polymer solution pore volume is introduced and then displaced by water. Surfactants are also injected or formed in situ to displace the crude oil. Mobility control in the CEOR is conducted in three steps:

- Fingering prevention of the chemical additives into an oil–water bank
- Mobility control between the polymer bank and the chemical slug
- Controlling the channeling of drive water with the polymer containing chemical

12.3.6 MISCIBLE GAS INJECTION

In this process, a slug of solvent is injected which shows either first contact miscibility or multiple contact miscibility behavior with oil. Since the viscosity of most solvents is much lower than the viscosity of oil or water, thus mobility ratio is usually quite undesirable. It results in some adverse effects including viscous fingering, channeling along high-permeability regions, and decreased sweep efficiency. Mobility control enhances the productivity of these miscible procedures. The WAG mechanism could also be applied for mobility control in miscible gas-injection systems. Mobility control may be accomplished by specifying a water–gas injection ratio which reduces gas bypass. However, the approach is dependent on synchronized flow of water and gas in a porous media as shown in Figure 12.5. Gas and water are injected as slugs practically. Slugs of water and gas for injection are designed in such a way that the two-phase flow is achieved in the combining region as gas tends to finger across the water slug. For linear displacement, WAG ratios could be calculated either using relative permeability results or empirically evaluated from field data analysis in field applications.

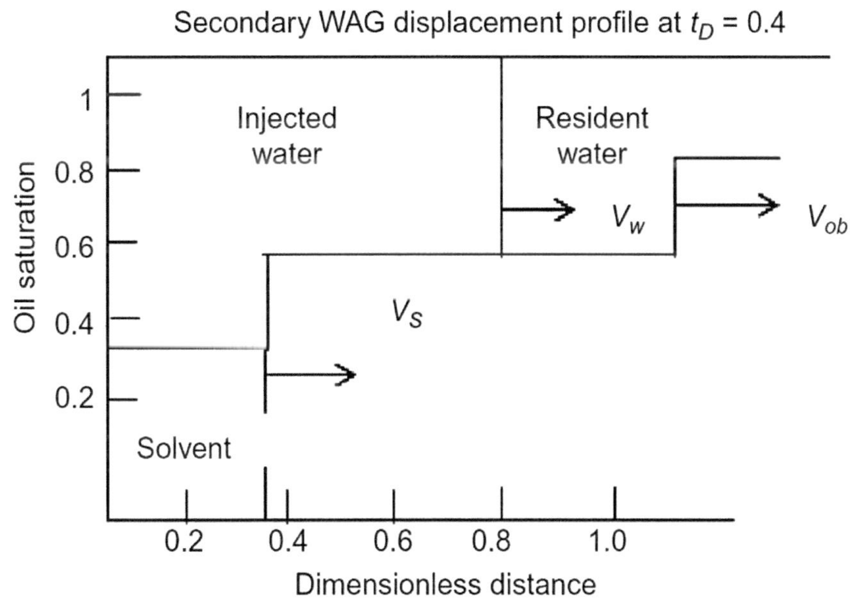

FIGURE 12.4 Mobility control with WAG process.
Source: Reprinted from Sheng, 2010. Copyright 2010. Elsevier.

Due to the fact that the gas viscosity is quite lower than the water or oil, the mobility ratio definitely will not be appropriate. Combined with minimizing the sweep efficiency, this element may also affect fingering and contribute to the mobility of gas in a region having high permeability. WAG procedure is usually applied to control the mobility of this technique. Mobility ratio is enhanced by deciding the adequate water/gas slug which reduces gas recycling. Water and gas are introduced as a slug, are combined inside the reservoir which results in enhanced injection performance.

12.3.7 STEAM FLOODING

Thermal recovery processes are significantly used to enhance the oil recovery from heavy-oil reservoirs. Steam injection is a thermal method in which steam is injected into heavy oil which reduces its viscosity and eases in flow (Al-Saedi et al., 2018; Li et al., 2017). It is the most frequently applied EOR method for producing heavy oils. During the steam injection process, the steam springs up toward the top of reservoir due to gravity segregation. Areal sweep is usually very high in such zones. Oil displacement is limited mainly to the top section of the reservoir, keeping the transition region of hot water just above the cold oil, as presented in Figure 12.6. Residual oil saturation can be as minimal as 5–10% in the steam-contacted zone. The mobility of steam in this area of steam displacement is higher, so the injection rate should be limited in order to prevent the unnecessary steam generation in the producers.

Vertical development of the steam region is slower compared to horizontal movements. Mobility control is required to minimize steam movement horizontally across the layout, thereby facilitating rapid vertical expansion of the steam zone. The efficiency of steam flooding can be increased by mobility control using additives like foam etc. to divert steam into high oil saturated zones (Al-Saedi et al., 2018). In situ foam production has the ability to be a mobility control method for miscible and steam injection systems. Foam is generated when a gaseous phase displaces a liquid solution having a surfactant inside a porous medium. It is a viscous, compressible gas dispersion in water isolated through a surfactant layer at the interface.

12.4 THEORETICAL INVESTIGATION

It is observed that the unit mobility ratio considering the traditional concept is not a rational criterion to differentiate "favorable" and "unfavorable" mobility control conditions. It has been observed that an improved criterion would be the unit mobility ratio, which is thought to be the ratio of injecting fluid mobility to the mobility of oil multiplied by the saturation of oil. An effort been made by researchers to validate the suggested concept from the consistency of displacement front has which yielded the following equation":

$$M_r = \frac{k_{wr} / \mu_u \overline{S_o}}{k_{ro}(S_w) / \mu_o} = \frac{\lambda_u}{\lambda_o} \overline{S_o} \tag{12.11}$$

In eq. 12.11, M_r represents the mobility ratio of the injection fluid (water/hybrid or polymer) to the displaced fluid (oil) in the supposed oil layer times the normalized movable oil saturation, S_o. Provided by the displacement front stability, M_r might be identical to or lower than 1. The new criterion for the mobility control in EOR processes will become as expressed as follows:

$$M_r = \frac{k_{wr} / \mu_u \overline{S_o}}{k_{ro}(S_w) / \mu_o} \leq 1 \tag{12.12}$$

Prats (1982) reported that a factor generally selected to determine the displacement front stability should be the ratio of the pressure gradient $\partial p/\partial n$ normal to and on the downstream side of the displacement front side compared to that the upstream side. Considering this concept, M_r will become as expressed as follows:

$$M_r = \frac{\left(\frac{\partial p}{\partial n}\right)_d}{\left(\frac{\partial p}{\partial n}\right)_u} \quad (12.13)$$

Here, M_r is showing the mobility ratio in all the above-mentioned equations. This represents the mobility ratio in the supposed oil layer or simply the product of typical mobility ratio and normalized oil saturation S_o. Based on the eq 12.13, the M_r is the downstream pressure gradient divided by the upstream pressure gradient in the presumed oil layer. The value needs to be equal to or less than 1. Practically, due to the unavailability of pressure gradients, eq 12.11 shall be applied for M_r.

Within the supposed water layer, in most cases the displacing fluid might be a polymer mixed in water or any other aqueous phase where the viscosity is greater than that of the water present. In case of thermal recovery, the case would be different as the displacement is absolutely stable Prats (1982) . This is because that small steam fingering occurs, if formed, have a tendency to lose heat at fairly high rates, eventually results in condensation and disappearing of the steam fingers (Sheng, 2010; Prats, 1982).

12.5 NUMERICAL INVESTIGATION

Due to heterogenous nature of the reservoir, it is very important to consider mobility impact on oil recovery factor in various formations including homogeneous and heterogenous (two-layered and random permeability).

12.5.1 HOMOGENEOUS FORMATION MOBILITY RATIO

Following the explanation of the mobility control utilizing the basic flow model, this part extends to a model having realistic water and oil relative permeability curves. For simulation studies, the parameters including connate water saturation, residual oil saturation, relative permeabilities (of both phases), Corey exponents of relative permeabilities and other necessary factors were considered. Figure 12.7 presents the simulation outputs of the data recovery factors upon single PV injection vs mobility ratio, that is characterized as the ratio of displacing fluid mobility divided by displaced fluid (oil) mobility multiplied by the normalized oil saturation, S_o (eq 12.11). This figure broadly demonstrates that if the mobility ratio is lower than 1, the recovery factors would not be affected toward the mobility ratio. Additionally, if the ratio is more than 1, the recovery factors will decrease steeply with the mobility ratio. Some simulated results are mentioned in Figures 12.5 and 12.6. These figures show the mobility ratio vs recovery factor by determining using eq. 12.11 and traditional method where the mobility is determined using an equation considering the injection fluid mobility divided by the total mobility.

12.5.2 LAYERED FORMATION MOBILITY RATIO

Combining the previous data of the homogenous model and a two-layer model having different top layer and bottom layer permeabilities, ratio of vertical permeability to horizontal permeability, and the total injection pore

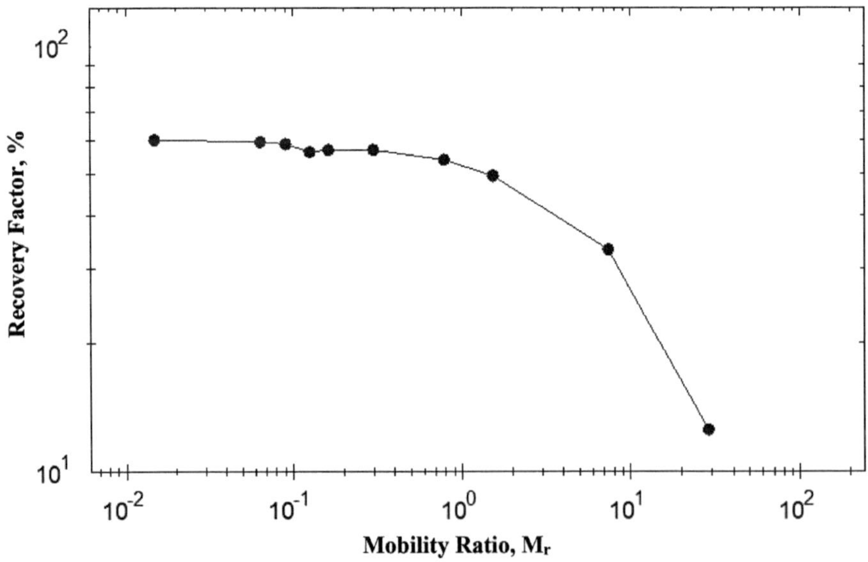

FIGURE 12.5 M_r defined by eq. 12.11 for homogenous formation.

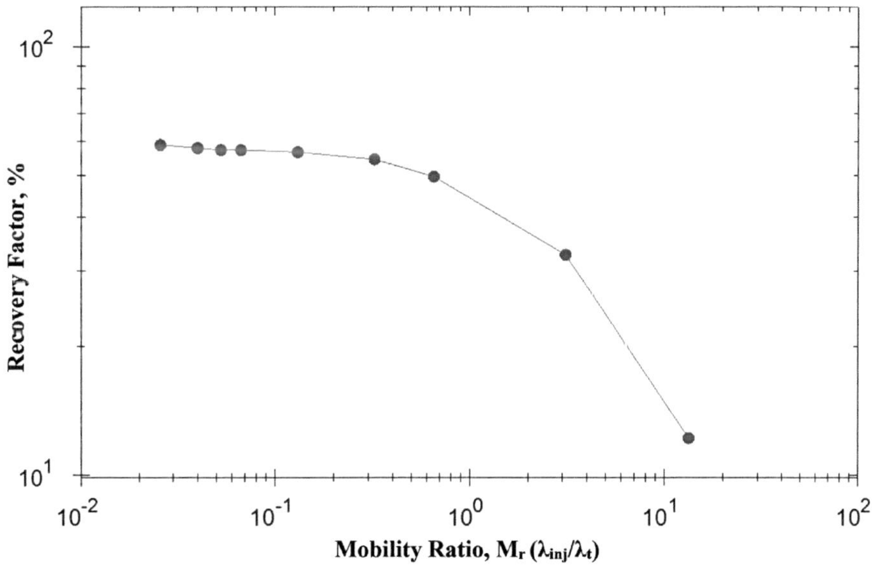

FIGURE 12.6 M_r defined by typical method for homogenous formation.

volume. The results showed in Figures 12.7–12.10demonstrates the recovery factors versus M_r and λ_{inj}/λ_t for different initial water saturations. From these results, it has been concluded that the mobility ratio defined as M_r given in eq 12.11 is the most suitable criterion than the typical one by considering the total mobility.

12.5.3 HETEROGENEOUS FORMATION MOBILITY RATIO

Taking the information within the heterogeneous model, the unique permeability distribution is developed utilizing the geostatistical software applications developed by Yang (1990) . The results presented in Figures 12.11 and 12.12 showed that the findings based on the homogeneous model are also valid for the heterogeneous model. It indicates that the unit mobility ratio determined by using eq 12.11 is a superior criterion compared to traditional one considering the total mobility.

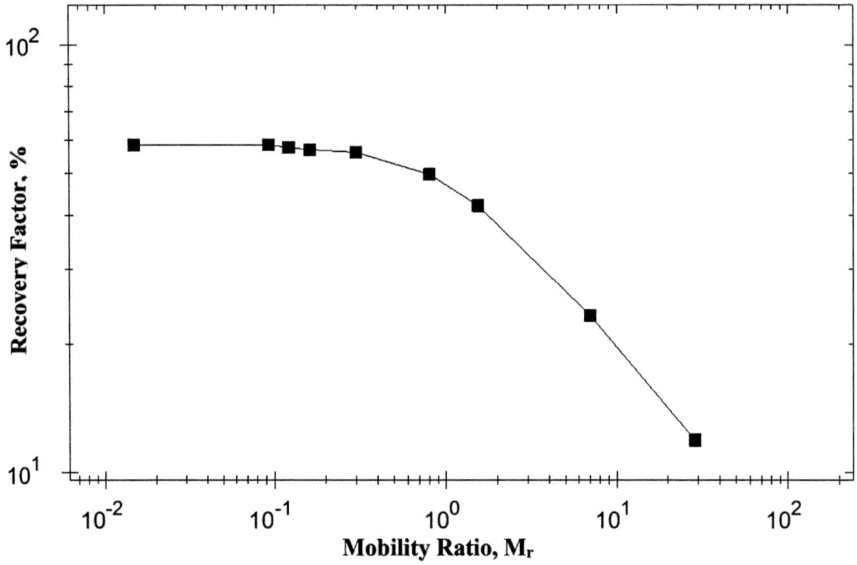

FIGURE 12.7 Mobility ratio for layered reservoir based on eq 12.11 at S_1.

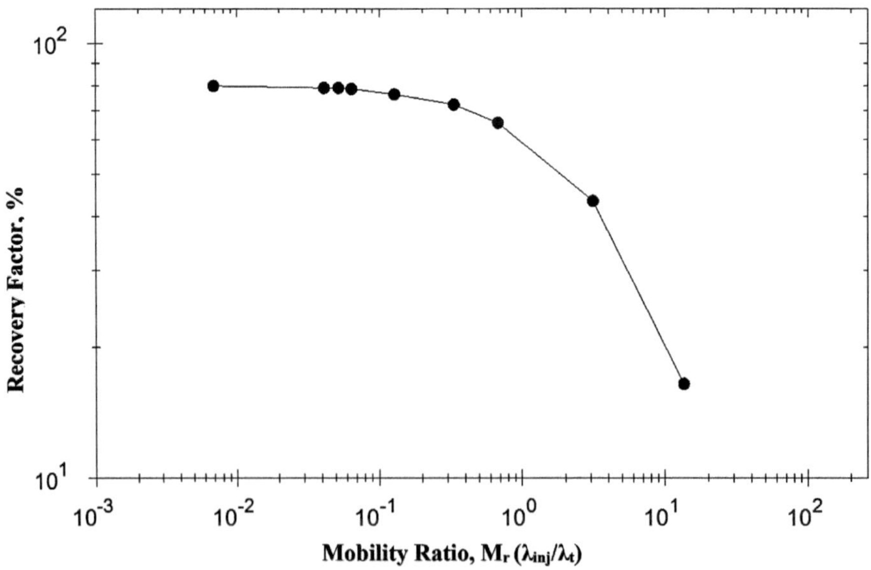

FIGURE 12.8 Mobility ratio for layered reservoir based on typical method at S_1.

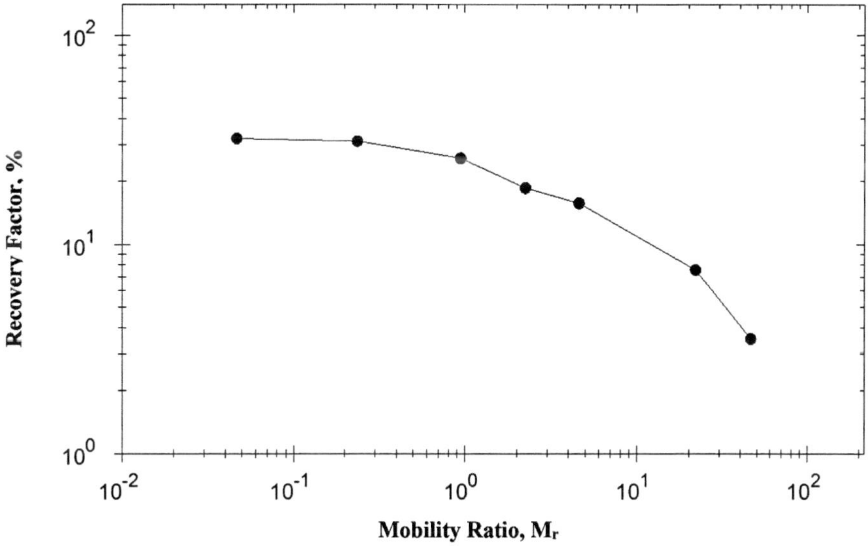

FIGURE 12.9 Mobility ratio for layered reservoir based on eq. 12.11 at S_2.

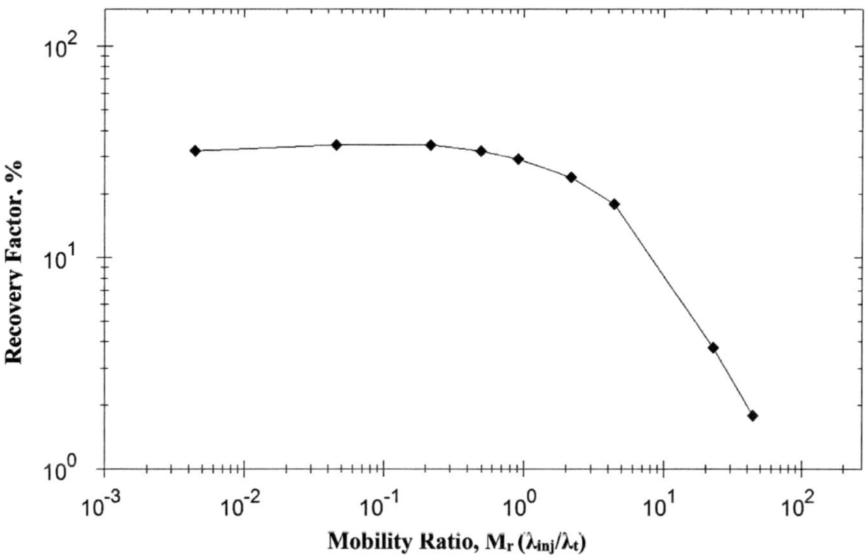

FIGURE 12.10 Mobility ratio for layered reservoir based on typical method at S_2.

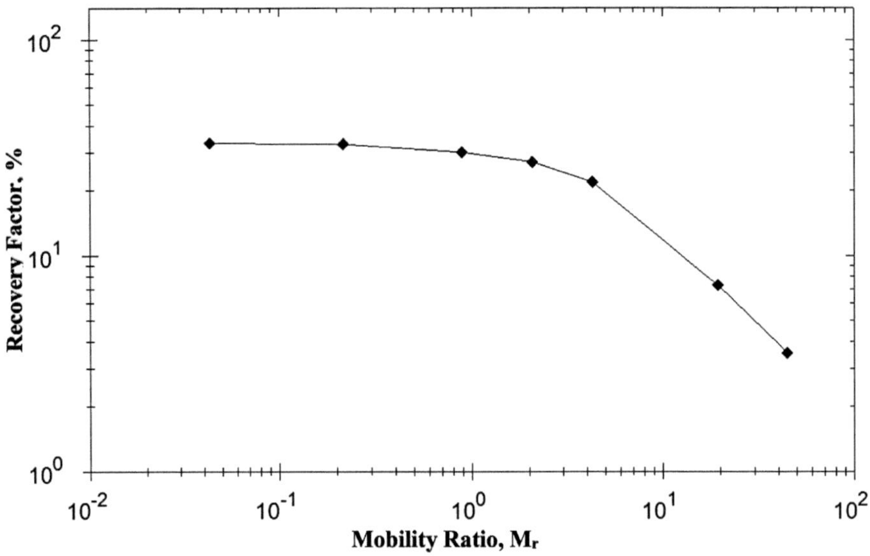

FIGURE 12.11 Mobility ratio defined eq 12.11.

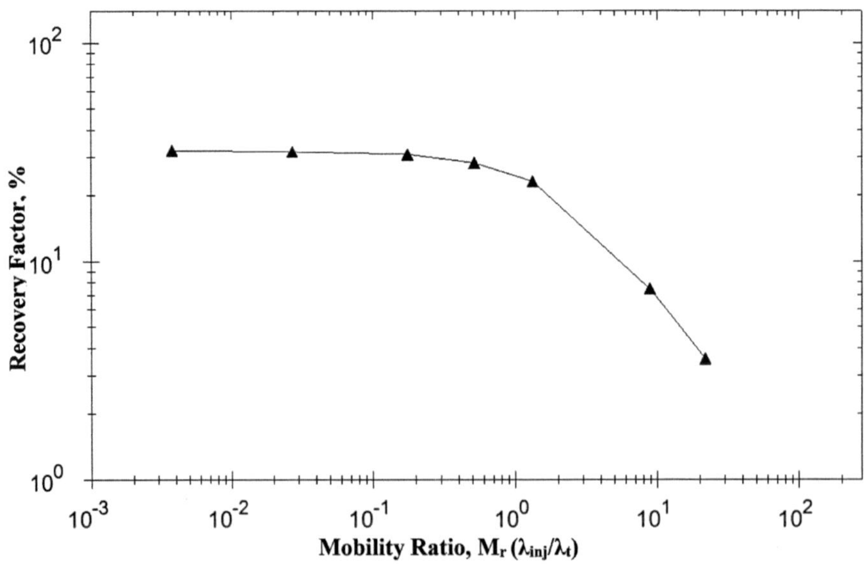

FIGURE 12.12 Mobility ratio defined typical method.

KEYWORDS

- **mobility**
- **mobility control**
- **mobility Ratio**
- **permeability**
- **viscosity**

REFERENCES

Alamooti, A. M.; Malekabadi, F. K. In *An Introduction to Enhanced Oil Recovery*; Fundamentals of Enhanced Oil and Gas Recovery from Conventional and Unconventional Reservoirs, 2018; p 1–40.

Al-Saedi, H. N.; et al. In *Coupling Low Salinity Water Flooding and Steam Flooding for Sandstone Reservoirs; Low Salinity-Alternating-Steam Flooding (LSASF)*; SPE Kingdom of Saudi Arabia Annual Technical Symposium and Exhibition; Society of Petroleum Engineers: Dammam, Saudi Arabia, 2018; p 10.

Bahadori, A. In *Fundamentals of Enhanced Oil and Gas Recovery from Conventional and Unconventional Reservoirs*; Gulf Professional Publishing, 2018.

Blevins, T.; Aseltine, R.; Kirk, R. Analysis of a Steam Drive Project, Inglewood Field, California. *J. Pet. Technol.* **1969,** *21* (09), 1141–1150.

Castro, R. H.; et al. Polymers for EOR Application in High Temperature and High Viscosity Oils: Rock–Fluid Behavior. *Energies* **2020,** *13* (22), 5944.

Ferreira, V. H.; Moreno, R. B. Experimental evaluation of low concentration scleroglucan biopolymer solution for enhanced oil recovery in carbonate. *Oil Gas Sci. Technol. Revue d'IFP Energies nouvelles* **2020,** *75*, 61.

Gbadamosi, A. O.; et al. An Overview of Chemical Enhanced Oil Recovery: Recent Advances and Prospects, In *International Nano Letters*, 2019; p.1–32.

Green, D. W.;Willhite, G. P. In *Enhanced Oil Recovery*; Henry, L., Doherty Memorial Fund of AIME, Society of Petroleum Engineers, 1998; vol 6.

Gogarty, W. B. Processes for the simultaneous displacement of petroleum and water in formations, Google Patents, 1969.

Gogarty, W.; Meabon, H.; Milton, Jr. H. Mobility Control Design for Miscible-Type Waterfloods Using Micellar Solutions. *J. Pet. Technol.* **1970,** *22* (02), 141–147.

Li, S.; Li, Z.; Sun, X. Effect of Flue Gas and n-hexane on Heavy Oil Properties in Steam Flooding Process. *Fuel* **2017,** *187*, 84–93.

Liang, K.; et al. Comparative Study on Enhancing Oil Recovery under High Temperature and High Salinity: Polysaccharides Versus Synthetic Polymer. *ACS Omega* **2019,** *4* (6), 10620–10628.

Prats, M. In *Thermal Recovery,* 1982.

Pu, W.; et al. A comprehensive review of polysaccharide biopolymers for enhanced oil recovery (EOR) from flask to field. *J. Ind. Eng. Chem.* **2018,** *61*, 1–11.

Sheng, J. Mobility Control Requirement in Enhanced Oil Recovery Processes, In *GEO 2010*; European Association of Geoscientists & Engineers, 2010.

Sheng, J. J. In *Modern Chemical Enhanced Oil Recovery: Theory and Practice*; Gulf Professional Publishing, 2010.

Sorbie, K. S. In *Polymer-Improved Oil Recovery*; Springer Science & Business Media, 2013.

Sun, Y.; Saleh, L.;Bai, B. Measurement and impact factors of polymer rheology in porous media, In *Rheology*; InTech, 2012; pp 187–202.

Sveistrup, M.; et al. Viability of Biopolymers for Enhanced Oil Recovery. *J. Dispers. Sci. Technol.* **2016,** *37* (8), 1160–1169.

Viñarta, S. C.; et al. Effects of Thermal, Alkaline and Ultrasonic Treatments on Scleroglucan Stability and Flow Behavior. *Carbohydr. Polym.* **2013,** *94* (1), 496–504.

Wang, D. Polymer Flooding Practice in Daqing, In *Enhanced Oil Recovery Field Case Studies*; Elsevier, 2013; p 83–116.

Xiao, J.; Qiao, G. Understanding the Problems and Challenges of Polymer Flooding Technique. *Oil Gas Res.* **2017,** *3* (126). DOI: 10.4172/2472-0518.1000126.

Yang, A. P. Stochastic Heterogeneity and Dispersion; Texas University, Austin, TX (United States), 1990.

Yen, W. S.; Coscia, A. T.; Kohen, S. I. Chapter 8, Polyacrylamides. In *Developments in Petroleum Science*; Donaldson, E. C., Chilingarian, G. V., Yen, T. F., Ed.; Elsevier, 1989; p 189–218.

Yen, W. S.; Coscia, A. T.; Kohen, S. I. *Polyacrylamides*, In *Enhanced Oil Recovery, II: Processes and Operations. Development in Petroleum Science*; Elsevier: Amsterdam, 1989; p 189–218.

CHAPTER 13

Ionic Liquid Applications in EOR

TAREK GANAT[1] and IMTIAZ ALI[2]

[1]*Department of Petroleum Engineering, Universiti Teknologi PETRONAS, Malaysia*

[2]*Department of Petroleum and Gas Engineering, BUITEMS, Quetta, Balochistan, Pakistan*

ABSTRACT

This chapter explores numerous applications of ionic liquids (ILs) in enhanced oil recovery (EOR) as an eco-friendly alternative to conventional chemical additives used in the petroleum industry. With a focus on reducing hazardous waste, ILs present unique properties such as non-volatility, low vapor pressure, high thermal and chemical stability, and tunability. The chapter details the synthesis of ILs, their classification into neutral, acidic, and basic types, and their potential impact on EOR processes. ILs show promise in modifying the properties of heavy crude oils, preventing asphaltene deposition, acting as demulsification agents, and enhancing the recovery of heavy oil from tar sands. The chapter also discusses the influence of ILs on critical properties like viscosity, thermal stability, density, and conductivity. Practical examples demonstrate the effectiveness of ILs in controlling interfacial tension, improving foam stability, and modifying wettability for better oil recovery. Despite their potential benefits, challenges such as environmental concerns, toxicity, and economic viability are addressed. The chapter finally emphasizes the evolving role of ILs in revolutionizing EOR processes and their potential contribution to a sustainable future in the petroleum industry.

Advancements in Chemical Enhanced Oil Recovery.
Tushar Sharma, Krishna Raghav Chaturvedi, Tarek Ganat, and Imtiaz Ali (Eds.)

13.1 INTRODUCTION

Environmental pollution is primarily caused by chemical industries which have boosted with multifold times in the past decades. It is the human's responsibility to design such processes that generate environment friendly and less hazardous chemicals waste. Ionic liquid is thought to be a source of hope to reduce such hazardous waste in future. These chemicals may substitute the harmful organic solvents and other chemicals. Ionic liquids (ILs) are primarily salts that vary from molten salts because they possess a melting point that ranges from −100 to 200°C (Turosung and Ghos, 2017). Generally, ILs comprise of organic cations and organic or inorganic polyatomic anions and are categorized as required (Xiao and Malhotra, 2005) These are found in molten salts state at room temperature. There are numerous identified and promising cations and anions, thus the quantity of ILs relative to organic solvents is very high and the development of a new ionic liquid is, therefore, very simple.

The central advantage of an ionic liquid is that it basically exerts no vapor pressure while presenting the advantages of conventional solvents that have high volatility. ILs are aprotic in nature and their properties including density, viscosity, reactivity, melting point, and hydrophobicity can be changed to meet the preferred conditions by changing in the cations and anions. Room temperature ionic liquids (RTILs) explicitly belong to such salts which form liquid electrolytes at or below room temperature. Their remarkable characteristics such as nonvolatility, exceptionally low vapor pressure, excellent thermal and chemical stability, high polarity, inert nature, minimal toxicity, and potential of modifying their cation and anion moieties to accomplish them task specified. These properties make ILs an explored reagent or solvent for separation and extraction processes.

The advancement of ionic liquid engineering applications began back in the 1990s when the first studies of continuous catalytic processes involving ILs and the first trials of ionic liquid processing were published. ILs are attracting a lot of recognition as a green chemical due to their special properties.

Chemical EOR methods have revealed great potential in oil recovery mechanisms. In CEOR processes, various chemicals are injected to enhance the oil production from the reservoir by modifying the characteristics of displacing or displaced fluids. Unfortunately, there are certain drawbacks to chemicals used such as massive costs, toxicities, and high adsorption ability (Fan et al., 2007). Due to the increased economics and environmental concerns coupled with the usage of traditional chemical EOR additives, ILs have gained substantial attention for EOR applications due to their green origin. ILs exhibit numerous practical characteristics including as thermal stability, solubility, high ionic conductivity wide electrochemical (conductivity) window, and reusability. In petroleum industry, the developments of emerging technology have been continuously explored in order to address numerous problems that have emerged during the displacement of oil from the reservoir to the processing facilities and further through the refining phase.

ILs have a very vast application in every field, but for our subject, this chapter is limited to its applications in EOR to replace some hazardous and toxic additives. Various applications of ILs in petroleum industry have been reported in the literature. For instance, ILs have been found very effective to modify properties of heavy crude oils such as API gravity and viscosity (Fan et al., 2009; Fan et al., 2007). ILs have also been used to prevent deposition of asphaltenes in the EOR process where the CO_2 flooding method is used. ILs behave to prevent asphaltene's precipitation by efficiently breaking the asphaltene association. Moreover, in some EOR techniques, ILs are utilized as demulsification agents due to their surface-active nature to break water/oil emulsion which further leads to enhance the oil recovery (Guzman-Lucero et al., 2010). Its efficiency to extract heavy oil from tar sands has also been examined by Painter et al. (2010) and Hezave et al. (2013) with a recovery efficiency of 90% with five folds IL recycling and reuse but no loss of efficiency.

ILs can be synthesized through several techniques by conducting two stages including the formation of required cation and anion exchange. In the first stage, the intended cation could be synthesized through protonation of the amine using acids or by amine quaternization with haloalkane and mixture heating. While in the second stage, anion exchange reactions may be done by treating halide salts with Lewis acids to create Lewis acid-based ILs, or anion metathesis (Singh and Savoy, 2020).

The primary disadvantage of ILs is its degradation that generates hydrogen fluoride (HF) and hydrogen chloride (HCl) that are detrimental for the environment. Moreover, the presence of halogen anions in ILs also limits their green nature.

13.2 IONIC LIQUIDS CLASSIFICATION

- **Cations**

  Generally, cations in ILs comprise of an organic structure bearing positive charge. Nitrogen or phosphorous containing organic ions are the most popular cations in ILs. The physical and chemical characteristics of ILs are significantly affected by the cations existing in them.

- **Anions**

  The anion present in ILs normally consists of weakly basic organic or inorganic compounds which are negatively charged. The most existing anions in ILs are nitrate, borate, sulfate, or acetate ions. Figures 13.1–13.2 and Table 13.1 show some common anions and cations of ILs.

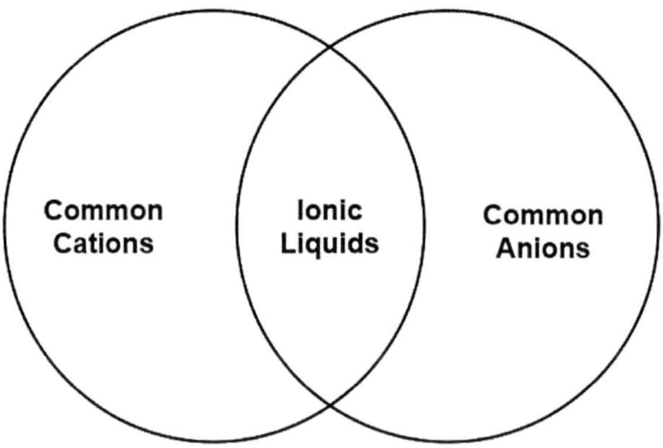

FIGURE 13.1 Ionic liquid cation and anions.

During the designing and selection of ILs, several characteristics are considered. Typically, the physical properties including viscosity, solubility, water miscibility, conductivity, and acidity or the basic properties of the component ions, are focused during the IL selection process.

- **Neutral Ionic Liquid**

  Due to weak electrostatic interactions of anions with the cations, results in low viscosity, and lower melting points. Examples of neutral type ILs include Hexafluorophosphate, bis(trifluoromethanesulfonyl) amide, and thiocyanate, etc. ILs with such anion formulations typically have greater thermal and electrochemical tolerance and are used like inert solvents in a broader range of processes.

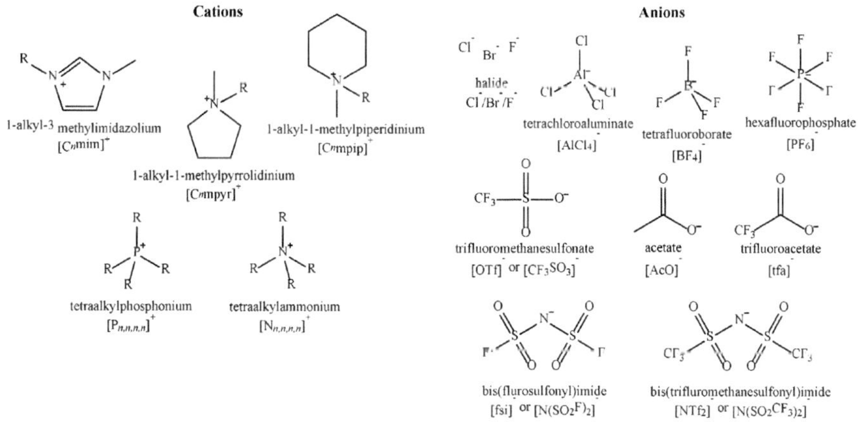

FIGURE 13.2 Typical cations and anions in ionic liquids.

TABLE 13.1 Common Cations and Anions.

Cations	Anions
Alkylammonium	Tetrafluroborate
N, N-dialkylimidazolium	Acetate
N-alkylpyridium	Nitrate
Alkylphosphonium	hexafluorphosphate
Pyrazolium	Methylsulfate
Oxazolium	Trifluoracetate
Triazolium	Chloride, bromide and iodide
Thiazolium	Hexafluroantimonate

- **Acidic Ionic Liquid**

  A significant category in the ionic liquid field is ILs with acidic properties. A low melting ionic salt with an acidic characteristic is termed as acidic ionic liquid. The acidic feature may be Brönsted, Lewis, or a mixture of both. The acidic functions or groups can either be in the cation, anion, or both. Bronsted or Lewis acidic ILs with acidic cation or acidic anion have a distinct significance since they have proton acidity and distinctive features of ILs at the same time. The inclusion of acidic Bronsted functional groups into ILs cations or anions greatly enhances their acidity and water solubility and makes a perfect promise for applying ILs as environmentally friendly catalysts in different reactions.

- **Basic Ionic Liquid**

  Among the ILs types, the basic ionic liquid is a category of essential functional ILs in the current developments. Basic ILs have not only the physical characteristics and potential benefits of typical ILs, but also the advantages of inorganic bases, such as balanced functionality in water and air. Basic ILs are easy to isolate, reuse, modify, non-corrosive, non-volatile, and could be blended with a lot of organic solvents. Basic ILs cannot only be applied like a solvent within the reaction environment but may also be used as a catalyst. Currently, it has played a significant part in several areas including catalytic reactions, adsorption, and separation. ILs further provide unique options for the production of eco-friendly basic catalysts according to the capabilities of integrating inorganic bases and ILs. The intensity of the cation, anion, or the composition of the cation and the anion, controls the acidity or basicity of reactive ILs (Greer et al., 2020).

- **Triazolium Salts**

  Various classes of ILs are available but among all, the most commonly used are those containing N-heterocyclic cations. *Imidazolium* salts are considered the real designer solvents and most prominent subclass in this field, and numerous types are commercially available. Their solvent properties including solubility, viscosity, and melting point can simply be modified in a vast range by changing the nitrogen atom substituents and also by varying the counter-ions. Owing to their promising characteristics, imidazolium-based ILs have many applications in several areas. However, they have some limitations such as they don't really function like an innocent solvent in highly basic conditions, because they can suffer from deprotonation at carbon 2 which leads to N-heterocyclic carbenes. 1,2,3-triazolium salts that lack an acidic ring carbon flanked by two N-atoms. The first exemplars of 1,2,3-triazolium salts have been reported since 1887.

13.3 PROPERTIES OF IONIC LIQUIDS

Properties of ionic liquid are entirely dependent on the substituents on the organic component and by the counterion. The physical and chemical features of ILs are often changed by the variation of ions in them. Following are the key properties of ILs.

- **Melting Point**

  The melting point is known to be a major property in ILs. Selection of cations and anions for ILs controls the melting point of ILs. The melting point of ILs reflects the lower end of the liquid range and thus determines the temperature range to be used as solvents. Due to the presence of a salt, ILs show reduced melting points due to lower lattice energy. Solid–liquid transition temperatures may be lower than ambient temperature and as low as −100°C. The effective approach for determining the melting temperature of ILs is using differential scanning calorimetry (DSC) technique.

- **Density**

  Similar to the melting point, density can also be affected with the cations and anions present in any ionic liquid. Generally, the density of ILs was observed between 1.0 and 1.35 g/cm^{-3}. Ionic liquid density can be determined by the weight of organic anion present in it. Density shows an inverse relationship with the weight of organic anions.

- **Thermal Stability**

  The majority of the available ILs are found thermally stable at 400?C or above. When ILs are compared with the organic solvents, they show better thermal stability. In thermal stability, the key contributor is the occurrence of anions in ionic liquid as compared to cations. Hydrophilicity of anions also affects the thermal properties of ILs. Thermogravimetric analysis (TGA) is used to evaluate the thermal stability of ILs with increase in temperatures. It is a thermal analysis method which examines the effect of temperature on mass loss of the provided sample in a controlled atmosphere.

- **Viscosity**

  Viscosity is an essential property of ILs for various applications. Typically, materials with more hydrogen bonding and van der Waals interactions possess higher viscosities (Ghandi, 2014). ILs were found in moderate range of viscosity as compared to other organic solvents. It is an advantage not having high viscous properties nor lowers to ease the mixing during synthesis and other reactions.

- **Volatility**

  A good number of ILs are non-volatile in nature, therefore, repeatability of the process is very high (Ghandi, 2014). Because of their low volatility, atmospheric pollution caused due to such chemicals is improbable (Amde et al., 2015). Organic solvents which are generally highly volatile, which make them to reproduce the process while ILs have solved this issue due to its non-volatile nature.

- **Conductivity**

  The conductivity of ILs changes with molecular weight and size of ions which is dependent on the available charge carriers and their mobility. Ion conductivity has an inverse relation with size of the cation. It means that ion conductivity declines by increasing the size of the cation. However, no noticeable trend has been found for the anions used. A Waldon plot (Fig. 13.3), showing equivalent conductivity against the log of the fluidity is a clear indication of the ionicity of ILs (Ghandi, 2014). Some general properties of ILs are given in Table 13.2.

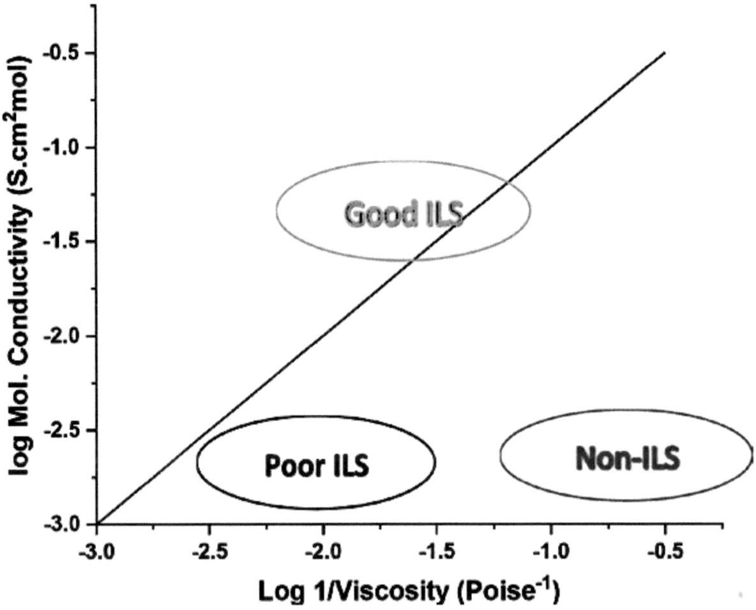

FIGURE 13.3 Walden plot of molar conductivity vs 1/viscosity.

TABLE 13.2 Typical ILs Properties (Gbadamosi et al., 2019; Johnson, 2007).

Viscosity	**Generally less than 100 cP , workable**
Polarity	Moderate
Specific conductivity	Less than 10 S cm^2mol^{-1}
Vapor pressure	Negligible
Thermal stability	High
Freezing point	Below 100C
Liquidus range	Above 200C

13.4 APPLICATIONS OF ILS IN PETROLEUM INDUSTRY

Numerous experiments have revealed the possible and effective use of ILs in petroleum industry for various applications such as surfactant, CO_2 capturing, drilling fluids additive, removal of hydrates, and shale stability. However, due to their high cost, it has not been used commercially. ILs are found the novel green surfactants in different chemical industries. Such ILs are also developing as an alternative for the surfactants utilized in petroleum industries. Figure 13.4 shows some of key applications of ILs in petroleum industry.

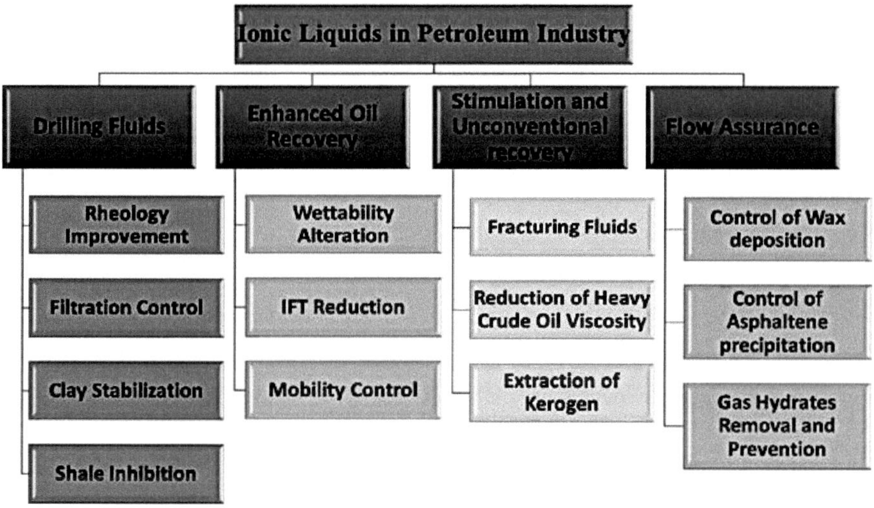

FIGURE 13.4 Application of ILs in petroleum industry.

13.5 SURFACE ACTIVITY OF IONIC LIQUIDS

In surfactant flooding, the selection of a suitable surfactant solution is very important because the surfactant must have the capability to minimize the low interfacial tension (IFT) between oil and water from 10–30 dynes.cm^{-1} to 10^{-3} dynes.cm^{-1}. This decrease in IFT will break up the oil into micro-droplets and will enable to be extricated from the void spaces by hydro-dynamic energy of the chasing water. Surfactants also assist in changing the wettability of reservoir rocks that could be an effective characteristic to improve oil recovery. Few co-surfactants including octanol are also employed to supplement the surfactant flooding process. Considering the substantial amounts of surfactants needed to displace commercial volumes of additional oil, the utilization of surfactants in most cases appears uneconomical.

Various solutions are available such as alkaline surfactant polymer flooding which reduces the cost up to some extent but still uneconomical. One of the solutions is to use new solvents and surfactant-IL blend as a replacement to typical surfactant injection into a reservoir. Conventional surfactants consist of a hydrophilic and a long-chain, non-polar hydrophobic elements, whereas in surfactant-ILs combination the surface behavior can be related to anions and cations or both.

Most of the ILs are found in liquid phase as well as non-volatile which is believed to be a better candidate over the conventional surfactants. Such blends can also be designed according to the reservoir conditions. Several ILs display a comparatively high viscosity that restricts them from fingering in the condition of an unacceptable mobility ratio. Additionally, IL blends do not need any co-surfactants, as they already have high cohesive forces that permit them to develop stable micelles with no requirement of supplementary additives.

Several studies have shown the effect of ILs in the improvements of oil recovery in CEOR such as tri-hexyl-(tetradecyl)-phosphonium-chloride was utilized as a surfactant by Lago et al. (2012). Its ability to perform as a surfactant was investigated to lower the interfacial tension between oil and water molecules. The surface activity of IL was analyzed through various experiments which showed better performance as surfactant as well as saline tolerance nature. Similarly, poly [1-hexadecyl-3-vinyl-imidazolium bromide]-a poly-ionic liquids (PILs) were tested as a chemical agent which showed significant salt tolerance behavior, wettability alteration properties, shear thinning effect, and prolonged thermally stabile behavior in the presence of monovalent and divalent ions. This confirms its applicability in high salinity and high temperature environments. Moreover, low interfacial tension (IFT) values were found at oil–water interfaces (Pillai and Mandal, 2020).

The surface properties of PIL can be calculated using surface tension data. The surface tension reduction effectiveness can be calculated using eq 13.1.

$$\Pi_p = Y_o - Y_p \tag{13.1}$$

Where Π_p represents the surface tension reduction, γ_o and γ_p are the surface tensions of pure solvent and at polymer, respectively. Smaller values of γ_p or higher values of Π_p indicate better capability to reduce surface tension of any solution. Figure 13.5 shows the impact of ILs on contact angle. The effect of water and other ILs can be seen clearly.

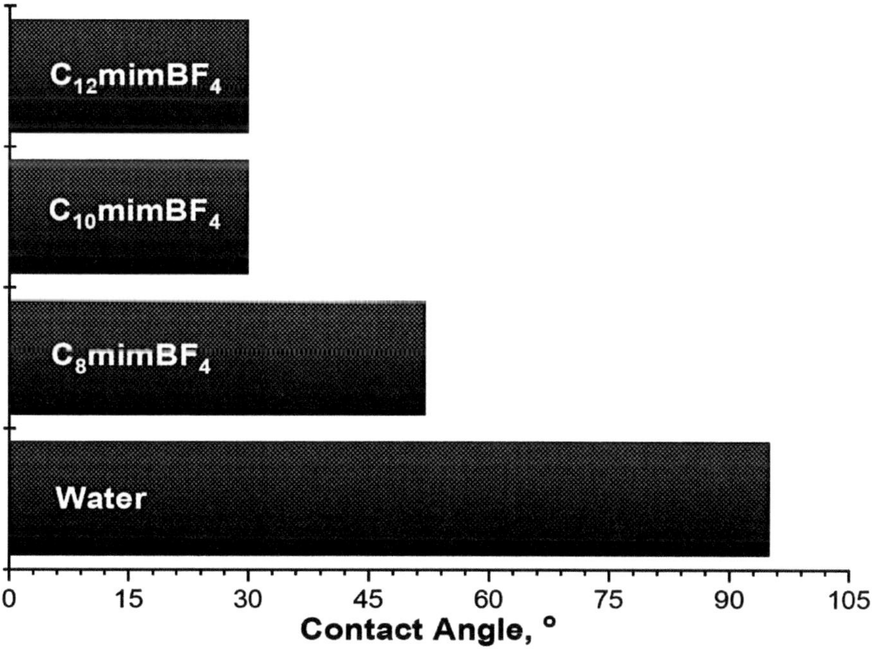

FIGURE 13.5 Wettability modification of rock surfaces by different types of ILs.

13.6 EFFECT OF IONIC LIQUIDS ON MOBILITY CONTROL

Some of the ILs have been found very effective by increasing the viscosity of injecting fluids. It has been experimentally observed that the viscosity of the injection fluid (water) increased with the addition of ILs along with other chemicals in the solution. It is very advantageous to be added in water injection process as it prevents the early water breakthrough at production well. The reason behind this comes from the concept of mobility ratio. It could really contribute to reduce the water–oil mobility ratio inside a reservoir, contributing to higher oil recovery. When the viscosity of the displacing fluid is increased, it will improve the oil recovery and will prevent the viscous fingering too. As like other properties, the viscosity of ILs is also dependent on the composition of anions, cations, and alkyl group present in ILs. But mainly viscosity is dependent on the alkyl group length.

Ionic liquid has a promising potential for enhancing surfactant-stabilized foam mobility control in CEOR procedures. Foam has been applied in EOR for mobility control to tackle several concerns arisen during gas injection. Foam stability has considered the major issue since it determines the efficacy of foam mobility control. Several chemical agents can be used to overcome the mentioned issue that produces synergistic effect in addition with surfactant. ILs are one of the agents which can support the surfactant efficiency at the interface. Hanamertani et al. (2018), for the first time introduced ionic liquid-based additives for improving surfactant-foam stability. Various experiments were performed at ambient as well as at high temperature conditions using Foamscan. The findings revealed that imidazolium and eutectic-based ILs were able to boost surfactant foam stability to differing degrees at high temperatures and had a more pronounced impact on foam stabilization at a surfactant/IL blend ratio of 60:40. The best change in foam stability expressed by the optimized formulation achieved 136% of that obtained without IL in the formulation. The foam stability results are given in Figure 13.6.

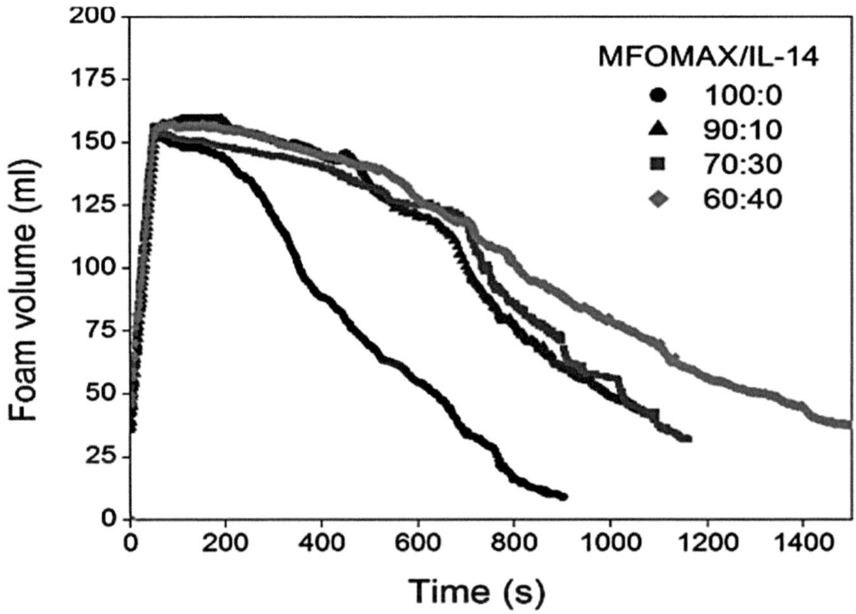

FIGURE 13.6 Foam decay at 90°C for optimized surfactant MFOMAX/IL-14

Source: Reprinted from Hanamertani et al., 2018. Copyright 2018. Elsevier.

Keeping in view the multifunctional nature of ILs, some ILs have been used to decrease the viscosity of heavy crude. Such effects have been seen while adding metal modified ILs in EOR injection fluid. A study showed a reduction of 44.05% by addition of 5% IL to the reaction system. The reaction mechanism is the reduction in sulfur contents in heavy oil which reduces the viscosity, molecular weight and asphaltenes content in the heavy crude oil (Fan et al., 2007; Sakthivel et al., 2015; Nares et al., 2007).

13.7 EFFECT OF IONIC LIQUIDS ON IFT

Most of the surfactants are employed in CEOR for reducing IFT which increases the capillary number by reducing IFT. For decreasing the saturation of oil in the reservoir, the IFT should be reduced to around 20–30 $m.Nm^{-1}$ and improving the capillary number to nearly 10^{-3} and 10^{-2} (Kamal et al., 2017; Omidi et al., 2020). Capillary number is a dimensionless quantity and is the ratio of the viscous forces to the capillary forces. Mathematically, it could be expressed as

$$NC = \frac{Viscous\ Forces}{Capillary\ Forces} \tag{13.2}$$

$$NC = \frac{V\mu}{\sigma Cos\theta} \tag{13.3}$$

Where NC represents the capillary number (dimensionless), V is the velocity (ms^{-1}), μ is the viscosity of the injecting fluid (cp), σ is the interfacial tension ($m.Nm^{-1}$), and θ is the contact angle (degrees).

Certain surface-active ILs have identical surface-active attributes like surfactants and are therefore believed to be appropriate agents for CEOR applications. Many ILs are utilized to investigate their surface activities by determining the IFT between oil and ionic liquid solutions.

It has been experimentally confirmed that surface activity of ILs can be enhanced with the increase in alkyl chain length. This is because of the increase in wan der Waals forces between oil and ILs due to its direct effect on each other. Hezave et al. (2013) experimentally observed the effect of imidazolium-based IL (1-dodecyl-3-methylimidazolium chloride) on the reduction of interfacial tension between water and oil. It was noticed that there was a decline in IFT from 39.98 $m.N.m^{-1}$ to 6.87 $m.N.m^{-1}$ at 5000 ppm concentration of IL. Remarkably, when IFT was tested again between saline and crude, there was a substantial reduction in IFT with just 100 ppm of IL. This decline in IFT with lower and higher IL concentrations in saline and distilled water can be explained by using the ions concept. It shows that the presence of ions in the water, either positive or negative, influences the performance of ILs.

ILs were found to be unable to conveniently organize themselves at the oil–water interface due to their high charge density resulting in molecular repulsion. Conversely, the existence of negative charges in saline water neutralizes the opposite surface charges of the cationic portion of IL. This results in easier aggregation of IL molecules only at oil-brine interfaces and then further decreases the IFT. Figure 13.7 shows the impact of different ILs concentration on the decline of IFT. Numerous researchers studied the effect of ILs on oil recovery in terms of IFT reduction. An imidazolium-based IL has been evaluated as a surface-active agent in carbonate rocks by Pillai and Mandal (2019). ILs modified rock surfaces more water-wet with reduction in contact angle even in the harsh conditions of salts and alkali. It was found that ILs concentration has a direct relationship with the adsorption on the carbonate surface till saturation point.

13.8 EFFECT OF TEMPERATURE AND SALTS ON ILs/CRUDE OIL IFT

Interaction between opposite ions induces electrostatic stability and increases IL molecular adsorption at the interface. Thus, IL molecules strongly occupy the brine/oil interface, which contributes to a higher decline in IFT (Bera and Belhaj, 2016).

The key advantages of ionic liquid in CEOR techniques are their alteration of wettability, IFT reduction, and stability at higher salinity and temperature conditions (Bera and Belhaj, 2016; Bera et al., 2020). Recent research has reported that PIL (poly ionic liquids) can improve oil recovery by up to 6% of OOIP from carbonate rocks (Alhussinan et al., 2018; Nandwani et al., 2017; Sakthivel et al., 2016). PILs are considered more advantageous because of their enhanced thermal stability, non-toxic nature, improved process ability, durability, and flexibility over that of ILs and other polymers (Abdullah et al., 2017). Compared to surfactants, laboratory experimentations have demonstrated that ILs are preferable substitutes since they minimize IFT at high temperatures and high salinity conditions. Furthermore, ILs have a remarkable tendency to self-organize throughout their aggregation behavior at the interface in such conditions.

Generally, high saline reservoirs comprise salts in the range of 10–15% by weight in the formation water (Zhao et al., 2006). Various studies showed the positive effects of ILs especially imidazolium-based ILs in case of saline environment (Velusamy et al., 2017). For instance, the effect of sodium chloride concentrations on IFT of ILs-oil system was studied by Sakthivel et al. (2014) at different concentrations of the ILs. It was noted that the ILs significantly reduced the IFT at salt concentration up to 100,000 ppm in most of the studies which is higher than reservoir brines. In some recent studies on salt tolerance of ILs, 30% of NaCl concentration showed stable results (Pillai et al., 2018). The values were found more improved than the typical surfactants used. However, in few cases, a negative effect was also noticed as compared to conventional ILs.

Experimental tests have confirmed that ILs can be successful additives for reduction in IFT as well as to modify the wettability to more water wet for favorable oil recovery. The presence of ILs in the injection fluid at a particular concentration and specific temperature, is expected to modify IFT behavior of a surfactant in saline environment.

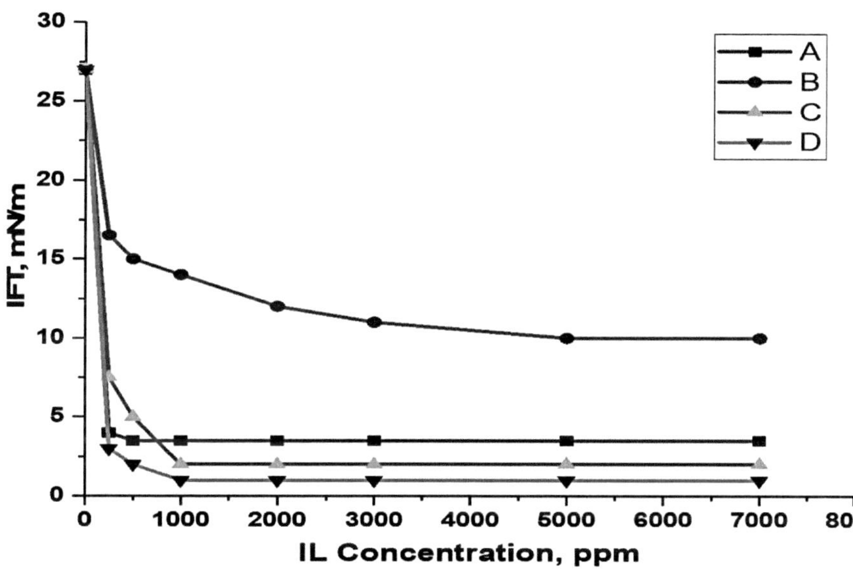

FIGURE 13.7 Effect of different Ionic liquids on IFT.

Source: Reprinted from Abdullah et al., 2017. Copyright 2017. Elsevier

13.9 EFFECT OF IONIC LIQUIDS ON THE SURFACTANT EFFICIENCY

The most efficient mechanism of surfactant flooding is IFT reduction, and research on this subject has been well established (Zeinolabedini Hezave et al., 2014). ILs alter the interfacial tension between water and oil by increasing the hydrophobic chain length. It is well known that the longer the length of hydrophobic chain, the high their adhesion to two different phases, owing to the distance of the two heads with different hydrophilic and hydrophobic characteristics. These two heads create more robust emulsions that displace the trapped oil (Hezave et al., 2013; Manshad et al., 2017; Zhou et al., 2018).

In CEOR processes, co-surfactants generally optimize the performance of surfactants at reservoir conditions. This can be alcohol or some synthetic surfactants. The addition of such agents enhances the salinity tolerance, thermal stability, wettability, and IFT reduction in some harsh conditions. ILs molecules and surfactant combination forms mixed micelles due to hydrophobic interactions, hydrogen bonding formation, and electrical repulsion stabilization between ionic heads of surfactant which further proceeds to variation of critical micelle concentration (CMC). Moreover, hydrophilic anions size can also decrease the CMC. As the hydration of the larger anions is quite challenging, hence they appear to be bound to the micelle surface, which is promising for micellization. Additionally, when ILs are added to the surfactant solution, the aggregation number and size could be increased due to decreased electrical repulsion between anionic head groups of surfactants. Below are few studies which show the importance of IL as potential surfactants in CEOR processes.

Nabipour et al. (2017), examined the effect of three different groups of surfactants commercial, conventional, and IL-based surfactants for oil recovery improvement. Core flooding tests were performed which revealed an 8% and 22% improvement in oil recovery efficiency based on initial oil in place (IOIP) and residual oil in place, respectively. Contact angle measurement confirmed the wettability modification of reservoir rock with the addition of these novel surfactants. Figure 13.8 shows the effect of two ILs on the oil recovery. The newly designed commercial surfactants and IL-based surfactants were found to withstand the higher saline environment, while the conventional surfactants lost their functionality due to high saline conditions. Moreover, with the increase in salinity, the reduction in IFT was observed.

In another study, the wettability modification and IFT reduction through IL-surfactants has been examined. Four ILs were selected. IFT, contact angle measurements were performed followed by core flooding tests on carbonate rocks. The core flooding tests with the selected IL showed 13% increase in oil recovery compared to brine flooding (Manshad et al., 2017).

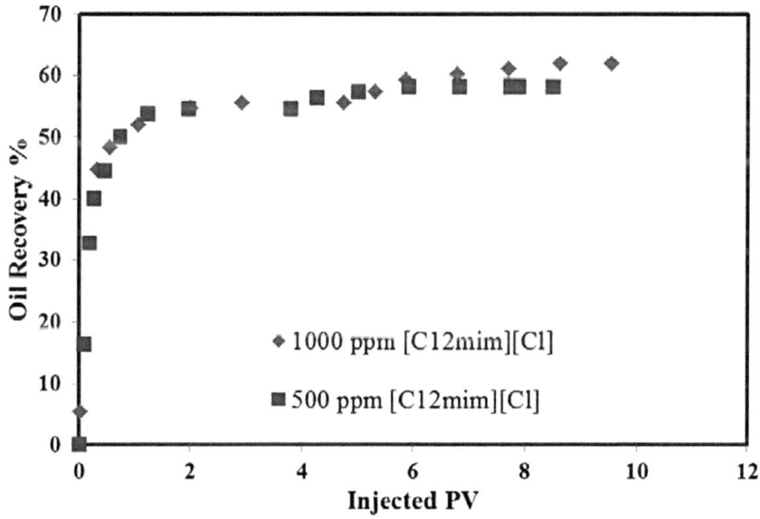

FIGURE 13.8 Oil recovery with different ionic liquid concentration.

Source: Reprinted from Nabipour et al., 2017. Copyright 2017. Elsevier.

Pillai and Mandal (2019) in their experimental research concluded that ILs may be used as an alternative to surfactants in chemical EOR processes, particularly for carbonate reservoirs. Figures 13.9–13.11 show the effect of varying IL concentration on the contact angle of different ILs.

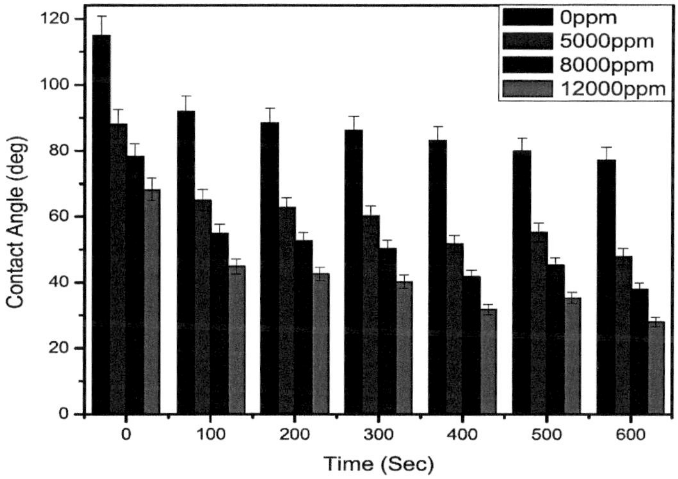

FIGURE 13.9 Effect of C_8mimBF_4 IL concentration on the contact angle.

Source: Reprinted from Pillai and Mandal, 2019. Copyright 2019. American Chemical Society.

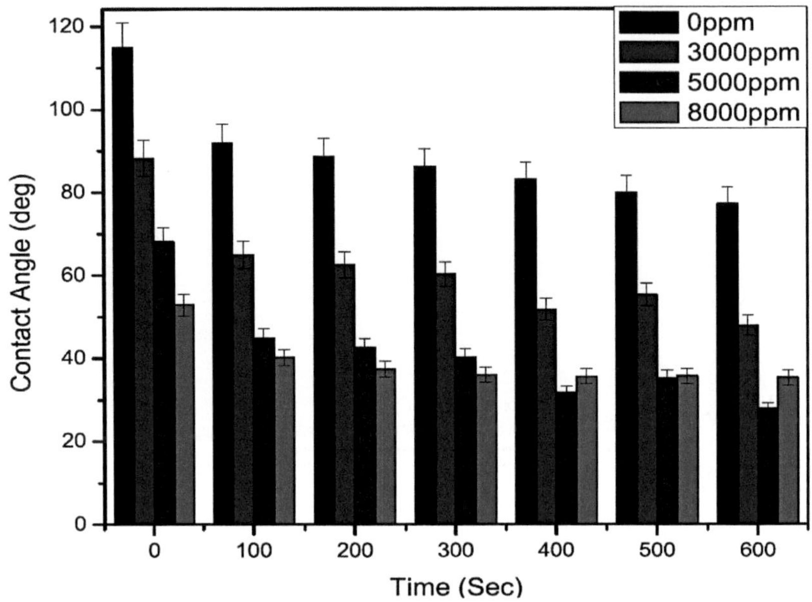

FIGURE 13.10 Effect of $C_{10}mimBF_4$ IL concentration on the contact angle.

Source: Reprinted from Pillai and Mandal, 2019. Copyright 2019. American Chemical Society.

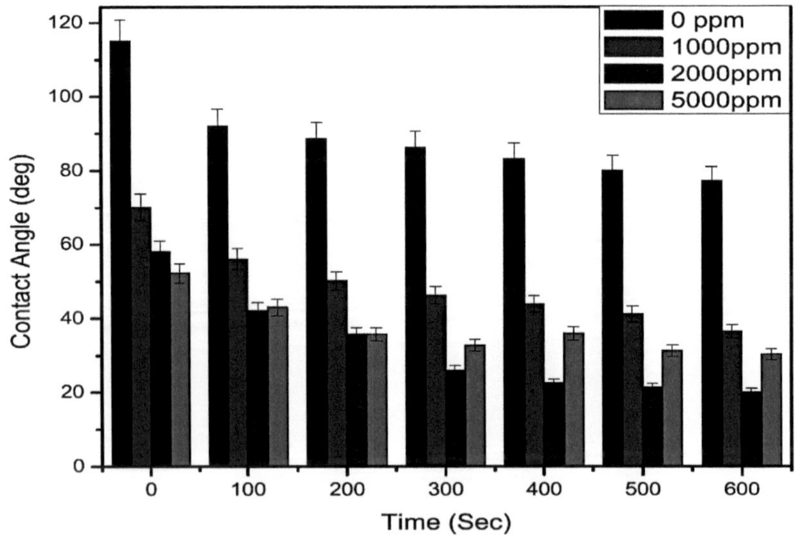

FIGURE 13.11 Effect of $C_{12}mimBF_4$ IL concentration on the contact angle

Source: Reprinted from Pillai and Mandal, 2019. Copyright 2019. American Chemical Society.

13.10 CHALLENGES IN EOR BY IONIC LIQUIDS

ILs have been believed as "green solvents" and accepted due to their multiple environmental applications. However, studies on their environmental fate highlighted their toxicity which puts a question on their greenness nature. The toxicity of ILs changes substantially across their type and provided conditions. Generally, it is suggested to create a database of environmentally friendly ILs based on the available data, which will play a vital role in the synthesis of environmentally friendly and degradable ILs. Additionally, the environmental processes of ILs in real environmental conditions should be studied because in real conditions the effect may be different than that of controlled laboratory conditions.

Another challenge to ILs in CEOR methods is their economics. Most of the ILs are not feasible to be applied in petroleum industry due to their high cost. Economic viability is one of the key considerations addressed in the CEOR processes. Additives used in such processes should be cost-effective so that the process can accomplish its economic viability.

Another important issue with some ILs is their release of HF and HCl in the presence of water which limits their applicability and green nature due to the harmful nature for environment. For some specific applications in petroleum industry, screening of ILs is also considered one of the key challenges to achieve the desired results.

Other challenges in the usage of ILs also include their toxic nature. Not all the ILs are environmentally friendly. The toxicity of ILs is majorly dependent on the cation type and the cation alkyl chain length. Literature showed that the toxicity of ILs increases with the increase in alkyl chain length.

All petroleum-related research areas will be benefited as soon as the spectrum of ILs synthesis on a huge scale is recognized. A number of researchers are trying to study and synthesize ILs which are specifically beneficial for petroleum industry. Since ILs based on imidazolium have acquired the greatest concern for application in EOR on the basis of characteristics of the reservoir rocks. Due to its recyclability with no substantial change in the physico-chemical properties, it can be used multiple times. ILs are expected to play a significant role in the petroleum industry in very near future (Bera et al., 2020).

KEYWORDS

- **ionic liquids (ILs)**
- **enhanced oil recovery (EOR)**
- **eco-friendly chemical**

REFERENCES

Abdullah, M. M.; et al. Synthesis of Novel Water Soluble Poly (Ionic Liquids) Based on Quaternary Ammonium Acrylamidomethyl Propane Sulfonate for Enhanced Oil Recovery. *J. Mol. Liquids* **2017,** *233*, 508–516.

Amde, M.; Liu, J. F.; Pang, L. Environmental Application, Fate, Effects, and Concerns of Ionic Liquids: A Review. *Environ. Sci. Technol.* **2015,** *49* (21), 12611–12627.

Alhussinan, S. N.; et al. *In Chemical Enhanced Oil Recovery With Water Soluble Poly Ionic Liquids in Carbonate Reservoirs,* SPE Kingdom of Saudi Arabia Annual Technical Symposium and Exhibition; Society of Petroleum Engineers, 2018.

Bera, A.; Belhaj, H. Ionic Liquids as Alternatives of Surfactants in Enhanced Oil Recovery—A State-of-the-Art Review. *J. Mol. Liquids* **2016,** *224*, 177–188.

Bera, A.; et al. Recent Advances in Ionic Liquids as Alternative to Surfactants/Chemicals for Application in Upstream Oil Industry. *J. Ind. Eng. Chem.* **2020,** *82*, 17–30.

Fan, H. F.; Li, Z. B.; Liang, T. Experimental Study on Using Ionic Liquids to Upgrade Heavy Oil. *J. Fuel Chem. Technol.* **2007,** *35* (1), 32–35.

Fan, Z. X.; Wang, T. F.; He, Y. H. Upgrading and Viscosity Reducing of Heavy Oils by [BMIM][AlCl4] Ionic Liquid. *J. Fuel Chem. Technol.* **2009,** *37* (6), 690–693.

Gbadamosi, A. O.; et al. An Overview of Chemical Enhanced Oil Recovery: Recent Advances and Prospects. *Int. Nano Lett.* **2019,** 1–32.

Ghandi, K. A Review of Ionic Liquids, their Limits and Applications. *Green Sustain. Chem.* **2014,** 2014.

Greer, A. J.; Jacquemin, J.; Hardacare, C. Industrial Applications of Ionic Liquids. *J. Mol.* **2020,** *25* (21), 5207. https://doi.org/10.3390/molecules25215207.

Guzman-Lucero, D.; et al. Ionic Liquids as Demulsifiers of Water-in-Crude Oil Emulsions: Study of the Microwave Effect. *Energy Fuels* **2010,** *24* (6), 3610–3615.

Hanamertani, A. S.; et al. The Use of Ionic Liquids as Additive to Stabilize Surfactant Foam for Mobility Control Application. *J. Pet. Sci. Eng.* **2018,** *167,* 192–201.

Hezave, A. Z.; et al. Investigating the Effect of Ionic Liquid (1-dodecyl-3-methylimidazolium Chloride ([C12mim] [Cl])) on the Water/Oil Interfacial Tension as a Novel Surfactant. *Colloids Surfaces A Physicochem. Eng. Aspects* **2013,** *421*, 63–71.

Johnson, K. E. What's an Ionic Liquid? *Interface Electrochem. Soc.* **2007,** *16* (1), 38–41.

Kamal, M. S.; Hussein, I. A.; Sultan, A. S. Review on Surfactant Flooding: Phase Behavior, Retention, IFT, and Field Applications. *Energy Fuels* **2017,** *31* (8), 7701–7720.

Lago, S.; et al. Enhanced Oil Recovery Using the Ionic Liquid Trihexyl (tetradecyl) Phosphonium Chloride: Phase Behaviour and Properties. *RSC Adv.* **2012,** *2* (25), 9392–9397.

Manshad, A. K.; et al. Wettability Alteration and Interfacial Tension (IFT) Reduction in Enhanced Oil Recovery (EOR) Process by Ionic Liquid Flooding. *J. Mol. Liquids* **2017,** *248*, 153–162.

Nabipour, M.; Ayatollahi, S.; Keshavarz, P. Application of Different Novel and Newly Designed Commercial Ionic Liquids and Surfactants for More Oil Recovery from an Iranian Oil Field. *J. Mol. Liquids* **2017,** *230*, 579–588.

Nandwani, S. K.; et al. Study on Interfacial Properties of Imidazolium Ionic Liquids as Surfactant and their Application in Enhanced Oil Recovery. *Colloids Surfaces A Physicochem. Eng. Aspects* **2017,** *516*, 383–393.

Nares, R.; et al. In *Upgrading Heavy and Extraheavy Crude Oil With Ionic Liquid*, International Oil Conference and Exhibition in Mexico; Society of Petroleum Engineers: Veracruz, Mexico, 2007; p 5.

Omidi, A.; et al. Smart- and Nano-Hybrid Chemical EOR Flooding Using Fe3O4/Eggshell Nanocomposites. *J. Mol. Liquids* **2020,** *316*, 113880.

Painter, P.; Williams, P.; Mannebach, E. Recovery of Bitumen from Oil or Tar Sands using Ionic Liquids. *Energy Fuels* **2010,** *24* (2), 1094–1098.

Pillai, P.; Kumar, A.; Mandal, A. Mechanistic Studies of Enhanced Oil Recovery by Imidazolium-Based Ionic Liquids as Novel Surfactants. *J. Ind. Eng. Chem.* **2018,** *63*, 262–274.

Pillai, P.; Mandal, A. Wettability Modification and Adsorption Characteristics of Imidazole-Based Ionic Liquid on Carbonate Rock: Implications for Enhanced Oil Recovery. *Energy Fuels* **2019,** *33* (2), 727–738.

Pillai, P.; Mandal, A. A Comprehensive Micro Scale Study of Poly-Ionic Liquid for Application in Enhanced Oil Recovery: Synthesis, Characterization and Evaluation of Physicochemical Properties. *J. Mol. Liquids* **2020,** *302*, 112553.

Sakthivel, S.; et al. Experimental Investigation on the Effect of Aliphatic Ionic Liquids on the Solubility of Heavy Crude Oil Using UV–Visible, Fourier Transform-Infrared, and 13C NMR Spectroscopy. *Energy Fuels* **2014,** *28* (9), 6151–6162.

Sakthivel, S.; et al. In *Nature Friendly Application of Ionic Liquids for Dissolution Enhancement of Heavy Crude Oil*, SPE Annual Technical Conference and Exhibition; Society of Petroleum Engineers: Houston, Texas, USA, 2015; p 12.

Sakthivel, S.; Gardas, R. L.; Sangwai, J. S. Effect of Alkyl Ammonium Ionic Liquids on the Interfacial Tension of the Crude Oil–Water System and Their Use for the Enhanced Oil Recovery Using Ionic Liquid-Polymer Flooding. *Energy Fuels* **2016,** *30* (3), 2514–2523.

Singh, S. K.; Savoy, A. W. Ionic Liquids Synthesis and Applications: An Overview. *J. Mol. Liquids* **2020,** *297*: 112038.

Turosung, S.; Ghosh, B. Application of ionic liquids in the upstream oil industry-a review. *Int. J. Petrochem. Res*. **2017,** *1* (1), 50–60.

Velusamy, S.; Sakthivel, S.; Sangwai, J. S. Effect of Imidazolium-Based Ionic Liquids on the Interfacial Tension of the Alkane–Water System and Its Influence on the Wettability Alteration of Quartz under Saline Conditions through Contact Angle Measurements. *Ind. Eng. Chem. Res.* **2017,** *56* (46), 13521–13534.

Xiao, Y.; Malhotra, S. V. Friedel-Crafts Acylation Reactions in Pyridinium Based Ionic Liquids. . *Organometall. Chem.* **2005,** *690* (15), 3609–3613.

Zeinolabedini Hezave, A.; et al. Mechanistic Investigation on Dynamic Interfacial Tension Between Crude Oil and Ionic Liquid using Mass Transfer Concept. *J. Dispers. Sci. Technol.* **2014,** *35* (10), 1483–1491.

Zhao, H.; et al. Synthesis, Characterization of Ag/MCM-41 and the Catalytic Performance for Liquid-phase Oxidation of Cyclohexane. *Catal. Lett.* **2006,** *108* (1), 49–54.

Zhou, H.; et al. Systematic Investigation of Ionic Liquid-Type Gemini Surfactants and their Abnormal Salt Effects on the Interfacial Tension of a Water/Model Oil System. *J. Mol. Liquids* **2018,** *249*, 33–39.

CHAPTER 14

Hybrid Chemical EOR Using Low-Salinity Water Flooding

TAREK GANAT[1] and IMTIAZ ALI[2]

[1]*Department of Petroleum Engineering, Universiti Teknologi PETRONAS, Malaysia*

[2]*Department of Petroleum and Gas Engineering, BUITEMS, Quetta, Balochistan, Pakistan*

ABSTRACT

This chapter investigates the evolving field of hybrid chemical enhanced oil recovery (EOR), focusing on the integration of low salinity water (LSW) flooding with various chemical agents to improve oil recovery. The chapter provides a comprehensive overview of the mechanisms behind LSW flooding, addressing challenges and limitations. It explores hybrid EOR methods such as LSW-polymer, LSW-surfactant, LSW-nanofluid, LSW-hot water, chemical alternating foam (CAF) flooding, and alkaline cosolvent polymer (ACP) EOR. The efficacy of these methods is demonstrated through laboratory experiments, field trials, and simulation studies. Emphasis is placed on LSW-Polymer hybrid EOR, showing its advantages in changing wettability, improving sweep efficiency, and minimizing polymer usage. Additionally, the chapter discusses LSW-Alkaline and LSW-Surfactant hybrid techniques, along with the emerging LSW-gas and LSW-nanofluid approaches. Practical insights into the advantages and challenge of each hybrid method contribute to a comprehensive understanding of advanced chemical EOR strategies for optimizing oil recovery.

Advancements in Chemical Enhanced Oil Recovery.
Tushar Sharma, Krishna Raghav Chaturvedi, Tarek Ganat, and Imtiaz Ali (Eds.)

14.1 INTRODUCTION

Water flooding is known to be the most commonly used means of increasing oil production rates. This method is using water as an injection source to increase oil production typically termed as secondary recovery. Generally, it extends after primary oil production, which produces oil from a reservoir using its natural energy including solution gas drive, gas cap, water drive, and combination drive, etc. This is accomplished by filling voids with water and moving the oil toward the production wells. It increases and maintains the reservoir pressure. The oil is displaced from the pore spaces and its efficiency is primarily dependent on oil viscosity and rock characteristics. Due to its various complications, it is not always considered the best technique for increasing oil recovery. The factors may include:

1. Injection water compatibility with the connate water of the reservoir
2. Injection water compatibility with the reservoir rocks
3. Injected water treatment to eliminate oxygen, bacteria, and unwanted contaminants.

Several recent studies revealed that the oil recovery improves if the salinity and ions concentration of the injecting water is modified (Teklu et al., 2016). The water injection with reduced salt concentrations or modified ion content induces a number of processes that change the wettability of sandstone and carbonate rocks. The linkage of polar components of crude oil to a carbonate or sandstone rock surface is impaired by both the salinity and the injection water composition. Normally, it influences the wettability of the rock surfaces. The polar parts of crude oil behave differently to the positively charged carbonate rock surface from the negatively charged quartz/clay in sandstone rocks, suggesting different mechanisms of chemical bonding reaction (AlAmeri, 2015).

Several strategies for low-salinity water flooding have been proposed in the literature, such as fine migration, rock dissolution, pH change, multicomponent ion exchange, and double layer expansion. However, the integration of these mechanisms is known to enhance the recovery of oil from sandstone and carbonate rocks.

Low-salinity water (LSW) flooding is an evolving and desirable EOR method for improving oil recovery. LSW injection as an EOR technique in sandstone reservoirs has been reported as early as 1967 (Bernard, 1967). A number of studies in previous years revealed the advantages of LSW both in the laboratory and on field scales (Dang et al., 2020). It has been employed in both sandstone and carbonate rocks to enhance the recovery of oil by various mechanisms. Figure 14.1 shows the improvement in recovery tested on both sandstone and carbonate rocks samples. Figure 14.2 shows the crude oil recovery curve for brine composition tests.

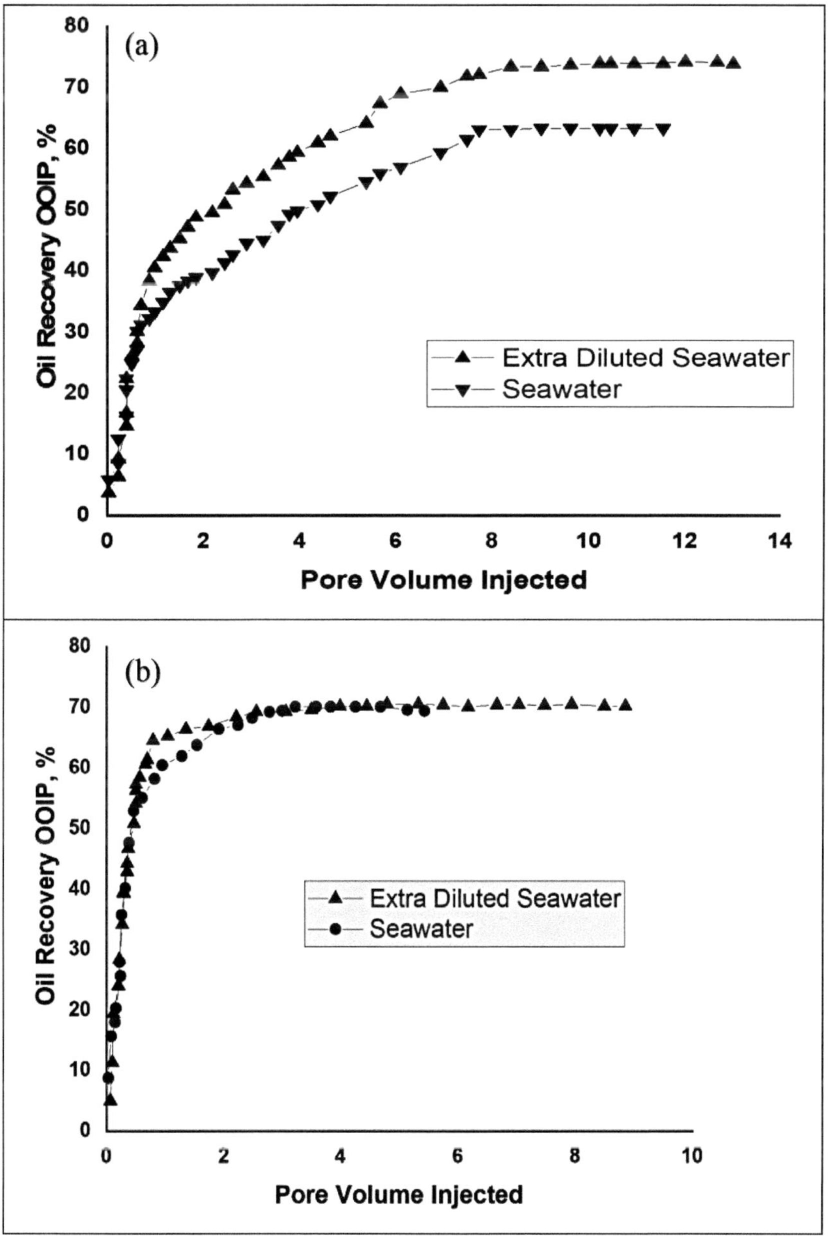

FIGURE 14.1 Core flooding of (a) sandstone and (b) carbonate rocks with various salinities

Source: Reprinted from Katende and Sagala, 2019. Copyright 2019. Elsevier.

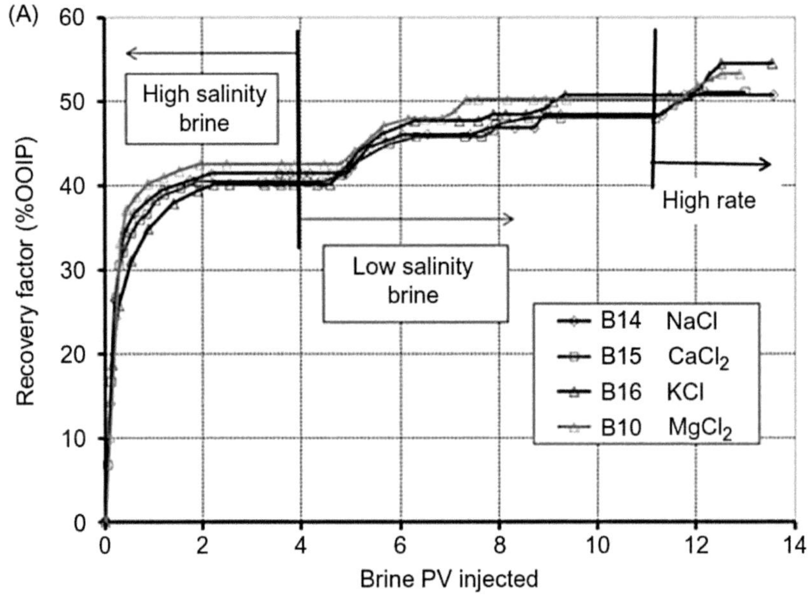

FIGURE 14.2 Recovery curves of crude oil for brine composition experiments.

Source: Reprinted from Austad, 2013. Copyright 2013. Elsevier

None of the studies confirmed the main driving mechanism responsible for improving oil recoveries in the low-salinity water flooding.

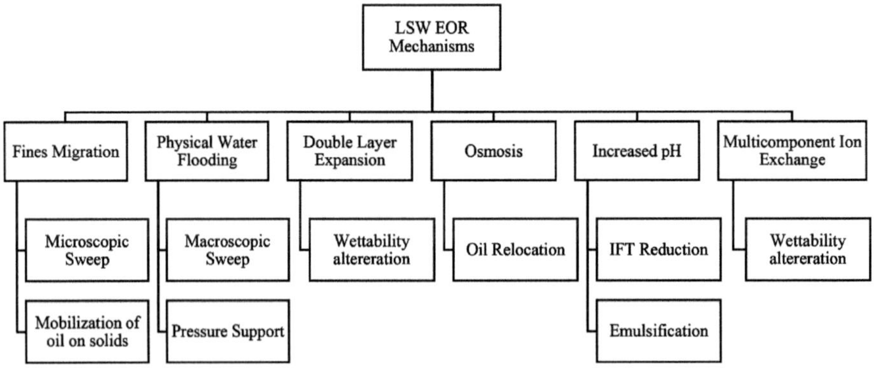

FIGURE 14.3 Low-salinity water flooding mechanisms.

Due to some limitations of LSW flooding such as formation damage and unstable nature due to its low-mobility ratio, a combination of other methods along with LSW has been developed called hybrid EOR methods. Its prime objective was to enhance crude/brine/rock (CBR) interactions which further leads to higher oil recovery. Oil trapping and fine migration are the two main shortcomings of conventional LSW flooding technique. This chapter includes the combination of various chemicals with LSW for the EOR purposes such as LSW-Polymer, LSW-Surfactant, LSW-Nanofluid, LSW-Hot water hybrid EOR techniques, Chemical Alternating Foam (CAF) Flooding, and Alkaline Cosolvent Polymer (ACP) EOR Technology. From the laboratory results, it is noticed that hybrid methods can contribute to higher oil recovery as compared to any other method used separately.

The focus of this chapter is to understand the wettability alteration due to LSW hybrid flooding. Previous experimental studies revealed that LSW flooding can change the shape and endpoints of water and oil relative permeability. Few factors such as formation water composition, displacing fluid composition, wettability conditions, and reservoir rock affect the LSW efficiency. In recent years, hybrid EOR methods have been introduced in wherein LSW is coupled with chemical flooding to boost the combined effect produced from both processes.

14.2 LSW-POLYMER HYBRID EOR

Polymer flooding is a popular and successful CEOR technique. It is often considered as a supportive technique to improve the performance of LSW flooding in a hybrid method. The main function of polymer injection into the reservoir in EOR is to improve mobility ratio and hence enhance the macroscopic sweep efficiency. In contrast, LSW technique improves the microscopic sweep efficiency by altering CBR wettability and CBR interactions. Thus, in LSW/polymer hybrid EOR technique, both advantages can be achieved with a higher oil recovery from the reservoir.

In literature, various injection systems were tested through different experimental studies such as LSW has been injected prior to polymer injection as a pre-flush or polymer has been injected before LSW. It was experimentally found that the LSW injection as a pre-flush is more effective than the polymer injection. The cause for this is the principle of wettability modification to more water wet by injection of low-salinity water prior to injection of polymer which changes the distribution of the residual oil saturation in the void spaces. Additionally, oil droplets are also separated and moved from small pores to medium and bigger sized pores. Therefore, the polymer injection recovers the oil more easily.

The key shortcoming of LSW flooding technique is its unstable nature due to its low-mobility ratio. Similarly, polymer cannot tolerate high salinity and its adsorption has a direct relationship with the salinity. The addition of polymer into low-salinity water can enhance sweep efficiency and displace oil droplets by modifying the wettability to more water wet. Valderhaug (2013) reported a 10% additional recovery by using LSW-Polymer hybrid flooding technique.

LSW-Polymer hybrid flooding can be employed to EOR in favorable conditions. Its effects have been experimentally seen by many researchers. Al-Qattan et al. (2018) experimentally compared the advantages of LSW and low-salinity polymer flooding on the Wara sandstone reservoir which is the world's second largest oilfield. The effectiveness of hybrid flooding was investigated in the laboratory scale as well as conducted a cycle of single-well chemical tracer (SWCT) tests in the Wara producing well. Both lab and field tests trials results showed reductions in residual oil saturation by using the LSW-polymer hybrid technique and proved this technique workable for tertiary oil recovery.

Another study conducted on the same technique resulted superior results by secondary low-salinity flooding in addition with polymers (Skauge and Shiran, 2013). AlSofi et al. (2016) examined the synergetic effect of LSW/polymer process in carbonate rocks by using heavy crude oil and high/low-salinity brines. Polymer injection followed by polymer-LSW after secondary waterflooding was resulted more than 24% additional recovery from OOIP. Lower salinity has an impact on several parameters of polymers including viscosity, injectivity, retention, and stability. Experiments confirmed that due to less ions in LSW, polymers chains have more repulsive interactions that caused the expanding of the polymer backbone. This expansion results in increased polymer viscosity in a solution that tends to pressure buildup and relatively lower polymer injectivity.

In addition to laboratory research, the simulation studies have also indicated the applications of the LSW-polymer hybrid technique. An analytical solution for the hybrid EOR process in sandstone formation was developed to describe the efficiency of this technique (Khorsandi et al., 2017). It was further simulated by Pourafshary and Moradpour (2019) which showed higher oil recovery as shown in Figure 14.5. This figure compares the recovery between the LSW and hybrid process. LSW-polymer flooding is resulted in higher recovery than the LSW. In another experimental study that compared the conventional polymer flood, low-salinity water flooding and low-salinity polymer floods on the intermediate water-wet cores. In all the flooding techniques, the significant recovery factor was obtained by deploying low-salinity polymers (Fig. 14.4).

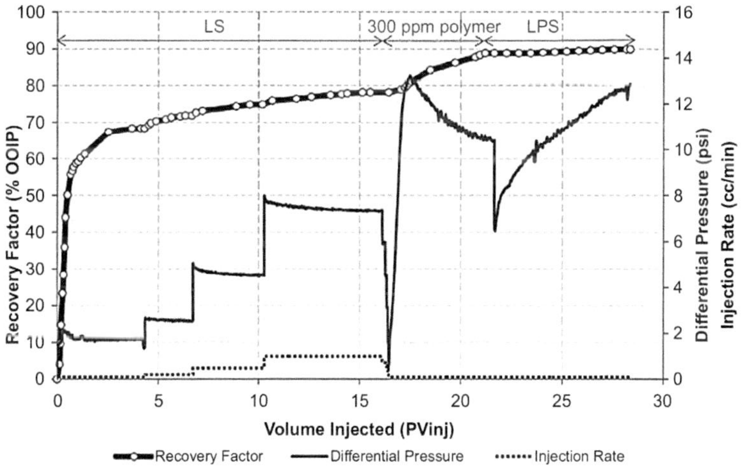

FIGURE 14.4 Comparison of conventional polymer, low salinity, and hybrid flood technique.
Source: Reprinted from Shaker Shiran and Skauge, 2013. Copyright 2013. Americal Chemical Society.

As discussed earlier, polymers alone cannot occupy the inaccessible PV; the water will remain after LSW that will be immovable in the pores, which lowers the channeling in the formation. Thus, the LSW-polymer technique can regulate the unstable front and channeling of low-viscosity LSW into the viscous oil. This ensures the hybrid approach, an extra useful technique for heavy oil recovery.

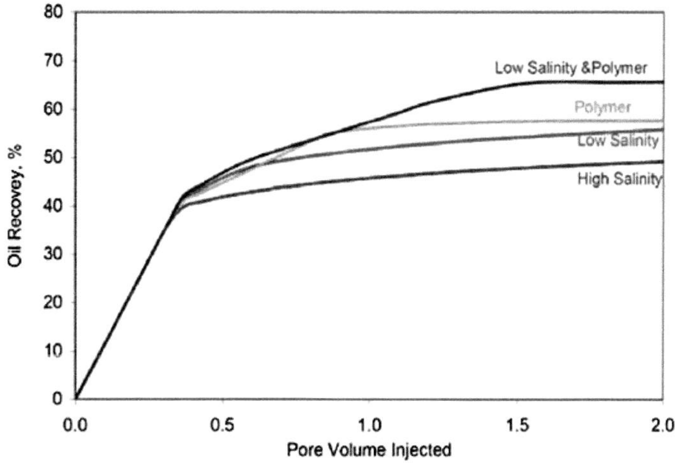

FIGURE 14.5 Oil recovery modeling by LSW, Polymer, and LSW-Polymer Hybrid EOR.
Source: Reprinted from Pourafshary and Moradpour, 2019. https://creativecommons.org/licenses/by/3.0/

The synergism also decreases polymer utilization that is a significant aspect. A lower polymer concentration is needed at lower salinity to attain the target viscosity, which makes the polymer flood process more cost-effective. It is considered economical because of low shipping, handling, and storage costs.

When conventional polymers are injected into the high-salinity water bearing reservoirs at various shear rates, the viscosity of polymers is lost due to the mechanical degradation and salinity conditions. Such effects can be seen in Figure 14.6.

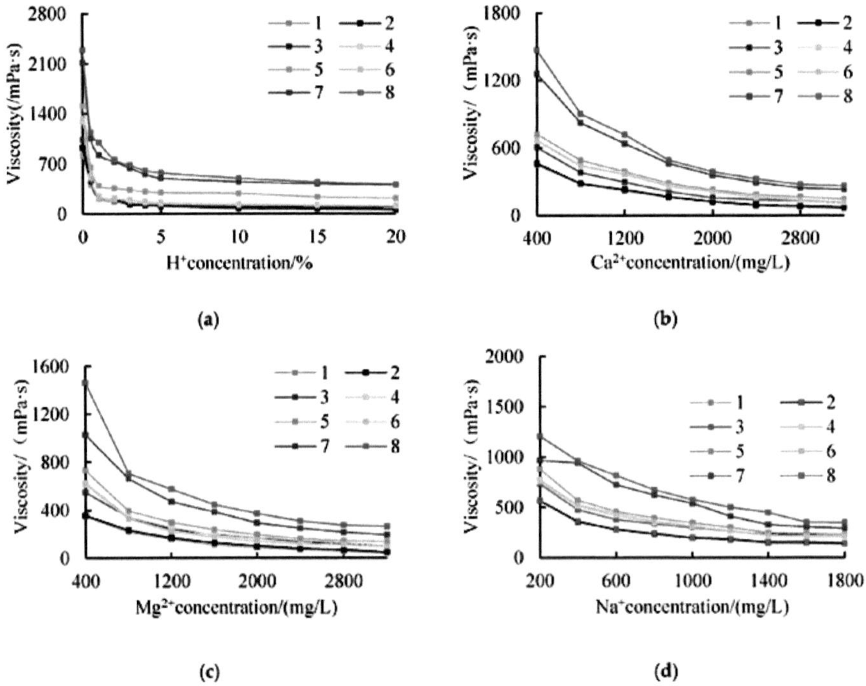

FIGURE 14.6 Polymers degradation at various share rates and salinity conditions.

14.2.1 LSW-POLYMER DESIGN PARAMETERS

Several parameters should be considered during designing a flooding process including initial wettability of the rock, brine composition, and concentration, etc. AlQuraishi et al. (2015) assessed the effectiveness and recovery mechanisms of LSW floods in sandstone and carbonate reservoirs.

The results indicated that the modification of wettability is the key process for increased oil recovery in both carbonate and sandstone rocks. In addition, the selection of appropriate time to move from typical waterflooding to the hybrid approach is also a factor which should be considered in design process. Few studies indicated that LSW-polymer process should be initiated with water cuts of less than or equal to 75% in order to increase the recovery of oil.

11.2.2 ADVANTAGES AND DISADVANTAGES OF LSW-POLYMER HYBRID EOR

The hybrid LSW-polymer technique has several benefits relative to the conventional polymer flooding, which incorporates seawater as the make-up brine. The major advantage of it is the synergetic influence, which enhances both wettability and mobility ratio. Resultantly, this approach improves both volumetric sweep and displacement efficiencies, thus enhances the oil production. The LSW-polymer hybrid EOR implementation is often made possible to inhibit clay swelling and minimize crucial problems. Moreover, owing to the intrinsic salinity-dependent viscosity of the polymer, it intends to utilize reduced dosing of polymers required to reach the target viscosity of the displacing fluid. The reduced salinity and hardness in the hybrid technique also inhibit the substantial chemical degradation of polymer. Chemical procurement, shipping, packaging, mixing, and hydration requirements are determined by the amount of polymer utilization. The smaller facilities needed for least polymer quantity with reduced capital expenditures (Lee and Lee, 2019).

Besides the advantages of hybrid polymer flooding, there exist some disadvantages. During the hybrid polymer flooding, de-stabilization in the initial equilibrium condition is caused in the reservoir and activates the re-equilibrium. Due to the re-equilibrium conditions, the clay cation exchange and low salinity in the solution could destabilize the clay and contribute to further clay swelling through a double layer development. Additionally, it causes mineral dissolution that expands the multivalent cations in the solution which cause the chemical degradation of polymer. The electrostatic interactions within polymers and the clay surfaces may also be caused by re-equilibrium. Thus it is very important to evaluate both positive and negative effects to make a complete economics evaluation of hybrid polymer process.

14.3 LSW–ALKALINE FLOOD HYBRID

Since higher pH raises ionic strength as well as the impacts of ionic strength or salinity are comparably less prevalent on alkaline flood performance. Very limited literature is available till date about LSW–alkaline hybrid technique in EOR. It is accomplished by combining both mechanisms. The synergetic effect of both processes has shown enhanced the oil recovery in few experimental studies.

A study investigated the effect of low salinity with alkaline injection concluded that the LSW–alkaline hybrid method showed less EOR potential when compared with low-salinity surfactant flooding. However, the economical factor of alkali agents has been found advantageous on the EOR implementation. Moreover, the reduced chemicals cost of alkali agent injection might prevent the adsorption of *in situ* generated surfactants (Shaddel and Tabatabae-Nejad, 2015). Suleimanov et al. (2017) evaluated the effects of this technique in terms of contact angle, IFT adsorption, and clay swelling. The results showed an indirect relation of salinity with the pH of solution. Furthermore, Both IFT and contact angle were also decreased. The IFT and contact angle measurements suggested that the acidic part of crude oil decreased the IFT at elevated pH conditions and modified the wettability. This solution also generated highly stable water-in-oil emulsion with lower bentonite swelling. The flooding tests showed an improved oil recovery by implementing the hybrid technique.

14.4 LSW–SURFACTANT HYBRID EOR

Low-salinity water injection separates the oil films/oil droplets from the rock surfaces resulting in the wettability alteration. Wettability alteration to water-wet is a well-known mechanism for the effectiveness of LSW. On the other hand, the main functions of surfactant injection are to reduce the interfacial tension (IFT) between oil droplets and brine and alter the rock's wettability. The resulted flooding method by combining the two approaches may potentially be a successful method by integrating the effects of both approaches, i.e., destabilizing the oil layer by LSW and IFT reduction using surfactant. This hybrid method results in additional oil recovery than LSW or surfactant flooding alone. Low-salinity water flooding provides a favorable medium for surfactant flooding and some of the residual oil could be solubilized by low-salinity surfactant. Applications of the LSW–surfactant hybrid flooding yields in fewer operational issues, lower surfactant consumption, and lower operational costs as compared to stand alone flooding techniques.

Numerous studies have described the effectiveness of LSW–surfactant methods. Olefin sulfonate as a surfactant was employed in core flooding experiments to examine the effectiveness of surfactants after LSW flooding in mixed wet sandstone rock samples. Results revealed that more than 90% of the OOIP was recovered when the surfactant was injected after the LSW flood. This recovery was much better when compared with the high-salinity water injection (Alagic and Skauge, 2010). The reason was the alteration of wettability of rocks to additional water wet which assisted to destabilize the oil droplets and reduce capillary forces which lead to make the displacement of oil easier. Mjøs (2014) described that the LSW–surfactant and polymers hybrid technique resulted in 11–32% recovery of remaining oil in place. The higher recovery was obtained from aged cores compared to unaged ones.

In some experiments, both processes, i.e., the modification of wettability and the reduction of interfacial tension in the hybrid procedure, resulted in a higher oil recovery than the predicted recovery from the capillary desaturation curves (CDCs). CDC describes the relationship of residual oil saturation (S_{or}) and capillary number. It is one of the basic curves in the recovery of oil, which demonstrates the flow conditions necessary for proper oil displacement in a porous media (Yeganeh et al., 2016).

Several experimental results showed the low-salinity environment that has brought numerous potential advantages including the improving solubility of surfactant and reduction of retention to surfactant EOR procedures.

14.5 LSW–GAS HYBRID EOR

The introduction of LSW–Gas hybrid EOR flooding has recently drawn the interest of numerous scientists. Using experimental and simulation techniques, numerous injection systems have been examined. Low-salinity water could be injected into both water alternating gas (WAG) and simultaneous water alternating gas (SWAG) methods. In addition, after execution of the LSW flooding, gas may be injected to enhance the overall recovery of oil. In this hybrid method, carbon dioxide is usually used as the injecting gas since it is inexpensive, has a lower minimum miscibility pressure (MMP), and offers environment friendly performance (Teklu et al., 2016). The gas solubility in water and the gas/oil interaction and total oil recovery are impaired by low-salinity water flooding. Various results have been reported by researchers about the advantages and disadvantages of the hybrid LSW–gas technique. Few researchers reported no enhancement in oil recovery while using this method, but some studies showed its positive effects on the oil recovery when compared with the conventional gas injection process.

In some studies, the solubility of carbon dioxide in saline environment showed improved results with minimized water salinity due to salting out process. Therefore, greater solubility of CO_2 results in smaller concentrations of free gas available to come into contact with oil and increase the mobility of oil, leading in reduction of oil recovery. However, several experimental studies resulted in enhancement in oil recovery with minimum water salinities. The greater solubility improves the diffusion of gas in water impacting the waterfront, the ratio of mobility, and the stability of the fluid injection. Enhancement in such factors contributes to lower injecting fluid fingering, which improves its interaction in the porous medium with already bypassed oil droplets. Hence, the waterfront geometry is modified with the dissolved gas in water that targets un-swept oil more simply compared to free gas.

The positive impact of hybrid LSW–gas technique has been seen both in sandstone and carbonate rocks by altering the mobility of displacing fluid. Numerous studies highlighted the importance of low salinity combined with carbon dioxide. The improved recovery was achieved due to the combined effect of both EOR techniques in a unique way. An experimental study on CO_2-WAG and CO_2-SWAG concluded that by decreasing water salinity, the oil recovery increased. This behavior is due to the increase in the solubility of carbon dioxide in water. The enhancement in oil recovery due to lower salinity reached up to 18% due to the improved displacement efficiency supported by CO_2 (Aleidan and Mamora, 2010). The comparison of 0 and 6 wt% oil and water in both schemes are given in Figures 14.7–14.10.

A similar study was conducted by Teklu et al. (2016) for improving oil recovery by injecting low-salinity water and CO_2 mixture on sandstone and carbonate rocks. From the experiments of contact angle and IFT, it was confirmed that significant wettability alteration and IFT reduction were accomplished with the hybrid EOR method. The recovery factor versus pore volume injection is given in Figure 14.11.

14.6 LSW-NANOFLUID HYBRID EOR TECHNIQUE

Nanomaterials are commonly used for a wide range of applications in EOR methods. Due to their small size and higher surface area, nanoparticles may

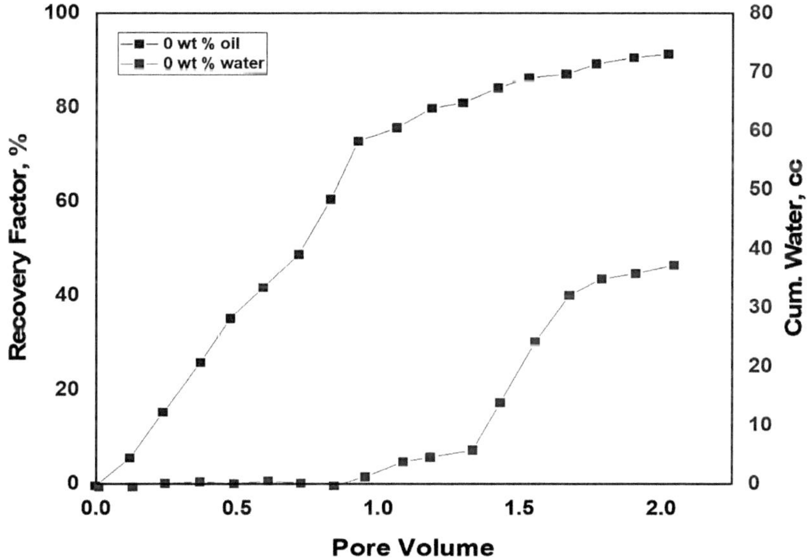

FIGURE 14.7 Recovery factor of WAG-CO_2 and cumulative water production at 0 wt% oil and water.

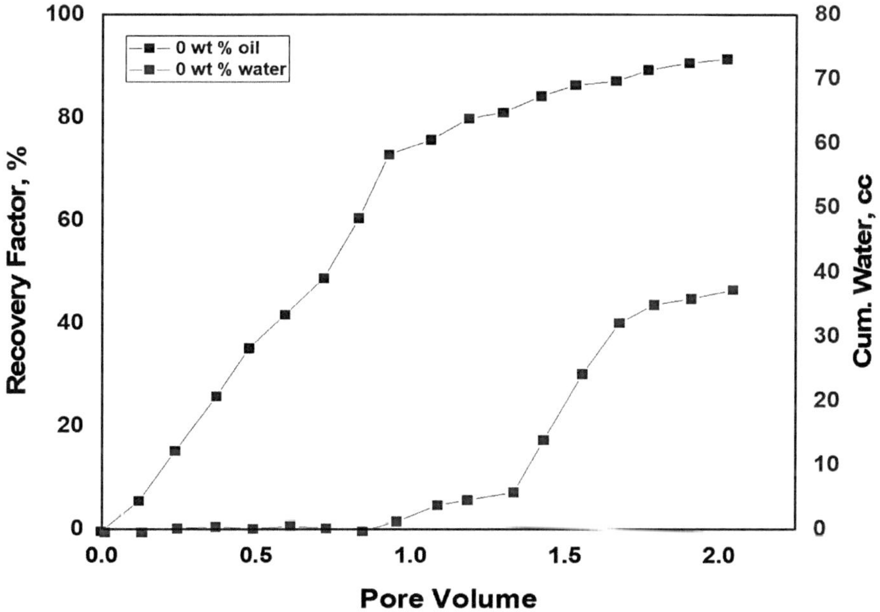

FIGURE 14.8 Recovery factor of WAG-CO_2 and cumulative water production at 6 wt% oil and water.

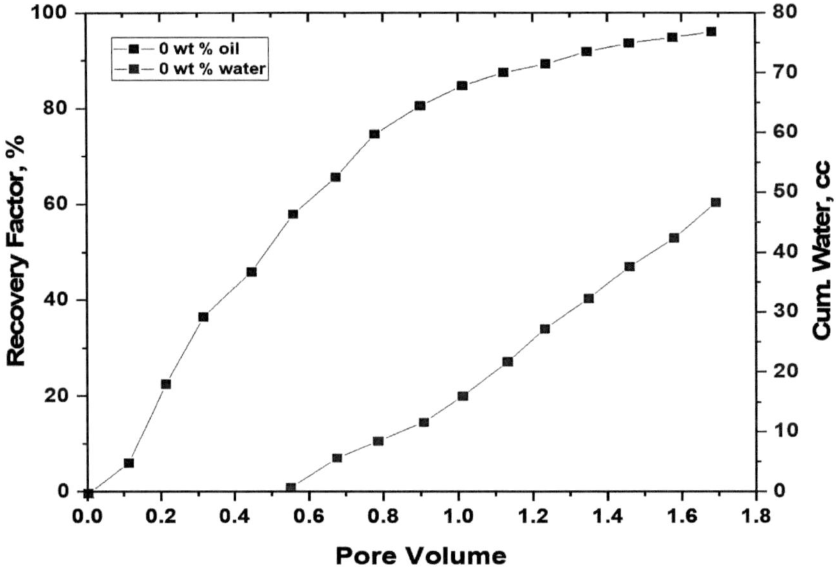

FIGURE 14.9 Recovery factor of SWAG-CO_2 and cumulative water production at 0 wt% oil and water.

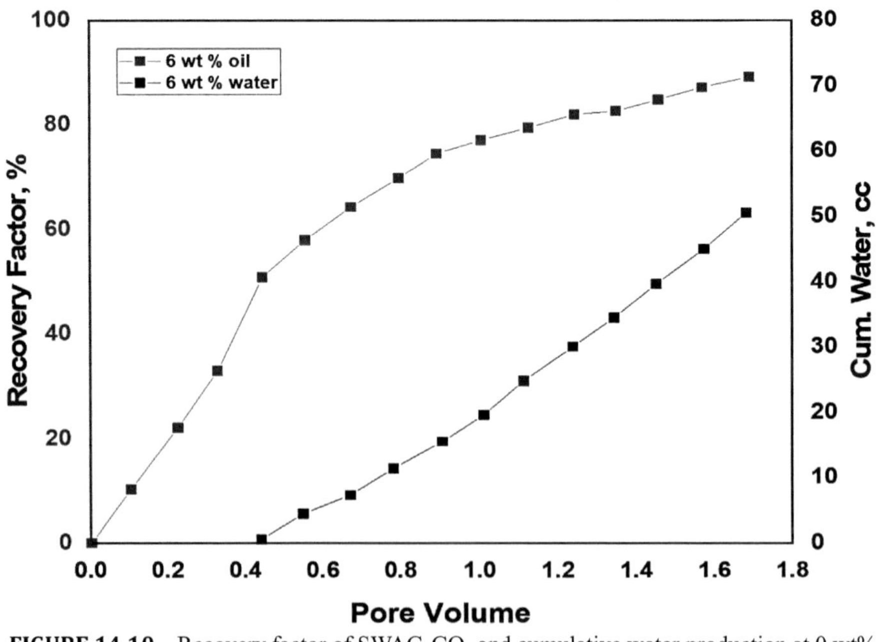

FIGURE 14.10 Recovery factor of SWAG-CO_2 and cumulative water production at 0 wt% oil and water.

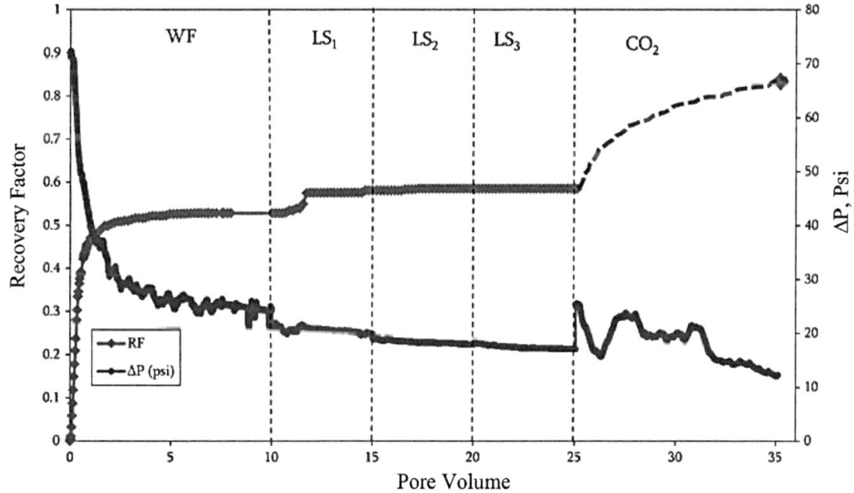

FIGURE 14.11 Recovery factor vs pressure difference

Source: Reprinted with permission from Teklu et al., 2016. Copyright 2016. Elsevier

flow through porous media easily. Some unique properties of NPs such as high-specific area, active surfaces, and higher chemical reactivity make NPs suitable additives in various applications (Hu et al., 2016; Zhang et al., 2015). Nanoparticles enhance the recovery of oil by the following mechanisms (Cheraghian and Hendraningrat, 2016; Kazemzadeh, et al., 2019; Khezrnejad, 2015; Zaid et al., 2014):

- Mobility ratio enhancement
- Improvement of the quality of the injection fluid
- Modification of wettability
- Modification of viscosity and density of injection fluid
- Enhancing rock–oil interactions
- Changing the heat transfer coefficient
- Reducing the possibilities of formation damage
- Reduction of interfacial tension
- *In situ* emulsification

Some of the most commonly used nanomaterials in EOR are TiO_2, SiO_2, MWCNT, ZnO, ZrO_2, Fe_3O_4, MgO, and γ-Al_2O_3 (AfzaliTabar et al., 2017; Jafari Daghlian Sofla et al., 2018; Kazemzadeh et al., 2019; Kumar Inturi et al., 2019; Nowrouzi et al., 2020; Tang et al., 2020). It is a well-known practice that when hydrophilic nanoparticles are injected into the reservoir, the wettability is altered to more water wet.

Even though LSW flooding has many positive effects on oil recovery but this technique imposes changes in terms of pH, temperature, and ionic strength in the reservoir. This further results to the destabilization of reservoir particles. When the injection fluid salinity is lowered than critical salinity, it causes formation damage due to the plugging of pores by fine migration into the pores. This improves the oil recovery factor by improving the mobility by plugging the high-permeable zones but can cause damage to the near wellbore region. Thus, it is required to reduce fine migration and prevent formation damage.

Several studies highlighted the positive effects of LSW in combination with nanofluids which assisted to minimize the detrimental effects of fine migration caused due to LSW flooding. Nanoparticles improve the attractive forces between fine particles and grain surfaces, especially by modifying the surface zeta potential of fine particles. If nanofluid is injected before the LSW flooding, it can decrease injection pressure drop and thus results into the less formation damage due to the particles invasion. This effect has been studied in sandstone core samples by Abhishek et al. (2018) through injecting the LSW–NP hybrid fluids. It was found that the fine migration was prevented by using silica NPs along with LSW.

This hybrid method also performs other functions in CEOR such as alteration of wettability and reduction of IFT. When hydrophilic NPs adsorb on the rock surface, water molecules occupy the surface and accumulate around the surface. This accumulation causes the rock more water wet and hence improves the oil recovery. A study conducted by Sadatshojaei et al. (2016) to investigate the simultaneous effects of LSW-SiO_2 NPs on oil-wet dolomite rock. Results revealed that LSW-NPs surprisingly altered the wettability of the dolomite rock. The comparative analysis showed the high efficiency of hybrid technique compared to LSW alone.

14.7 LSW–HOT WATER HYBRID EOR TECHNIQUE

Thermal EOR methods are generally applied for the recovery of heavy crude oils. The primary function of thermal recovery is to decrease the viscosity of oil and displace oil toward the production well. Many complications are associated with thermal methods such as large supply of high-quality freshwater and heat losses through injection lines and wellbores.

LSW–Hot water hybrid technique is a novel method to reduce the oil viscosity and aid in alteration of wettability of the rock for achieving higher recovery. Various experiments have been performed to understand the phenomenon of this method. A research conducted on sand pack and core flood test designed to examine the efficiency of this emerging method in terms of recovery. The results confirmed a substantial improvement in oil recovery due to low-salinity hot water injection. The low salinity in addition with thermal recovery can recover 25% more oil as compared to hot water flooding (Abass and Fahmi, 2013).

14.8 CHEMICAL ALTERNATING FOAM (CAF) FLOODING

Foam flooding is a technique by injecting foam into an oil reservoir. Foam is a distinct form of colloidal dispersion, generally referred to a process in which a gas phase is dispersed in a liquid phase. The foams can be injected in either pre-generated or not. The key functions of the foam flooding in CEOR are to reduce the gas relative permeability in a reservoir and mobility control. In the meantime, the high-permeability zone is also blocked which substantially reduce the heterogeneity of a reservoir. Due to better rheological and flowing characteristics, foam has been used in drilling, stimulation, and EOR methods.

Owing to some benefits of foam flooding, some drawbacks also exist in the foam flooding. The most critical disadvantage of foam flooding is the foam durability and stability (Schramm and Novosad, 1990). The foam stability can be affected by several factors such as electric layers repulsion, high saline and high-temperature environment, capillary suction, and dispersion force attraction, etc.

Chemical-alternating foam (CAF) flooding is an emerging method utilized to overcome the shortcomings of the typical foam flooding including inadequate quantity of *in situ* foam, surfactant retention, and serious foam instability. Different additives have been used to overcome the mentioned disadvantages of foam flooding and still research is underway to improve its efficiency. Xu and Saeedi (2017) compared the conventional foam flooding and chemical alternating foam flooding with same composition of chemicals. Results depicted that for CAF flooding, the oil recovery was noted about 10–15% more than that of normal foam flooding. CAF flooding also resulted in a significant volume of incremental oil with 98% water cut while the typical foam flooding attained at 60% water cut with the injection of 1.8 PV. Additionally, CAF flooding acquired better residual resistance factor (RRF) at high temperature. RRF is the ratio of relative brine permeability before and after EOR application. It demonstrates the ability of the EOR process to reduce permeability in porous media. If the brine is flowing at a same rate before and after EOR process, then RRF can be determined using eq. 14.1:

$$RRF = \frac{K_w}{K_f} \tag{14.1}$$

$$RRF = \frac{\Delta P_2}{\Delta P_1} \tag{14.2}$$

where K_w and K_f represent the brine relative permeability before and after flooding, respectively. ΔP_1 and ΔP_2 represent the pressure drops of the brine flow before and after the process, respectively. Figure 14.12 illustrates the difference between typical foam and CAF flooding.

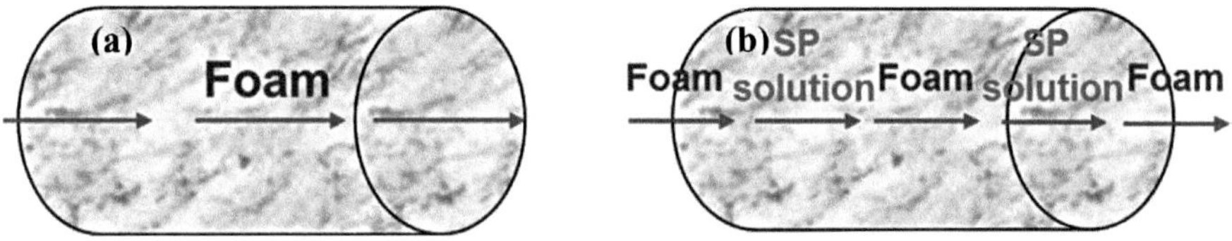

FIGURE 14.12 (a) Conventional foam flooding and (b) CAF flooding

Source: Reprinted from Xu and Saeedi, 2017. https://creativecommons.org/licenses/by/4.0/

Kutay and Schramm (2004) experimentally studied the effectiveness of the polymer-enhanced foam (PEF) in terms of foam stability and durability. It was noted that foam durability can be significantly enhanced by optimizing viscosity of the bulk phase and lowering the rate of liquid drainage. Silicon-based NPs were employed to stabilize the foam and was also found very successful (Worthen et al., 2013). The implementation of chemically enhanced water alternating gas (CWAG) floods was also investigated using the CMG STAR simulator to visualize the performance of the water alternating gas (WAG) process (Majidaie et al., 2012). In this new approach, a chemical slug composed of alkaline, surfactant, and polymer was introduced during WAG process. The objective was to decrease the IFT and enhance the mobility ratio. The chemical slug was followed by water then gas slug and other gases and chemicals. The combined effect of all the mentioned additives reduced IFT and viscosity of crude oil with further improvement in the mobility.

14.9 ALKALINE COSOLVENT POLYMER (ACP) EOR PROCESS

Owing to some limitation of the previous CEOR methods, researchers are constantly and actively exploring advanced and feasible chemical EOR strategies capable of achieving higher oil recovery without the detriment effects of additives. Such effects including high-pressure gradients and phase trapping using alkaline polymer (AP) flooding, poor phase behavior due to alkaline concentration that cause in ultralow IFT which is considered limited for practical purposes and ASP formulations find it hard to pass stringent tests of aqueous stability.

More recently, relative to traditional EOR approaches, ACP flooding has been developed as a novel chemical EOR of great prospect and capability. In this method, low concentrations of cheap light cosolvents are combined with AP solutions. By this way, the increase in the efficacy of the recovery of oil is achieved with two essential ways (Gbadamosi et al., 2019).

1. Formation of low-viscous microemulsions due to the presence of cosolvents
2. Improvement in the phase behavior due to the occurrence of light cosolvent in the ternary combination.

ACP flooding performs similarly as ASP while being simpler and more robust with the availability of polymers for mobility control.

KEYWORDS

- **hybrid chemical EOR**
- **low salinity water flooding**
- **EOR mechanisms**
- **LSW-polymer hybrid**
- **advanced chemical EOR**
- **oil recovery strategies**
- **hybrid EOR techniques**

REFERENCES

Abass, E.; Fahmi, A. In *Experimental Investigation of Low Salinity Hot Water Injection to Enhance the Recovery of Heavy Oil Reservoirs*, North Africa Technical Conference and Exhibition; Society of Petroleum Engineers: Cairo, Egypt, 2013; p 10.

Abhishek, R.; Hamouda, A. A.; Murzin, I. Adsorption of Silica Nanoparticles and Its Synergistic Effect on Fluid/Rock Interactions During Low Salinity Flooding in Sandstones. *Colloids Surf. A: Physicochem. Eng. Asp.* **2018,** *555,* 397–406.

AfzaliTabar, M.; et al., Preference of Multi-Walled Carbon Nanotube (MWCNT) to Single-Walled Carbon Nanotube (SWCNT) and Activated Carbon for Preparing Silica Nanohybrid Pickering Emulsion for Chemical Enhanced Oil Recovery (C-EOR). *J. Solid State Chem.* **2017,** *245,* 164–173.

Alagic, E.; Skauge, A. Combined Low Salinity Brine Injection and Surfactant Flooding in Mixed−Wet Sandstone Cores. *Energy Fuels* **2010,** *24* (6), 3551–3559.

AlAmeri, W. In *Low-Salinity Waterflooding in a Low-Permeability Carbonate Formation of a Giant Middle East Field*, Colorado School of Mines. Arthur Lakes Library, 2015.

Aleidan, A. A.; Mamora D. D. In *SWACO2 and WACO2 Efficiency Improvement in Carbonate Cores by Lowering Water Salinity, Canadian Unconventional Resources and International Petroleum Conference*; Society of Petroleum Engineers, 2010.

Al-Qattan, A.; et al. In *Low Salinity Waterflood and Low Salinity Polymer Injection in the Wara Reservoir of the Greater Burgan Field*, SPE EOR Conference at Oil and Gas West Asia. Society of Petroleum Engineers, 2018.

AlQuraishi, A. A.; AlHussinan, S. N.; AlYami, H. Q. In *Efficiency and Recovery Mechanisms of Low Salinity Water Flooding in Sandstone and Carbonate Reservoirs*, Offshore Mediterranean Conference and Exhibition; Offshore Mediterranean Conference: Ravenna, Italy. 2015; p 14.

AlSofi, A. M.; et al. In *SmartWater Synergy with Chemical EOR for a Slightly Viscous Arabian Heavy Reservoir*, SPE Heavy Oil Conference and Exhibition; Society of Petroleum Engineers, 2016.

Bernard, G. G. Effect of Floodwater Salinity on Recovery of Oil from Cores Containing Clays. In *SPE California Regional Meeting*. Society of Petroleum Engineers, 1967.

Cheraghian, G.; Hendraningrat, L. A Review on Applications of Nanotechnology in the Enhanced Oil Recovery Part A: Effects of Nanoparticles on Interfacial Tension. *Int. Nano Lett.* **2016,** *6* (2), 129–138.

Dang, C.; et al. AI Based Mechanistic Modeling and Probabilistic Forecasting of Hybrid Low Salinity Chemical Flooding. *Fuel* **2020,** *261,* 116445.

Gbadamosi, A. O.; et al. An Overview of Chemical Enhanced Oil Recovery: Recent Advances and Prospects. *Int. Nano Lett.* **2019,** 1–32.

Hu, Z.; et al. Nanoparticle-Assisted Water-Flooding in Berea Sandstones. *Energy Fuels* **2016,** *30* (4), 2791–2804.

Jafari Daghlian Sofla, S.; James, L. A.; Zhang, Y. Insight into the Stability of Hydrophilic Silica Nanoparticles in Seawater for Enhanced Oil Recovery Implications. *Fuel* **2018,** *216,* 559–571.

Katende, A.; Sagala, F. A Critical Review of Low Salinity Water Flooding: Mechanism, Laboratory and Field Application. *J. Mol. Liq.* **2019,** *278,* 627–649.

Kazemzadeh, Y.; et al. Review on Application of Nanoparticles for EOR Purposes: A Critical Review of the Opportunities and Challenges. *Chin. J. Chem. Eng.* **2019,** *27* (2), 237–246.

Khezrnejad, A. Nanofluid Enhanced Oil Recovery—Mobility Ratio, Surface Chemistry, or Both? **2015.**

Khorsandi, S.; Qiao, C.; Johns, R. T. Displacement Efficiency for Low-Salinity Polymer Flooding Including Wettability Alteration. *SPE J.* **2017**, *22* (2), 417–430.

Kumar Inturi, S.; Ranjan Dash, G.; Kodavaty, J. Role of Silica Nano Particles in Altering Rheological Properties of Drilling Fluid in Enhanced Oil Recovery. *Mater. Today: Proc.* **2019,** *17,* 354–361.

Kutay, S. M.; Schramm, L. L. Structure/Performance Relationships for Surfactant and Polymer Stabilized Foams in Porous Media. *J. Can. Pet. Technol.* **2004,** *43* (2), 10.

Lee, K.; Lee, J. Hybrid Chemical EOR Using Low-Salinity and Smart Waterflood. In *Hybrid Enhanced Oil Recovery using Smart Waterflooding*; Lee, K. S., Lee, J. H., Eds.; 2019; pp 65–110.

Majidaie, S.; et al. In *A Simulation Study of Chemically Enhanced Water Alternating Gas (CWAG) Injection*, SPE EOR Conference at Oil and Gas West Asia; Society of Petroleum Engineers: Muscat, Oman, 2012; p 9.

Mjøs, J. E. S. *Low Salinity Waterflood in Combination with Surfactant/Polymer: Effect of Brine Composition*. The University of Bergen, 2014.

Nowrouzi, I.; Khaksar Manshad, A.; Mohammadi, A. H. Effects of TiO_2, MgO and γ-Al_2O_3 Nano-Particles on Wettability Alteration and Oil Production Under Carbonated Nano-Fluid Imbibition in Carbonate Oil Reservoirs. *Fuel* **2020,** *259,* 116110.

Pourafshary, P.; Moradpour, N. Hybrid EOR Methods Utilizing Low-Salinity Water. In *Enhanced Oil Recovery Processes-New Technologies*; IntechOpen, 2019.

Sadatshojaei, E.; et al., Effects of Low-Salinity Water Coupled with Silica Nanoparticles on Wettability Alteration of Dolomite at Reservoir Temperature. *Pet. Sci. Technol.* **2016,** *34* (15), 1345–1351.

Schramm, L. L.; Novosad, J. J. Micro-Visualization of Foam Interactions with a Crude Oil. *Colloids Surf.* **1990,** *46* (1), 21–43.

Shaddel, S.; Tabatabae-Nejad, S. Alkali/Surfactant Improved Low-Salinity Waterflooding. *Transp. Porous Media* **2015,** *106* (3), 621–642.

Shaker Shiran, B.; Skauge, A. Enhanced Oil Recovery (EOR) by Combined Low Salinity Water/Polymer Flooding. *Energy Fuels* **2013**, *27* (3), 1223–1235.

Sheng, J. J. *Enhanced Oil Recovery Field Case Studies*. Gulf Professional Publishing, 2013.

Skauge, A.; Shiran, B. Low Salinity Polymer Flooding. In *IOR 2013-17th European Symposium on Improved Oil Recovery*; European Association of Geoscientists & Engineers, 2013.

Sui, Y.; et al. Synthesis and Mechanism Study of Temperature- and Salt-Resistant Amphoteric Polyacrylamide with MAPTAC and DTAB as Monomers. *Processes* **2022,** *10* (8), 1666.

Suleimanov, B. A.; et al. In *Low Salinity and Low Hardness Alkali Water as Displacement Agent for Secondary and Tertiary Flooding in Sandstones*, SPE Annual Caspian Technical Conference and Exhibition; Society of Petroleum Engineers, 2017.

Tang, X.; et al. Regulation of Polymerizable Modification Degree of Nano-SiO_2 and the Effects on Performance of Composite Microsphere for Conformance Control. *Colloids Surf. A: Physicochem. Eng. Asp.* **2020,** *585,* 124100.

Teklu, T. W.; et al. Low-Salinity Water-Alternating-CO_2 EOR. *J. Pet. Sci. Eng.* **2016,** *142,* 101–118.

Valderhaug, O. M. *Investigating EOR for SS by Low Salinity Water*. University of Stavanger: Norway, 2013.

Worthen, A. J.; et al. Carbon Dioxide-in-Water Foams Stabilized with Nanoparticles and Surfactant Acting in Synergy. *AIChE J.* **2013,** *59* (9), 3490–3501.

Xu, X.; Saeedi, A. Evaluation and Optimization Study on a Hybrid EOR Technique Named as Chemical-Alternating-Foam Floods. *Oil Gas Sci. Technol. – Rev. IFP Energies Nouvelles* **2017,** *72* (1), 1.

Yeganeh, M.; et al. In *Capillary Desaturation Curve Fundamentals*, SPE Improved Oil Recovery Conference; Society of Petroleum Engineers, 2016.

Zaid, H. M.; et al. Effect of Morphology of Aluminium Oxide Nanoparticles on Viscosity and Interfacial Tension (IFT) and the Recovery Efficiency in Enhanced Oil Recovery (EOR). *AIP Conf. Proc.* **2014,** *1621* (1), 705–710.

Zhang, T.; et al. Investigation of Nanoparticle Adsorption During Transport in Porous Media. *SPE J.* **2015,** *20* (4), 667–677.

CHAPTER 15

Polymer Rheology in Porous Media

TAREK GANAT[1] and IMTIAZ ALI[2]

[1]*Department of Petroleum Engineering, Universiti Teknologi PETRONAS, Malaysia*

[2]*Department of Petroleum and Gas Engineering, BUITEMS, Quetta, Balochistan, Pakistan*

ABSTRACT

This chapter explores the complex interaction of polymer rheology in porous media, particularly within petroleum engineering discipline. Following an introduction explaining the microscopic confinement of oil within reservoir pores, the chapter investigates the essential characteristics of non-Newtonian polymer solutions passing through porous media. The discussion covers polymer viscosity, retention mechanisms (adsorption, mechanical entrapment, and hydrodynamic retention), and permeability reduction, emphasizing their pivotal roles in enhanced oil recovery processes. The impact of polymer stability, including mechanical, chemical, and biological degradation, is examined, highlighting its critical importance in maintaining polymer efficacy during enhanced oil recovery operations. The chapter also presents various models for porous media, describing the complexities of fluid movement and emphasizing the limitations of conventional capillary bundle models. A comprehensive examination of Darcy's law and fluid flow, as well as the behavior of pseudo-plastic fluids in porous media, further enriches the reader's understanding. In-situ rheology of polymers, both pseudoplastic and viscoelastic, is explored with a focus on the complex influences of porous media geometry, mechanical degradation, and tortuosity. The theoretical approaches to polymer rheology in porous media are

Advancements in Chemical Enhanced Oil Recovery.
Tushar Sharma, Krishna Raghav Chaturvedi, Tarek Ganat, and Imtiaz Ali (Eds.)

discussed, highlighting the challenges and recent advancements in numerical modeling, particularly in pore network modeling approaches.

15.1 INTRODUCTION

After the primary and secondary oil production stages, most of the oil remains in the reservoir. From a microscopic perspective, the remaining oil is observed to be confined in the pores, surrounded by another fluid. This confining process depends on the rock's wettability, pore geometry, and interfacial tension (IFT). These factors are responsible for the trapping and mobilization of oil across a porous medium.

The movement of polymers through porous media has a variety of applications. The most important use of such polymers in the petroleum industry is in EOR processes, where polymer solutions are injected to control mobility by increasing the viscosity of displacing fluids. Two central polymers employed in EOR are biopolymers and synthetic polymers. Xanthan gum is most widely used as a biopolymer, and partially hydrolyzed polyacrylamide (HPAM) is considered a synthetic polymer. The flow of such polymers through a porous medium causes some interesting rheological effects.

This chapter describes the essential characteristics of the rheological behavior of non-Newtonian polymer solutions flowing through porous media.

15.2 FLOW BEHAVIOR OF POLYMERS

This part covers the fundamentals of polymer viscosity, retention, and permeability reduction in porous media.

15.2.1 POLYMER VISCOSITY

In general terms, a fluid's viscosity relates to how "thick" the fluid is; for instance, honey is "thicker" as compared to water. It shows that honey viscosity is higher than water viscosity. According to the Navier–Stokes equation, the viscosity of any fluid is termed as its resistance to shear, as given by eq 15.1.

$$Shear\ Stress(\tau) = \frac{Force}{Area} = \frac{F}{A} \tag{15.1}$$

Considering the rate of deformation of a fluid, the above equation can be generalized as:

$$\tau = -\mu\left(\frac{dV}{dr}\right) = \mu\gamma \tag{15.2}$$

where τ is shear stress, μ represents viscosity, and γ is the shear rate. Newtonian fluids are those in which the proportionality between τ and (dV/dr) in eq 15.2 is constant over a wide range of deformations. For non-Newtonian fluids, the viscosity is not constant at various shear rates. Polymer solutions always exhibit non-Newtonian behavior at appropriately higher polymer concentrations. Shear stress and shear rate relations for non-Newtonian fluids are expressed as follows:

$$\tau = -\eta\left(\frac{dV}{dr}\right) = \eta\gamma\gamma \tag{15.3}$$

where η is the viscosity function which is shear rate dependent.

It is challenging to determine the viscosity of a polymeric mixture in porous media directly. It could be computed using Darcy's equation by conducting core flood experiments. As polymer viscosity is dependent on shear rate, several flooding tests would be required to obtain the viscosities at different shear rates. Due to the core flooding test's economics for multiple runs, a model is necessary to estimate polymer viscosities at different shear rates in porous media. Practically, it is not so simple to do so; instead, one needs to conduct a few core flood experiments for the calibration of the model.

For the estimation of polymer solution viscosities in a porous medium, several steps need to be followed, including the conversion of flow rates into shear rates, plotting bulk viscosity against different shear rates, core porosity and permeability determination, core flooding, and then using some equations to finally estimate the polymer solution viscosity. The power law model is generally considered a better rheological model for non-Newtonian fluids, which is given below.

$$\eta = K\gamma^{n-1} \tag{15.4}$$

15.2.2 POLYMER RETENTION

When a polymer solution flows through a porous medium, there are three major mechanisms responsible for polymer retention: polymer adsorption, mechanical entrapment, and hydrodynamic retention.

The technical and economic performance of a polymer flooding program is deeply influenced by polymer retention. Polymer retention is inversely proportional to the recovered oil volume per pound of injected polymer. It has three different mechanisms, including mechanical trapping, adsorption, and hydrodynamic retention. Mechanical trapping and hydrodynamic retention are the two main mechanisms in polymer retention. Mechanical trapping arises when larger polymer molecules flow through small pore throats (Huh et al., 1990). It is the key mechanism responsible for polymer retention in porous media. Adsorption implies the adhesion of atoms, ions, or molecules from a gas, liquid, or dissolved solid to any surface. In a porous medium, it is the interaction between the polymer molecules and the rock surfaces. This keeps the polymer molecules bound to the rock surface either by physical adsorption (van der Waals) or hydrogen bonding. The main objective of polymer adsorption is to change the interaction between the surfaces, which in turn enhances the dispersion of particles, flocculation processes, and surface properties (Michael et al., 2018). Hydrodynamic retention is generally ignored in field applications due to its low impact on total retention and its low flow velocity. Polymer retention can be affected by some factors, including permeability, the polymer's molecular weight, clay content, and wettability. Permeability and oil wetness showed an inverse relationship, while polymer MW, clay content, and the existence of crude oil showed a direct relationship with polymer retention.

Heterogeneity in terms of permeability has a significant role in the adsorption of polymers. It has been observed that polymer flooding in moderately heterogeneous fields showed better performance as compared to heterogeneous fields (Choi et al., 2015). Polymer retention has an inverse relationship with rock permeability.

Before starting a polymer flood, polymer retention must be thoroughly studied, determined, and analyzed. For any proposed polymer project, it is typically estimated by performing a flooding test on representative reservoir rock samples with the provided reservoir conditions.

15.2.3 REDUCTION OF PERMEABILITY

Permeability reduction is mainly caused by polymer flow through reservoir matrix rock. It is primarily initiated by the polymer adsorption mechanism. A factor representing the reduction of permeability is called the resistance factor. It can be determined using the following equation:

$$F_{kr} = \frac{K_{wf}}{K_{pf}} \tag{15.5}$$

where F_{kr} is the permeability reduction, K_{wf} represents the permeability of rock when water flows, and K_{pf} is the permeability of rock when polymer solution flows.

Permeability reduction is an irremediable mechanism, and thus it remains at the reservoir post polymer waterflooding. This post permeability reduction is called the residual permeability reduction factor. It is the proportion of permeability of rock to water before and after polymer flow, as given below.

$$F_{krr} = \frac{K_{rb}}{K_{ra}} \tag{15.6}$$

where F_{krr} is the residual permeability reduction factor and K_{rb} and K_{ra} are the permeabilities of rock to water before and after polymer flow, respectively.

Different studies have found that the polymer's molecular weight has a significant impact on the reduction of the residual permeability factor. In cases of similar molecular weight, this factor will be higher if the polymer has a broad molecular weight distribution (Pang et al., 1998). Higher MW polymers will increase recovery due to high F_{krr}, but the molecular weight must be restricted by formation permeability. It has also been noted that the permeability reduction factor increases with increasing injection velocity and decreasing temperature (Huang et al., 1998; Sheng, 2013). Figure 15.1 shows the molecular weight distribution for a common HPAM polymer sample used in polymer flooding. The high molecular weight tail of the MW distribution is very noticeable, as shown in Figure 15.1.

15.3 POLYMER STABILITY

It is absolutely important that the polymer properties do not degrade significantly when used in enhanced oil recovery. Degradation is generally the change of polymer characteristics from their original state. The change may

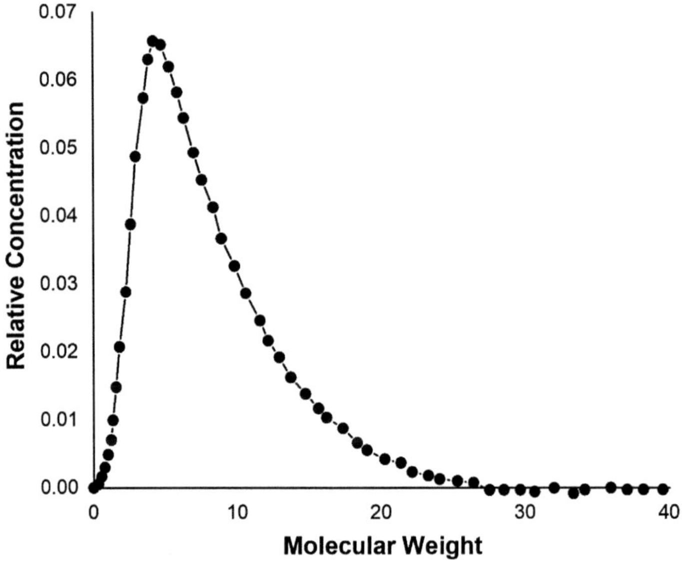

FIGURE 15.1 Molecular weight distribution of a typical HPAM polymer used in polymer floods.

either be advantageous or not desirable. Therefore, degradation is recognized as a generic term for all sorts of reactions in a polymer (Kourde-Hanafi et al., 2017; Maya et al., 2007). These processes can lead to a significant decline in the polymeric material's quality, that is, poor mechanical or electrical properties, before finally losing its functionality.

The important characteristic to consider in this case is usually the viscosity of the polymer solution, although the potential of the polymer to decrease the permeability of the exposed rocks of the reservoir can even be of certain relevance for some polymers. Polymer degradation applies to any mechanism that destroys the macromolecule's molecular structure, and the major degradation routes of interest in applications for oil recovery are as follows:

1. Mechanical degradation
2. Chemical degradation
3. Biological degradation

When the polymer molecules are exposed to high shear rates near the wall, they degrade for a short time. This degradation is termed mechanical degradation or sometimes shear degradation. In chemical degradation, the deterioration of polymer molecules occurs due to the contaminants' attachment or hydrolysis. The microbial deterioration of polymer molecules attributable to bacteria is called biological degradation. Polymers undergo breakdown of their polymer attachments in biological tissues through hydrolytic or enzymatic lysis, thereby producing waste metabolites. This is critical if appropriate biocides are not available or are at lower temperatures. It could arise either during the preinjection preparation of the polymer or in the cold areas of the reservoir. The physical and chemical characteristics of a polymer are very important for biodegradation. Degradation is commonly recognized as an inevitable process (Sazali et al., 2020; Sorbie, 1991; Goonoo et al., 2015; Glaser, 2019).

In petroleum reservoirs, polymers are injected into the reservoir for the improvement of the mobility ratio, which enhances the oil recovery. Owing to some problems associated with such polymers due to high-temperature conditions, the polymers thermally degrade. Thermal degradation refers to the degradation of chemical compounds caused by heat into smaller components that do not recombine upon cooling. Several studies have reported on the thermal degradation of polymers in the presence of higher reservoir temperatures.

The mechanical degradation of HPAM can be minimized when the polymer solution is subjected to a certain amount of mechanical degradation before injection into the reservoir. This procedure is called the polymer preshearing process. Mechanical degradation of polymers has been studied by Al-Shakry et al. (2018) to examine the influence of mechanical degradation and filtration on HPAM rheology. They conclude that at higher flow velocities, core flooding experiments revealed a prevailing shear thickening behavior due to successive contraction and expansion flows in pores. At low shear rates, no substantial change in in situ viscosity was found, and shear thickening was decreased when presheared. Results also showed that continual deterioration resulted, indicating that the presumption that the maximum point of shear determines mechanical degradation in a porous medium does not carry valid conditions for all fields.

15.4 MODELS OF POROUS MEDIA

There is certainly no questioning that fluid movement in porous media is a major subject in several fields. Heterogeneity-dominated systems over a vast range of scales, such as porous media, are still challenging in modeling initiatives (Battiato et al., 2019; Wei et al., 2020; Battiato et al., 2019). In this section, various models will be discussed, focusing on porous media.

15.4.1 *EXPERIMENTAL EXAMINATION OF PORE STRUCTURE*

A valuable working framework for illustrating flow in porous material has become the continuum methodology to portray a porous medium, in which little consideration is paid to the complex pore structure. An example of this type of method is Darcy's law for a single-phase fluid's pressure drop and flow rate behavior.

Such procedures were reviewed and termed phenomenological models. Nevertheless, such methods did not assist in illustrating various observations dependent on fluid behavior at the microscopic scale (Sorbie, 1991). Therefore, many efforts have recently been made to describe the phenomena with respect to the microscopic structure as precisely as possible. Direct microscopic analysis is a perfect reference point for modeling porous media at the pore scale. To accomplish this, various strategies have been developed to analyze pore geometries. The detailed pore structures can be viewed through electron micrography techniques. The resulting micrographs are of expanding significance for the examination of pore structures. The information can help develop an appropriate mathematical model for pore space. This thorough examination provides valuable parameters for suitable microscopic mathematical models.

The most significant factors controlling the macroscopic reservoir performance and fluid movement in tight sandstones are microscopic pore throat structures. Microscopic pore structures, such as pore geometry, pore size distribution, and pore connectivity, are of considerable significance for the maintenance and enhancement of oil recovery (Lai et al., 2018). Pore morphological structure and connectivity are also very important for correlating porosity and permeability data (Desbois et al., 2011).

Computer-assisted evaluation of thin segments of core samples is very useful for pore size distribution. Mercury porosimetry is considered the most widely used experimental approach and an extremely useful characterization technique for porous materials because it obtains information that affects the pore size distribution. It is an analytical approach being used to examine different scientific attributes of a material's porous nature, including pore size distribution, surface area, pores diameter, and total pore volume, or porosity. Different pore sizes can be investigated using this approach (Giesche, 2006).

By comparing both mercury porosimetry and image analysis, it has been reported by various authors that mercury porosimetry is a satisfactory method to characterize pore size distributions as compared to other techniques (Giesche, 2006; Dullien and Dhawan, 1974).

15.4.2 CAPILLARY BUNDLE MODELS

Theoretical models are very important to represent any porous medium. These models can be treated mathematically to predict behavior under different conditions.

Various approaches are available to model and characterize a porous material. The first technique is the continuum approach, where the porous structure is interpreted as a bulk, utilizing bulk parameters, such as diffusion coefficients and permeability, to measure the physical behavior of the system. Such parameters are computed through assumptions and approximations attributable to phenomenological equations. The above assumptions and approximations restrict the efficacy of such models to very limited systems with physical and chemical characteristics comparable to those of the modeled ones.

Another approach is the capillary bundle model (CBM), in which the porous medium is supposed to have aligned and separate capillaries, as shown in Figure 15.2. The basic principle in CBMs is to consider the porous medium like a stack of capillaries. In the earliest capillary model, the porous medium was considered a stack of uniform radius capillaries. In CBMs, pressure balance amongst the capillaries is specified, and capillary forces have been used. This attribute keeps the concept totally different from the conventional models of tube bundles, in which fluids are independent of one another in various capillaries (Dong et al., 2005). CBM is limited to a maximum porosity of 78.5%, as the capillary diameters do not make it feasible to put the capillaries much closer to each other.

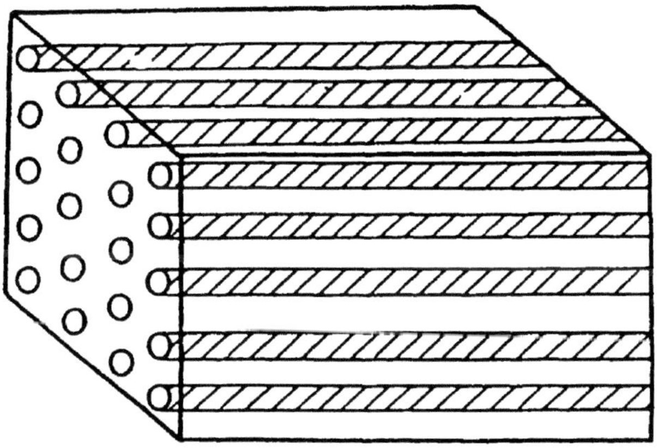

FIGURE 15.2 Capillary bundle model.

Based on the mentioned assumptions, the below relation has been developed, which shows the relationship between porosity, permeability, and pore radius.

$$k = \frac{\phi R^2}{8} \tag{15.7}$$

where R represents the pore radius, ϕ and k are the porosity and permeability, respectively.

In this model, the R value is considered an average pore radius when applied to the real cores. Another limitation of this model is that it considers unidirectional permeability, and capillaries are considered parallel without any orthogonal flow among the capillaries. However, real rocks show anisotropic behavior in terms of permeability.

The first revision made to the mentioned model was to reduce the permeability three times by putting one-third of the capillaries in each of the three spatial dimensions. Equation 15.7 will become

$$k = \frac{1}{3}\left(\frac{\phi R^2}{8}\right) \tag{15.8}$$

This model has been further modified by considering that the entire capillaries permitting flow in a specific direction are parallel to that direction but differ in pore diameter. The pore size distribution term αR has been added, and the modified equation becomes

$$k = \frac{\phi}{24}\int_0^\infty R^2 (R)\, dR \tag{15.9}$$

In another modification, the "shift factor" C has been added to the equation. It is a dimensionless quantity described in terms of the model's parameters, which demonstrates the influence of the deviations from the straight capillaries. It is related to tortuosity (T), which expresses diffusion and fluid flow in a porous medium.

$$k = \frac{\phi R^2}{24C} \tag{15.10}$$

Another model has been developed by Thanh et al. (2018) using the concept of the capillary tube model for the "streaming potential coupling coefficient (SPCC)" over saturated conditions, and the model was also validated using published results. SPCC is a parameter that controls the amount of coupling between the fluid flow and the electrical flow (Luong and Sprik, 2014).

All the capillary models do not show any agreement with the experimental results due to some key limitations discussed below.

- Tortuosity and complex geometry of the porous medium have been ignored in all the models.
- A unidirectional flow that does not show any interconnectivity among the pores.
- The assumption of a single radius neglects the cross-sectional area variation, which results in the contraction–expansion nature of non-Newtonian fluids in porous media.

An experimental study showed the capillary model's failure by studying the rheological characteristics of polymer solutions in porous media. The pressure drop was significantly higher than that indicated by capillary models, especially at the lower Carreau power index values (Duda et al., 1983).

15.4.3 DARCY'S LAW AND FLUID FLOW

In a broad range of science and engineering applications, fluid flow through porous media is considered very important. The fluid flow applications are based on an experimental law proposed by Henry Darcy in 1856. It is based on the experimental results conducted on water flow through sandbeds, which form the basis of hydrogeology, a branch of earth sciences. This law expresses a linear relationship between flow rate and pressure drop, as given below.

$$k = \frac{\mu Q L}{A \Delta P} \tag{15.11}$$

where k is the permeability, μ is the viscosity of a Newtonian fluid, Q is fluid flow, L is the sample length, A is the cross-sectional area, and ΔP is the pressure drop.

From the above expression, the permeability can be easily measured by conducting experiments on a rock sample where flow rates and pressure drops are measured.

Here, the term Q/A represents the Darcy velocity, which can be expressed as:

$$v = \frac{Q}{A} \tag{15.12}$$

For non-Newtonian fluids, eq 15.11 has been modified by expressing the macroscopic parameter (μ_a) as follows:

$$k = \frac{\mu_a Q L}{A \Delta P} \tag{15.13}$$

In the above equation, generally, the term ΔP will not be a linear function of flow rate. Moreover, the term μ_a refers to the polymeric fluid's observed macroscopic rheology in a porous medium. The factor μ_a (apparent viscosity) is used to illustrate the observed macroscopic rheology of the polymeric fluid in a porous medium.

In all the above-mentioned models, the complex topological nature of porous media has been ignored, which generally comprises an interconnected series of pores. Due to the complex geometry of porous matrix with different throat segments and capillary pores, an analytical approach sometimes does not work. Thus, different numerical models for Newtonian and non-Newtonian fluids have been developed (Sorbie et al., 1989).

15.5 PSEUDOPLASTIC FLUIDS FLOW IN POROUS MEDIA

This section covers the behavior of pseudoplastic (shear-thinning) fluids in porous media. Due to the economic conditions and unavailability (in most cases), the representative cores are difficult to obtain, test for various shear rates, and then correlate with the reservoir shear rates through the existing rheological models.

15.5.1 IN-SITU RHEOLOGY

The viscosity of polymers is strongly dependent on shear rate, which is typically measured with a rheometer. During the test, the polymer solution is subjected to various shear rates. Viscosity against each shear rate is measured. Figure 15.3 shows the relationship between viscosity and shear rate. This is called bulk viscosity.

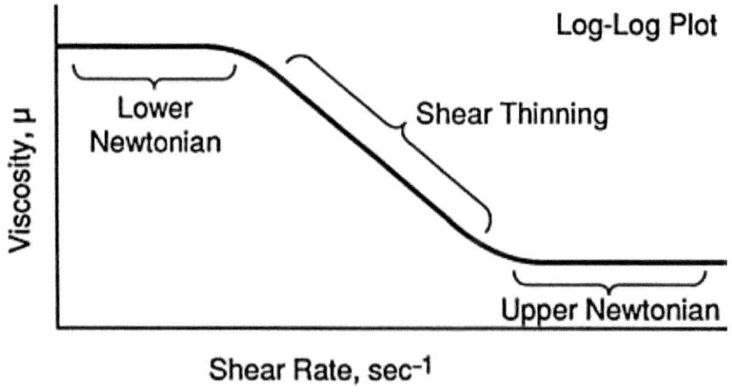

FIGURE 15.3 Rheology of pseudoplastic fluids.

In the case of polymers flowing through porous media, the behavior of the flow entirely changes due to the following reasons (Skauge et al., 2018):

- The complex geometry of porous media
- Mechanical degradation of polymers
- Tortuosity effect

The flow of both biopolymers and synthetic polymers exhibits a pseudoplastic behavior in viscometric flow, such as that experienced in a capillary (Green and Willhite, 1998). One of the simplest empirical models to explain the viscosity versus shear rate behavior of such fluids is the power law model.

$$\eta = K\gamma^{n-1} \tag{15.14}$$

where η is the viscosity, γ is the shear rate, and K and n are constants.

Let us take an example of xanthan: a rigid and viscous biopolymer generally shows Newtonian behavior throughout a spectrum of low shear rates. Thus, for a wider shear rate range, some other models can be used, such as the Carreau equation, which is considered more suitable for its flow through porous media (Chauveteau, 1982). This model is based on shear rheology, which has been shown to be appropriate for viscous biopolymers exhibiting shear-thinning behavior in bulk and porous media.

For in-situ rheology determination, the equations of power law and Carreau can still be applied because of XG shear-thinning behavior across a broad range of flow rates in a porous medium. However, the shear rate should be substituted with an "in-situ flow rate" or "equivalent" shear rate due to the variation of parameters in the simple models. On the other hand, HPAM, a synthetic viscoelastic polymer, reveals a different behavior at high flow rates as the apparent viscosity rises very high above a critical flow rate in porous media. This behavior is considered viscoelastic behavior (Green and Willhite, 1998; Sheng, 2010).

15.6 IN-SITU RHEOLOGY OF XANTHAN

From the published data about the in-situ rheology of xanthan, it has been reported that it has a purely pseudoplastic behavior, and in a few cases, a lower Newtonian plateau was seen. From literature data, it has been summarized that xanthan-based polymer solutions exhibit shear-thinning behavior (Ghoumrassi-Barr and Aliouche, 2016). The shear-thinning behavior can be portrayed across flow rates through a power law model. However, in some conditions, a Newtonian plateau may be obtained at relatively low flow rates.

Additionally, the flow behavior index (n) was either observed similar to or higher in some cases in porous media. Similarly, the model overestimated the in-situ viscosity based on purely bulk quantities (Sorbie, 2013).

15.7 IN-SITU RHEOLOGY OF VISCOELASTIC FLUIDS

Elastic behavior correlates with the "memory" of any material in that it tends to restore the material to its original configuration.

Generally, the polymers used for heavy oil recovery have a higher molecular weight that exhibits viscoelasticity and induces apparent viscosity thickening that cannot be estimated using the shear-thinning models discussed above. At high rates, the viscoelastic impacts were believed to reduce the mobility of synthetic polymer HPAM in porous media. Such effects were correlated with the Deborah number. It is a dimensionless quantity applied under particular flow conditions to describe the fluidity of materials. Viscoelastic effects have been stated to be predominant when the Deborah number is in the range between 0.1 and 10. Additionally, this number increases with increased flow rates. High rates are generally observed near the wellbore where injection of the polymer solution is taking place. Viscous polymer solutions show shear-thinning behavior with higher injectivity. Viscoelastic polymer solutions reveal thickening behavior after a critical flow rate in the porous medium (Yuan, 1981; Magbagbeola, 2008).

Thickening results in an increase in viscosity with reference to shear rate. Conversely, it shows shear thinning in the bulk shear field. Polymer solutions that reveal shear-thinning behavior in the shear field show thickening behavior in the extensional field. This was ascribed to the viscoelastic nature of HPAM (Barnes et al., 1989). The molecular weight of polymers with flexible and linear chain structures also has an important effect on viscoelasticity (Azad and Trivedi, 2019).

15.8 THEORETICAL APPROACH OF POLYMER RHEOLOGY

Owing to the complex geometry of porous media, the solution of existing models is a difficult task for both Newtonian and non-Newtonian fluids. Currently, some numerical solutions exist for non-Newtonian fluids that are so simple that they cannot be applied to porous media. If the fluid's type is selected, then both Newtonian and inelastic fluids can be explained through the existing models. However, there are still issues with the numerical solutions of such viscoelastic fluids. The problems multiply when the elastic effects powerfully influence the flows at higher flow rates. Computational fluid mechanics of non-Newtonian fluids can be used to calculate the flow rate and pressure drop behavior in some simple channels. These calculations can further be used to model porous media.

To analyze the in-situ rheology of polymer solutions, it is required to develop empirical relationships for different quantities associated with the flows. Various attempts have been made to examine the in-situ rheology both analytically and numerically. Numerous numerical studies were performed, but very limited success was achieved in relating the in-situ viscosity to the bulk viscosity by considering polymer and porous media characteristics. Such models were developed considering the non-Newtonian flow across capillary stacks, which simplified the complex geometry of porous media.

A pore network modeling approach has been suggested to avoid the simplification of porous media as discussed in capillary models. These models were applied to study non-Newtonian fluids that show shear-thinning behavior. In these models, porous media were visualized as integrated bundles with idealized geometries. The pore bodies are linked through small pore throats. Numerous researchers applied this approach and found agreement between analytical and experimental results (Lopez et al., 2003; Balhoff and Thompson, 2006; Balhoff and Thompson, 2006).

KEYWORDS

- **polymer rheology**
- **porous media**
- **enhanced oil recovery**
- **non-Newtonian fluids**
- **pore network modeling**

REFERENCES

Al-Shakry, B.; et al. Impact of Mechanical Degradation on Polymer Injectivity in Porous Media. *Polymers* **2018,** *10* (7), 742.

Azad, M. S.; Trivedi, J. J. Novel Viscoelastic Model for Predicting the Synthetic Polymer's Viscoelastic Behavior in Porous Media Using Direct Extensional Rheological Measurements. *Fuel* **2019,** *235*, 218–226.

Balhoff, M. T.; Thompson, K. E. A Macroscopic Model for Shear-Thinning Flow in Packed Beds Based on Network Modeling. *Chem. Eng. Sci.* **2006,** *61* (2), 698–719.

Barnes, H. A.; Hutton, J. F.; Walters, K. *An Introduction to Rheology*, vol. 3; Elsevier, 1989.

Battiato, I.; et al. Theory and Applications of Macroscale Models in Porous Media. *Transp. Porous Media* **2019,** *130* (1), 5–76.

Chauveteau, G. Rodlike Polymer Solution Flow Through Fine Pores: Influence of Pore Size on Rheological Behavior. *J. Rheol.* **1982,** *26* (2), 111–142.

Choi, B.; Choi, J.; Lee, K. S. Integrated Analysis of Permeability Reduction Caused by Polymer Retention for Better Understanding Polymer Transport. *J. Chem.* **2015,** *2015*, 291905.

Desbois, G.; et al. High-Resolution 3D Fabric and Porosity Model in a Tight Gas Sandstone Reservoir: A New Approach to Investigate Microstructures from mm-to nm-Scale Combining Argon Beam Cross-Sectioning and SEM Imaging. *J. Pet. Sci. Eng.* **2011,** *78* (2), 243–257.

Dong, M.; et al. Immiscible Displacement in the Interacting Capillary Bundle Model Part I. Development of Interacting Capillary Bundle Model. *Transp. Porous Media* **2005,** *59* (1), 1–18.

Duda, J. L.; Hong, S. A. Klaus, E. E. Flow of Polymer Solutions in Porous Media: Inadequacy of the Capillary Model. *Ind. Eng. Chem. Fundam.* **1983,** *22* (3), 299–305.

Dullien, F. A. L.; Dhawan, G. K. Characterization of Pore Structure by a Combination of Quantitative Photomicrography and Mercury Porosimetry. *J. Colloid Interface Sci.* **1974,** *47* (2), 337–349.

Ghoumrassi-Barr, S.; Aliouche, D. A Rheological Study of Xanthan Polymer for Enhanced Oil Recovery. *J. Macromol. Sci. B* **2016,** *55* (8), 793–809.

Giesche, H. Mercury Porosimetry: A General (Practical) Overview. *Part. Part. Syst. Charact.* **2006,** *23* (1), 9–19.

Glaser, J. A. Biological Degradation of Polymers in the Environment. In *Plastics in the Environment.*IntechOpen, 2019.

Goonoo, N.; Bhaw-Luximon, A.; Jhurry, D. Biodegradable Polymer Blends: Miscibility, Physicochemical Properties and Biological Response of Scaffolds. *Polym. Int.* **2015,** *64* (10), 1289–1302.

Green, D. W.; Willhite, G. P. *Enhanced Oil Recovery*, vol. 6; Henry L. Doherty Memorial Fund of AIME, Society of Petroleum Engineers, 1998.

Huang, P.; et al. *Prediction of HPAM permeability reduction factor.* in *Chemical Flooding Symposium—Research Results During the Eighth Five-Year Period (1991À1995).* 1998.

Huh, C.; Lange, E. A.; Cannella, W. J. Polymer Retention in Porous Media. In *SPE/DOE Enhanced Oil Recovery Symposium.* Society of Petroleum Engineers: Tulsa, Oklahoma, 1990; p 20.

Kourde-Hanafi, Y.; et al. Influence of PVP Content on Degradation of PES/PVP Membranes: Insights from Characterization of Membranes with Controlled Composition. *J. Membr. Sci.* **2017,** *533,* 261–269.

Lai, J.; et al. A Review on Pore Structure Characterization in Tight Sandstones. *Earth Sci. Rev.* **2018,** *177,* 436–457.

Lopez, X.; Valvatne, P. H.; Blunt, M. J. Predictive Network Modeling of Single-Phase Non-Newtonian Flow in Porous Media. *J. Colloid Interface Sci.* **2003,** *264* (1), 256–265.

Luong, D. T.; Sprik, R. Examination of a Theoretical Model of Streaming Potential Coupling Coefficient. *Int. J. Geophys.* **2014,** *2014*, 471819.

Magbagbeola, O. A. Quantification of the Viscoelastic Behavior of High Molecular Weight Polymers Used for Chemical Enhanced Oil Recovery. 2008.

Maya, E.; et al. Chemical Modification of Copolyimides with Bulky Pendent Groups: Effect of Modification on Solubility and Thermal Stability. *Polym. Degrad. Stabil.* **2007**, *92* (12), 2294–99.

Michael, F. M.; et al.; 2 - Surface Modification Techniques of Biodegradable and Biocompatible Polymers. In *Biodegradable and Biocompatible Polymer Composites*; Shimpi, N. G., Eds.; Woodhead Publishing, 2018; pp 33–54.

Pang, Z.; et al. In *Effect of Polymer Molecular Weight on Residual Permeability Reduction Factor,* Proceedings of the Symposium on Chemical Flooding Research Results During the Eighth Five-Year Period (1991–1995), 1998.

Sazali, N.; et al. In *Degradation and Stability of Polymer: A Mini Review*, IOP Conference Series: Materials Science and Engineering; IOP Publishing, 2020.

Seright, R.; Maerker, J.; Holzwarth, G. Mechanical Degradation of Polyacrylamides Induced by Flow Through Porous-Media. In *Abstracts of Papers of the American Chemical Society*. Amer Chemical Soc 1155 16th St, NW: Washington, DC, 1981; 20036.

Sheng, J. J. *Modern Chemical Enhanced Oil Recovery: Theory and Practice*. Gulf Professional Publishing, 2010.

Sheng, J. J. Polymer Flooding—Fundamentals and Field Cases. In *Enhanced Oil Recovery Field Case Studies*. Elsevier, 2013; pp 63–82.

Skauge, A.; et al. *Polymer Flow in Porous Media*, 2018.

Sochi, T.; Blunt, M. J. Pore-Scale Network Modeling of Ellis and Herschel–Bulkley Fluids. *J. Pet. Sci. Eng.* **2008,** *60* (2), 105–124.

Sorbie, K. *Polymer Improved Oil Recovery.* Blackie and Son Ltd: Glasgow and London, 1991.

Sorbie, K. S. *Polymer-Improved Oil Recovery*. Springer Science & Business Media, 2013.

Sorbie, K. S.; Clifford, P. J.; Jones, E. R. W. The Rheology of Pseudoplastic Fluids in Porous Media Using Network Modeling. *J. Colloid Interface Sci.* **1989,** *130* (2), 508–534.

Sorbie, K. *Polymer-Improved Oil Recovery, 115 Glasgow.* Blackie & Son: Scotland, 1991; pp 126–163.

Thanh, L. D.; et al. A Fractal Model for Streaming Potential Coefficient in Porous Media. *Geophys. Prospect.* **2018,** *66* (4), 753–766.

Wei, Q.; Zhou, H.; Yang, S. Non-Darcy Flow Models in Porous Media via Atangana-Baleanu Derivative. *Chaos Solitons Fractals* **2020,** *141,* 110335.

Yuan, M.-K. A Rheological Study of Polymer and Microemulsion in Porous Media. 1981.

Sorbie, K. *Polymer-Improved Oil Recovery*, 115 Glasgow. Blackie & Son: Scotland, 1991; pp 126–163.

CHAPTER 16

Reservoir Simulation of Chemical EOR

KRISHNA RAGHAV CHATURVEDI, AYUSH JOSHI, ANSHI JAIN, and TUSHAR SHARMA

Department of Petroleum Engineering and Geological Sciences, Rajiv Gandhi Institute of Petroleum Technology, Jais, Amethi, India

ABSTRACT

Reservoir simulation is a comparatively inexpensive solution which allows researchers to work at a field scale and simulate the field relevant conditions, within a lab. Reservoir simulation helps in the development of new fields by estimating the number of wells required, the optimal well completion, the present and future requirement for artificial lifts along with other surface/subsurface production facilities, and the expected oil, water, and gas production rates from the field as well as it may also be utilized for mature fields to determine the most effective enhanced oil recovery (EOR) method. Given that chemical EOR is an expensive undertaking, reservoir simulation and modeling becomes crucial for ensuring that planned chemical EOR pilots are successful, and that no reservoir parameter compromises the viability of oil production

16.1 INTRODUCTION

The process of developing and operating a model that closely resembles the behavior of a real reservoir is referred to as simulation of petroleum reservoir performance. A model might be physical (e.g., a laboratory sandpack) or mathematical (Kechut et al., 2011; Yang et al., 2017). A mathematical model is a collection of equations that, when certain assumptions are made, describe

Advancements in Chemical Enhanced Oil Recovery.
Tushar Sharma, Krishna Raghav Chaturvedi, Tarek Ganat, and Imtiaz Ali (Eds.)

the physical processes that occur in a reservoir. While the model is devoid of a reservoir, its behavior simulates and assumes the look of the real thing (Jerauld et al., 2008). In general, simulation is both a theoretical or physical representation of an operation and a replication of its systems or processes in real life. Reservoir simulation is a branch of petroleum engineering that uses computer modeling of porous media to approximate the dynamics of fluids. Its objective is to forecast field performance under a variety of different manufacturing procedures. Reservoir simulation is based on well-known engineering equations; reservoir engineering was previously computed using fundamental mathematical models. Although reservoir modeling is not a new concept in the industry, it has become more efficient than ever due to the improved capabilities of modern technology. Due to their competence, efficiency, and efficacy, several engineers were competent in the concept and its growth. Simulating a field's performance (e.g., oil recovery) under one or more production plans is the objective of simulation. While a field can only be created once at a high cost, a model may be built or run numerous times over a short period of time at a low cost. Observing model results that represent a variety of potential producing scenarios aids in finding the reservoir's optimal set of producing conditions (Ding et al., 2022).

The techniques for reservoir simulation range from intuitive and judgmental engineering to intricate mathematical models that require the use of digital computers (Rios et al., 2020). The choice is not whether to replicate or not, but which instrument or method to use. By giving property values and characterizing the reservoir's average time, pressure, porosity, and other characteristics under a solution gas drive, material balance equation (MBE) techniques are used to assess if a reservoir is homogeneous and to forecast its production behavior. The tank model, sometimes referred to as "the foundation of reservoir simulators," is a depiction of a reservoir model. It is a zero-dimensional model, as is the case with the oil MBE equation, since the difference between the initial and final oil entering the reservoir equals cumulative net withdrawal. By assuming no oil enters the reservoir, the tank model determines the reservoir's average values for all of its properties. Due to lithological variation, the tank model cannot well represent and average out sandbars as a whole, but it can adequately represent and average out their constituents. Due to the complexity introduced by fluid expansion, the MBE equation and Darcy's law are used to describe each component's behavior in this example (Hao et al., 2008; Kök and Bal, 2019). Because this one-of-a-kind model includes several components in one direction and just one component in the other two dimensions, it is a one-dimensional simulator. Other models have been increased and changed to account for the fact that lithologies, such as two- and three-dimensional simulators, differ in several dimensions. Additionally, before picking a simulator, a detailed understanding of the reservoir's characteristics, as well as extensive data collection and testing, are required (Gyan et al., 2019; Liu et al., 2014). On the other hand, the two-dimensional simulator is the most often used; it supports three typical geometries: horizontal (x-y), vertical (x-z), and radial (r-z). Models have been classified into a variety of categories according to their kind, including:

- Black-oil
- Compositional
- Thermal
- Generalized
- IMPES (implicit, sequential, and adaptive implicit)
- Single-porosity or dual-porosity

These classifications offer a confusing context for discussing models; some refer to the application (e.g., thermal), others to the model formulation (e.g., implicit), and still others to a reservoir formation component (e.g., dual-porosity). While there are exceptions, the historical trend has been and continues to be toward the generalized model, which includes all of the above categories as well as others (Li et al., 2019). It is the broadest model, which incorporates the vast majority of models now in use or in development. Corporations in the oil and gas industry use reservoir simulation models to help in the development of new fields. Models are frequently used in developed regions to aid with investment choices that need production projections. Due to the time and effort required to develop and maintain a trustworthy and accurate model of an area, models are frequently generated only when important investment decisions are at risk. The improvement of simulation software is a continuous process, which has reduced the time necessary to construct a model (Fig. 16.1). Furthermore, rather than high-end workstations, models may be run on low-cost personal computers (Brady and Krumhansl, 2012; Sun et al., 2022).

16.2 USES OF RESERVOIR SIMULATION

Models may assist in the development of new fields by estimating the number of wells required, the optimal well completion, the present and future requirement for artificial lift, and the expected oil, water, and gas production (Sun et al., 2022). By combining hydraulic fracturing and ongoing reservoir management, models can assist in maximizing oil recovery. Additionally, wells that are highly divergent or horizontal can be shown. Hydraulic fracturing may be produced using specialized software, and the associated productivity gains can be included in the field model. Reinjecting generated gas or injecting water into an aquifer can also be used to research future advances in oil recovery via pressure control. Floods have a significant effect on oil displacement, which is intensively explored using reservoir modeling. The field must possess the necessary characteristics to efficiently implement enhanced oil recovery (EOR) processes. This evaluation can be aided by model studies. EOR techniques include miscible displacement using natural gas, CO_2, or nitrogen, as well as chemical inundation (polymer, alkaline, surfactant, or a combination of these).

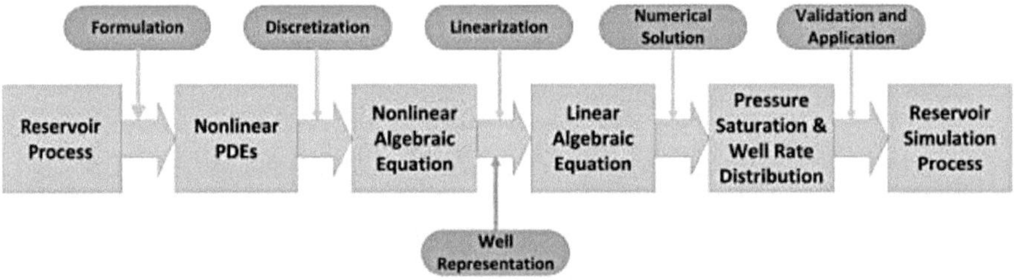

FIGURE 16.1 Typical flow path observed during a reservoir simulation job.

Source: Reprinted with permission from Deb et al., 2017. Copyright 2017. Elsevier.

To model these processes, simulation software must possess certain characteristics. In some miscible applications, "smearing" of the floodfront, also known as numerical dispersion, may be a concern. It is usual to employ reservoir modeling to explore the possibility of increasing oil output from heavy oil sources. Oil recovery is boosted by reducing the viscosity of the oil using steam or hot water. Two typical methods are steam soaks (injecting steam and then producing oil from the same well) and steam floods (separate steam injectors and oil producers). These processes need the use of exact property simulators to account for heat transmission to existing fluids and the formation, as well as future property changes and heat losses external to the formation (Khan et al., 2016; Pal and Mandal, 2020). The reservoir modeling tool is the outcome of an unprecedented level of collaboration spanning geosciences, reservoir and production engineering, petrophysics, computer science, and economics. Reservoir modeling is a technique that is frequently used in the oil and gas industry to extend and create new technologies, recoveries, and well-site infill. A detailed overview of the reservoir simulation process has been provided as follows:

- **Geophysical model**

  The geologic model (geocellular model) is created to represent the reservoir numerically or mathematically. It analyzes seismic well structure data, petrophysical data from wells with documented depositional parameters, and a three-dimensional volume model (porosity, permeability, and water saturation). It is composed of around 50 million cells distributed in a wide complex reservoir, allowing the industry to simulate an equal number of cells of this size in a reservoir simulator.

- **Gridding**

  The geological model is comprehensive, with lithology, facies, major faults, and pinchouts all included. As a consequence, the model is vertically and spatially upscaled, allowing for the development of novel methodologies for predicting transmissibilities, such as local flow-based methods and global flow-based methods. Using generic fault information and projected transmissibilities, upscaled models of faults and pinchouts were built. Faults have become an increasingly significant component of reservoir data and modeling.

- **Schematic diagrams**

  Cartesian or corner-point structured grids are employed in complex geometry reservoir modeling research. The grid-based 3D system enables a more detailed assessment of the placement of infill wells and water/gas estimates.

- **Wells that are not traditional**

  In the petroleum industry, unconventional wells are used to boost production rates and recoveries. Numerous strategies are being employed by the company to replicate unconventional wells and assist the well in maintaining a steady inflow.

- **Consolidation of several reservoirs and reservoir models**

  By linking several wells from disparate fields (e.g., subsurface and surface fields) and modeling their integration, reservoir management may be enhanced, leading to increased recovery and development.

- **Using larger models**

  The maximum model size has increased in response to the phenomenal growth of the computer hardware sector. As a result, the simulation industry underwent a sea change.

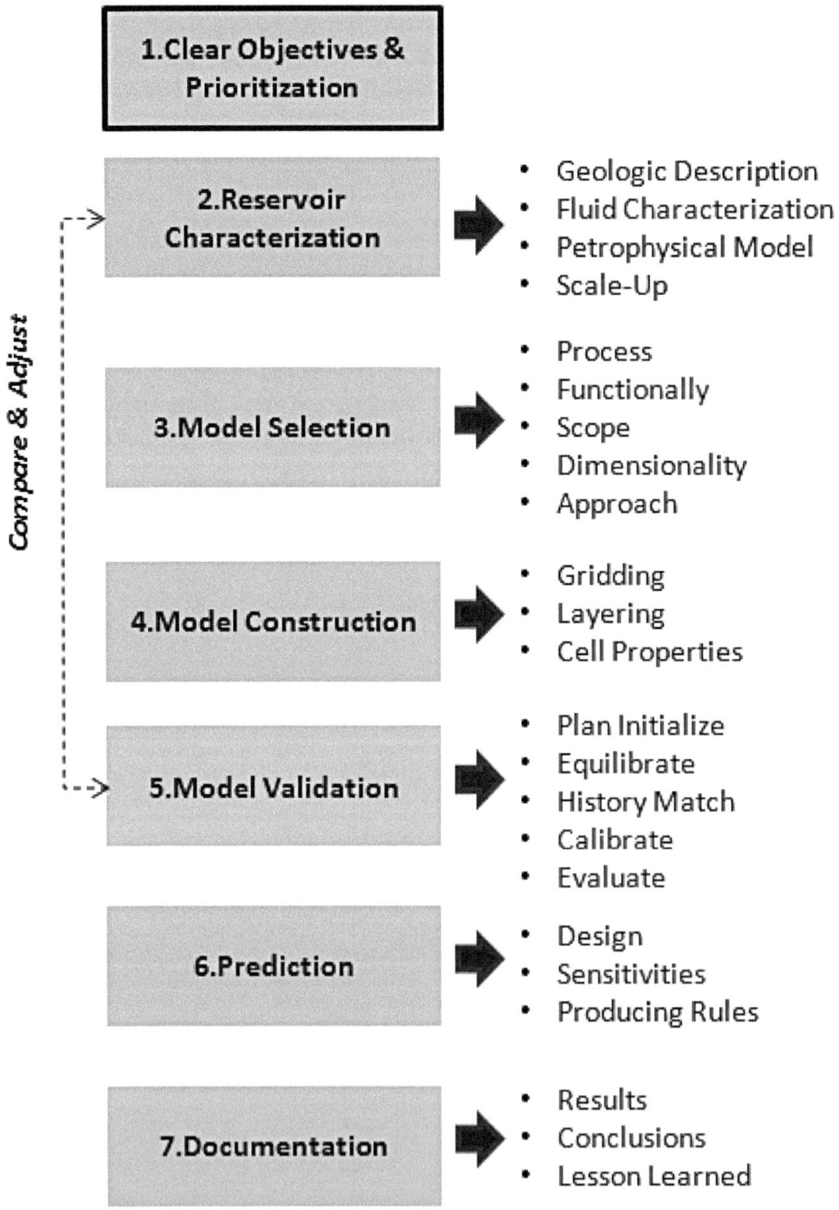

FIGURE 16.2 Typical overview of the steps involved in a reservoir simulation study. Image available on Oil and Gas portal ("Reservoir Simulation Fundamentals," n.d.).

- **Subsurface modeling**

  A structural model built using subsurface modeling accurately represents complex three-dimensional structures. Merging new datasets with existing reservoir and geological data is possible. This application provides operators with an understanding of the subsurface, which aids in production planning, reserve estimation, and well location.

- **Geomechanics**

  "Geomechanics technologies employ novel methods for designing multistage fracturing and forecasting future production from unconventional resources." In reservoir modeling, rock mechanics are connected to fluid flows via two variables: medium permeability change and porosity. By minimizing high-hazard situations, software alleviates rock mechanical worries and hence improves well planning.

- **Fracturing hydraulic**

  A structured grid was constructed to imitate the effects of production over time in a range of hydraulically fractured wells. The proper software in reservoir modeling offers a clear image of the distribution and geometry of complicated hydraulic fractures, allowing for the selection of the most effective sites and knowledge of fracture solutions.

- **Characterization and analysis of fluids**

  Engineers may use accurate PVT modeling and analysis to swiftly characterize fluid phases; it has a fast computation speed, a variety of modeling equations, multiple data-gathering methodologies, high-quality lines at predetermined intervals, and periods to compute fluid parameters.

- **Conducting historical data analysis and forecasting production**

  The process of generating reservoir simulation models is preserved and utilized to replicate the performance of a field under similar prior system circumstances. The computed-to-stated performance ratio allows for a more exact assessment of the steps required to reduce the error percentage and improve future estimations.

16.3 COMMERCIAL SOFTWARE USED FOR RESERVOIR SIMULATION

CMG: CMG offers the IMEX black oil simulator, GEM composition simulator, and STARS thermal composition simulator. CMG-STARS is a powerful process reservoir simulator that can simulate polymer flooding, surfactant flooding, steam flooding, and in-situ combustion (Maurya et al., 2017; Trivedi and Babadagli, 2005). CMG-STARS operates in two modes: fully implicit and adaptive implicit. Following that, the distinct properties of polymer and surfactant flooding will be investigated. The viscosities of polymer flooding, permeability reduction, adsorption, and rheology will be compared to those of UTCHEM. There is no microemulsion (ME) phase since the surfactant model is based on Type I (water and oil). The reduction in IFT may be shown as a function of surfactant concentration (Chaturvedi et al., 2021).

Supercomputer clusters can run IMEX, GEM, and STARS, as well as the Rock Flow Dynamics Navigator.

ECLIPSE: This simulator was created by Schlumberger Exploration Consultants Limited and is now owned by GeoQuest. It uses corner-point grids to solve the fluid model equations (black oil). ECLIPSE is composed of two modules: ECLIPSE100, which represents the black oil model, and ECLIPSE300, which represents the compositional equation of state, which will be eliminated from this study. ECLIPSE100 is a three-phase, three-dimensional general-purpose black oil simulator that can simulate a variety of chemical EOR processes such as polymer and surfactant flooding (Foroozesh et al., 2016; Sadeed et al., 2018). ECLIPSE300, on the other hand, operates in completely implicit, IMPES, and adaptive implicit (AIM) modes, whereas ECLIPSE100 simulates the effect of changing a polymer's salt content on its viscosity. In a completely mixed polymer solution, ECLIPSE100 evaluates viscosity as a function of polymer concentration. The influence of temperature on polymer viscosity, on the other hand, has not been examined. This study will compare the features of polymer flooding to those of UTCHEM, including viscosity, permeability reduction, adsorption, and rheology. The surfactant model, which is based on Type I surfactants, does not include a distinct ME phase. Surfactant viscosity, on the other hand, is demonstrated by varying water viscosity in proportion to ME viscosity (Chaturvedi and Sharma, 2021).

The MATLAB Reservoir Simulation Toolbox (MRST) introduces novel reservoir modeling ideas and methodologies for unstructured grids. These strategies and ideas apply to both large and complex systems.

Reveal (Petroleum Experts, 2012) is a full-field reservoir simulator that can simulate surfactant phase behavior and mobility control via polymer or gel choices. The surfactant module, like the one in UTCHEM, enables the construction of different phase behaviors in response to salt (Type I, Type II, and Type III). Using near-wellbore shear-thinning behavior, Reveal models polymer and polymer-gel dynamics. Reveal allows you to reduce permeability, increase the volume of inaccessible holes, gel the polymer, add a cross-linker, and degrade. It also has a foam model for enhancing gas-phase viscosity, which is very useful in heavy oil reservoirs ("REVEAL," n.d.).

VIP, Landmark's reservoir modeling package, allows thermal injection of hot water and steam, as well as polymer flooding in the black oil model (Lager et al., 2008). It is a numerical simulation method for anticipating pressure destitution over time and calculating saturation poverty. It is extremely beneficial for educational purposes.

The Open Porous Media (OPM) program seeks to develop methods and tools for modeling fluid flow in porous media (Rasmussen et al., 2021).

UTCHEM is a three-dimensional multiphase, multicomponent chemical composition simulator that can simulate a variety of chemical EOR processes (Khalilinezhad et al., 2019; Sharma and Mohanty, 2018). Complex phase behavior and chemical interactions may be accounted for by the simulator. The simulator provides improved numerical precision and dispersion control methods, as well as the ability to build up to four phases (gas, aqueous, oleic, and ME). A ME is a combination of water, oil, surfactant, and cosurfactant that may form a single thermodynamically stable phase under particular temperature, pressure, and salinity conditions. It can simulate, among other things, wettability, capillary pressures, up to four-phase relative permeabilities as a function of trapping number, full tensor dispersion, molecular diffusion, adsorption, chemical reactions that occur during high pH alkaline flooding, and nonequilibrium mass transfer between phases.

16.4 RESERVOIR SIMULATION OF CHEMICAL EOR

In recent years, several studies have explored the use of reservoir simulation techniques for the design, development, and testing of chemical EOR methods. A highly detailed and well-investigated study that listed the reservoir simulators as per their performance in the simulation of chemical EOR was reported by Goudarzi et al. (2016). Traditional finite difference simulators dominate both conceptually and practically in reservoir simulation. The Darcy approximation is based on three fundamental concepts in classical fluid dynamics (FD) modeling: mass conservation, isothermal fluid phase behavior, and the Darcy approximation for fluid flow through porous media. Thermal simulators (which are widely used in heavy crude oil applications) add to the list by allowing reservoir temperatures to change.

The following numerical algorithms and methodologies are used by modern simulators:

- The majority of modern FD simulation systems can generate three-dimensional representations suitable for use in full-field or single-well models.
- Numerous conceptual models, such as cross-sections and two-dimensional radial grid models, are approximated in two dimensions.
- In principle, finite difference models provide an accurate representation of reservoir geometry through the use of organized and more complex unstructured grids. Numerous models take advantage of local grid changes to accurately reflect the effects of multiphase flow near the wellbore. This "fine meshing" of wellbores is crucial for resolving reservoir issues like water and gas coning. Apart from the normal neighbor-to-neighbor connections, these models must anticipate the transmissibilities of intercell flow for nonadjacent layers.
- Natural fracture simulation (also known as dual porosity and dual permeability) is a complex feature that simulates hydrocarbons trapped in dense matrix blocks. There is movement between the dense matrix blocks and their surrounding porous fracture networks, as well as between the blocks and the wells. The black oil model takes the development of the field's hydrocarbon content into account. Compositional modeling is a more sophisticated approach in which the PVT characteristics of oil and gas phases are fitted to a state equation using a collection of components (EOS). After that, the simulator uses the fitted EOS equation to track phase and component movement in real time in the field.

Conventionally, simulation models estimate the saturation change of three phases (oil, water, and gas) and the pressure of each phase in each cell at each time step. Gas is liberated from oil when pressure lowers, as reservoir depletion studies reveal. When pressures are increased as a result of water or gas injection, the gas reemerges in the oil phase. A developed field modeling project typically involves "history matching," in which estimated values are compared to historical field output and pressures. Field by field, the model parameters are changed until a satisfactory match is reached, which is generally the case for all wells. Water conservation, water-oil ratios, and gas-oil ratios are all connected in general (Kohata et al., 2017; Yuan et al., 2019).

A successful pilot is important to the EOR process; it is the stage during which laboratory solutions are assessed on a larger scale, and it is frequently a make-or-break moment (Alcorn et al., 2019; Kohata et al., 2017).

The development of a successful pilot program begins in the laboratory, where field variables like water composition and reservoir temperature necessitate formulation tweaks (Moreau et al., 2010). However, the reservoir's geology and architecture must also be addressed. Effective execution of an EOR chemical project needs the use of reservoir characterization techniques aimed at identifying the reservoir's and fluid's reactivity to the injection of any sort of chemical. Extrapolating core flood data to the reservoir scale usually produces inaccurate results (e.g., when calculating average fluid saturations).

Geologists and reservoir engineers must comprehend the reservoir's flow obstacles and critical conduits. A pilot zone representative of the field's characteristics will be established and, to the extent possible, free of identifying features like faults or fractures that might skew the findings. While this is not always possible, a pilot zone should have a production and injection history; this sets a baseline and provides insight into the reservoir's FD. Tracers are occasionally used to assist in comprehending such movements.

Reservoir modeling should be used to develop the injection scenario and analyze the reservoir reaction. A static and dynamic three-dimensional model is created using geological, geophysical, and production data. Chemical EOR techniques often need a reevaluation of novel rock types with different chemical reactions that would be missed by conventional recovery, as well as an assessment of the sweeping mechanisms of the chemical slugs injected into the reservoir heterogeneous media. By initially controlling the reservoir model and fluid behavior, the EOR project's cornerstone is formed by combining historical production data using history-matching techniques. The whole reservoir characterization process is overseen by a multidisciplinary team of geologists, petrophysicists, chemists, and reservoir engineers. An expansive study on the same topic has been published by Douarche and coworkers and is strongly recommended for anyone following chemical EOR reservoir simulation (Douarche et al., 2012). A rock-type definition (Easytrace) and a geocellular model were created using the Openflow/Cobraflow package (these softwares are part of the Eclipse package discussed earlier). The simulations can be conducted mostly using Pumaflow, a multipurpose reservoir simulator. It includes a number of EOR alternatives for chemical enhanced recovery, such as adsorption, IFT modification, ion exchange, and polymer combinations in a single or dual medium (Ayache et al., 2018; Moreau et al., 2010). Coalbed methane (CBM) production modeling is a recent use of reservoir simulation. This program requires the use of a specially constructed CBM emulator (Yan et al., 2019). In addition to normal fractured (fissured) formation data, CBM modeling incorporates gas content data at beginning pressure, sorption isotherms, diffusion coefficients, and parameters to estimate changes in absolute permeability owing to pore-pressure depletion and gas desorption. Once in the field, it is vital to continuously monitor the injection and output rates of the well. Water samples must be obtained regularly to check for surfactant and/or polymer breakthroughs, and observation wells are occasionally employed. Field data must be compared to anticipated reservoir models, and any variations from the forecast must be addressed quickly to allow remedial actions to be implemented. Later in the field, the pilot's history will be matched to enable additional discoveries and the modeling of other flood propagation scenarios (Kechut et al., 2011; Mishra and Pandey, 2017; Sinan et al., 2020).

16.5 CONCLUSION

Chemical EOR is an exceedingly expensive endeavor, needing millions of dollars in laboratory research and development before the first field application phase. Chemical EOR has very little margin for error in producing well since poorly implemented EOR technologies have shutoff pores, resulting in significant production loss, even by a large margin. Thus, reservoir modeling is crucial for ensuring that scheduled field tests are successful and that no reservoir parameter compromises the viability of oil production.

KEYWORDS

- **reservoir simulation**
- **partial differential equation**
- **material balance equation**
- **black-oil**
- **fluid dynamics**

REFERENCES

Alcorn, Z. P.; Fredriksen, S. B.; Sharma, M.; Rognmo, A. U.; Føyen, T. L.; Fernø, M. A.; Graue, A. An Integrated Carbon-Dioxide-Foam Enhanced-Oil-Recovery Pilot Program with Combined Carbon Capture, Utilization, and Storage in an Onshore Texas Heterogeneous Carbonate Field. *SPE Reserv. Eval. Eng.* **2019,** *22,* 1449–1466.

Ayache, S. V.; Lamoureux-Var, V.; Michel, P.; Preux, C.; Reservoir Simulation of Hydrogen Sulfide Production During a Steam-Assisted-Gravity-Drainage Process by Use of a New Sulfur-Based Compositional Kinetic Model. *SPE J.* **2018,** *22,* 080–093.

Brady, P. V.; Krumhansl, J. L.A Surface Complexation Model of Oil-Brine-Sandstone Interfaces at 100°C: Low Salinity Waterflooding. *J. Pet. Sci. Eng.* **2012,** *81,* 171–176.

Chaturvedi, K. R.; Narukulla, R.; Amani, M.; Sharma, T. Experimental Investigations to Evaluate Surfactant Role on Absorption Capacity of Nanofluid for CO_2 Utilization in Sustainable Crude Mobilization. *Energy* **2021,** *225,* 120321.

Chaturvedi, K. R.; Sharma, T. Rheological Analysis and EOR Potential of Surfactant Treated Single-Step Silica Nanofluid at High Temperature and Salinity. *J. Pet. Sci. Eng.* **2021,** *196,* 107704.

Deb, P. K.; Akter, F.; Imtiaz, S. A.; Hossain, M. E. .Nonlinearity and Solution Techniques in Reservoir Simulation: A Review. *J. Nat. Gas Sci. Eng.* **2017,** *46,* 845–864.

Ding, B.; Dong, M.; Chen, Z.; Kantzas, A. Enhanced Oil Recovery by Emulsion Injection in Heterogeneous Heavy Oil Reservoirs: Experiments, Modeling and Reservoir Simulation. *J. Pet. Sci. Eng.* **2022,** *209,* 109882.

Douarche, F.; Rousseau, D.; Bazin, B.; Tabary, R.; Moreau, P.; Morvan, M. Modeling Chemical EOR Processes: Some Illustrations from Lab to Reservoir Scale. *Oil Gas Sci. Technol. – Rev. d'IFP Energies Nouv.* **2012,** *67,* 983–997.

Foroozesh, J.; Jamiolahmady, M.; Sohrabi, M. Mathematical Modeling of Carbonated Water Injection for EOR and CO_2 Storage with a Focus on Mass Transfer Kinetics. *Fuel* **2016,** *174,* 325–332.

Goudarzi, A.; Delshad, M.; Sepehrnoori, K. A Chemical EOR Benchmark Study of Different Reservoir Simulators. *Comput. Geosci.* **2016,** *94,* 96–109.

Gyan, P. S.; Xie, C.; Brantson, E. T.; Atuahene, S. Computer Modeling and Simulation for Undersaturated Primary Drive Recovery Mechanism: *Adv. Mech. Eng.* **2019,** *11,* 1–18.

Hao, W.; Wang, J.; Fan, S.; Hao, W. Evaluation and Analysis Method for Natural Gas Hydrate Storage and Transportation Processes. *Energy Convers. Manag.* **2008,** *49,* 2546–2553.

Jerauld, G. R.; Lin, C. Y.; Webb, K. J.; Seccombe, J. C. Modeling Low-Salinity Waterflooding. *SPE Reserv. Eval. Eng.* **2008,** *11,* 1000–1012.

Kechut, N. I.; Sohrabi, M.; Jamiolahmady, M. In *Experimental and Numerical Evaluation of Carbonated Water Injection (CWI) for Improved Oil Recovery and CO_2 Storage*, 73rd European Association of Geoscientists and Engineers Conference and Exhibition 2011: Unconventional Resources and the Role of Technology. Incorporating SPE EUROPEC 2011. Society of Petroleum Engineers, 2011; pp 2831–2842.

Khalilinezhad, S. S.; Hashemi, A.; Mobaraki, S.; Zakavi, M.; Jarrahian, K. Experimental Analysis and Numerical Modeling of Polymer Flooding in Heavy Oil Recovery Enhancement: A Pore-Level Investigation. Arab. *J. Sci. Eng.* **2019,** *44,* 10447–10465.

Khan, M. Y.; Kohata, A.; Patel, H.; Syed, F. I.; Al Sowaidi, A. K. In *Water Alternating Gas WAG Optimization Using Tapered WAG Technique for a Giant Offshore Middle East Oil Field*, Society of Petroleum Engineers—Abu Dhabi International Petroleum Exhibition and Conference, 2016. Society of Petroleum Engineers, 2016.

Kohata, A.; Willingham, T.; Chen, P.; Al Sowaidi, A. *Evaluation of Miscible Water Alternating Gas WAG Injection by Comparing with Lesson Learnt from an Immiscible Gas Injection Pilot*, Society of Petroleum Engineers—SPE Abu Dhabi International Petroleum Exhibition and Conference, 2017. Society of Petroleum Engineers, 2017.

Kök, M. V.; Bal, B. Effects of Silica Nanoparticles on the Performance of Water-Based Drilling Fluids. *J. Pet. Sci. Eng.* **2019,** *180,* 605–614.

Lager, A.; Webb, K. J.; Collins, I. R.; Richmond, D. M. LoSal™ Enhanced Oil Recovery: Evidence of Enhanced Oil Recovery at the Reservoir Scale. *Proc. - SPE Symp. Improv. Oil Recover.* **2008,** *3,* 1313–1324.

Li, C.; Yang, J.; Ye, J.; Zhou, J.; Zhang, R.; Feng, Q. Semi-Analytical Modeling of Advanced Water Injection in Low Permeability Reservoirs Under Non-Darcy-Flow Condition. *Adv. Mech. Eng.* **2019,** *11*. DOI: 10.1177/1687814019846767

Liu, T.; Diwu, P.; Liu, R.; Jiang, L.; Jiang, B. Fast Algorithm of Numerical Solutions for Strong Nonlinear Partial Differential Equations. *Adv. Mech. Eng.* **2014,** *2014*. DOI: 10.1155/2014/936490

Maurya, N. K.; Kushwaha, P.; Mandal, A. Studies on Interfacial and Rheological Properties of Water Soluble Polymer Grafted Nanoparticle for Application in Enhanced Oil Recovery. *J. Taiwan Inst. Chem. Eng.* **2017,** *70,* 319–330.

Mishra, S.; Pandey, A. Modelling the Impact of Reservoir CO_2 on Polymer Flood Performance in Aishwariya Field. In *ADIPEC*. SPE, 2017. DOI: 10.2118/186880-MS

Moreau, P.; Morvan, M.; Rivoal, P.; Bazin, B.; Douarche, F.; Argillier, J. F.; Tabary, R. An Integrated Workflow for Chemical EOR Pilot Design. In *SPE - DOE Improved Oil Recovery Symposium Proceedings*. OnePetro, 2010; pp 912–922.

Pal, N.; Mandal, A. Numerical Simulation of Enhanced Oil Recovery Studies for Aqueous Gemini Surfactant-Polymer-Nanoparticle Systems. *AIChE J.* **2020**. DOI: 10.1002/aic.17020

Rasmussen, A. F.; Sandve, T. H.; Bao, K.; Lauser, A.; Hove, J.; Skaflestad, B.; Klöfkorn, R.; Blatt, M.; Rustad, A. B.; Sævareid, O.; Lie, K. A.; Thune, A. The Open Porous Media Flow Reservoir Simulator. *Comput. Math. Appl.* **2021,** *81,* 159–185.

Reservoir Simulation Fundamentals [WWW Document], n.d. Oil Gas Portal. http://www.oil-gasportal.com/reservoir-simulation/fundamentals/ (accessed Jan 14, 2022).

REVEAL [WWW Document], n.d. https://www.petex.com/products/ipm-suite/reveal/ (accessed Jan 14, 2022).

Rios, V. S.; Avansi, G. D.; Schiozer, D. J.; Practical Workflow to Improve Numerical Performance in Time-Consuming Reservoir Simulation Models Using Submodels and Shorter Period of Time. *J. Pet. Sci. Eng.* **2020,** *195,* 107547.

Sadeed, A.; Tariq, Z.; Nasir Janjua, A.; Asad, A.; Enamul Hossain, M. Smart Water Flooding: An Economic Evaluation and Optimization. In *Society of Petroleum Engineers - SPE Kingdom of Saudi Arabia Annual Technical Symposium and Exhibition 2018*, SATS 2018. Society of Petroleum Engineers, 2018. DOI: 10.2118/192330-ms

Sharma, H.; Mohanty, K. K. An Experimental and Modeling Study to Investigate Brine-Rock Interactions During Low Salinity Water Flooding in Carbonates. *J. Pet. Sci. Eng.* **2018,** *165,* 1021–1039.

Sinan, S.; Glover, P. W. J.; Lorinczi, P. Modelling the Impact of Anisotropy on Hydrocarbon Production in Heterogeneous Reservoirs. *Transp. Porous Media* **2020,** *133,* 413–436.

Sun, K.; Liu, H.; Wang, J.; Wei, X.; Ma, L.; Kang, Z.; Zhang, Y. Three-Dimensional Physical Simulation of Water Huff-n-Puff in a Tight Oil Reservoir with Stimulated Reservoir Volume. *J. Pet. Sci. Eng.* **2022,** *208,* 109212.

Trivedi, J. J.; Babadagli, T. In *CO_2 and Flue Gas Sequestration During Tertiary Oil Recovery: Optimal Injection Strategies and Importance of Operational Parameters*, Canadian International Petroleum Conference 2005, CIPC 2005. Petroleum Society of Canada (PETSOC), 2005.

Yan, J.; Lu, Y. Y.; Zhong, D. L.; Zou, Z. L.; Li, J. B. Enhanced Methane Recovery from Low-Concentration Coalbed Methane by Gas Hydrate Formation in Graphite Nanofluids. *Energy* **2019,** *180,* 728–736.

Yang, W.; Wang, T.; Fan, Z.; Miao, Q.; Deng, Z.; Zhu, Y. Foams Stabilized by in Situ-Modified Nanoparticles and Anionic Surfactants for Enhanced Oil Recovery. *Energy Fuels* **2017,** *31,* 4721–4730.

Yuan, Y.; Yan, W.; Qi, Z.; Luo, T.; Fan, W. Study on Property Evaluation and Numerical Simulation of Air-Foam Flooding in Low Permeability Reservoir. *Energy Sources, Part A Recover. Util. Environ. Eff.* **2019**. DOI: 10.1080/15567036.2019.1674427

CHAPTER 17

Chemical Enhanced Oil Recovery in Unconventional Reservoirs

SANKARI HAZARIKA and ANNAPURNA BARUAH

Department of Petroleum Engineering and Earth Science, University of Petroleum and Energy Studies, Dehradun, India

ABSTRACT

A significant amount of oil and gas volumes are locked in the "unconventional resources" like shale, coal bed methane reservoirs and tight oil/gas reserves of low/very low permeability. The unique geophysical and chemical properties of these reserves add an additional layer of technical sophistication during the implementation of the chemical enhanced oil recovery (EOR) methods. This chapter discusses about the challenges in unconventional reservoirs, including super close grids and high conductivity, requires rigorous testing and experiments. Surfactant flooding in unconventional reservoirs aims to lower IFT and improve wettability, which is crucial for recovering residual oil. The technology allows the transformation from oil-wet to water-wet, facilitating enhanced oil recovery. Alkaline flooding enhances oil recovery through emulsification. This method, cost-effective but prone to corrosion and scaling, has shown success in reducing residual oil saturation. Recent trends in CEOR involve nanotechnology, the application of ionic liquids, and advancements in existing techniques. Subsequently, this chapter discusses about the Gemini surfactant, which has significant potential in EOR due to its high interfacial action, low critical micelle concentration, and wettability alteration benefits. This chapter also delves into challenges related to gas injection in CO_2 flooding, emphasizing the role of foam in improving mobility control. Furthermore, the discussion extends to hydraulic fracturing, specifically in unconventional reservoirs,

Advancements in Chemical Enhanced Oil Recovery.
Tushar Sharma, Krishna Raghav Chaturvedi, Tarek Ganat, and Imtiaz Ali (Eds.)

where waterless fracture methods are considered as alternatives. Finally, this chapter concludes with a case study on the Permian Basin, examining the design and execution of a CEOR Huff-n-Puff trial in the Midland Basin. The conclusion highlights the economic implications of CEOR strategies in shale oil development, particularly in improving oil recovery through consistent wettability alteration. It emphasizes the role of factors such as salinity, hardness, oil characteristics, reservoir mineralogy, and natural fractures in the success of chemical applications. The potential of various CEOR methods, including Huff-n-Puff and trial completions, is acknowledged, showcasing increased oil recovery and improved overall production results.

17.1 INTRODUCTION

Enhanced oil recovery is a process to increase the recovery rate after the depletion of primary production (see Fig. 17.1) (Belhaj et al., 2020). It is also called “tertiary recovery,” which is a cycle for separating oil that has not been recovered through the essential or auxiliary oil recovery methods and to utilize and improve the overall oil and gas extraction process with different EOR methodologies (Gbadamosi et al., 2018; Hanamertani et al., 2018). At the time of the oil recovery process, the whole oil displacement is a combination of pore scale and volumetric sweep efficiency (Gbadamosi et al., 2019). It means that increased oil recovery is greatly adept at the nanoscale range (Gbadamosi et al., 2019; Olajire, 2014). The formulated and used EOR strategies are significantly organized into thermal and nonthermal EOR techniques (Thomas, 2007).

Enhanced oil recovery is mainly categorized into three stages: chemical injection, gas injection, and thermal injection. Basically, for gas injection, nitrogen or carbon dioxide (CO_2), flammable gases, indicate almost 60% of EOR creation in the US (2020). The thermal infusion process represents 40% of EOR creation in the US and California (2020). Chemical injection, which can include the usage of since quite a while ago tied particles called polymers to grow the reasonability of waterfloods, speaks to around 1% of EOR creation in the US. Nowadays, there is a quick lessening solicitation of conventional oil resources, so we need to investigate unconventional oil and gas increments.

The unconventional reservoir has low porosity and ultralow permeability and a very complex geographical feature and structure, which contrasts with the traditional reservoir. It consequently applied some innovation to break the formation with sand and water under high pressure and temperature and to permit gas to stream. Likewise, low-permeability reservoirs contain a critical and developing part of the world's oil reserves; however, their exploitation is frequently connected with helpless recovery even after a waterflood. In such cases, chemical enhanced oil recovery (CEOR) is regularly considered (Speight, 2017).

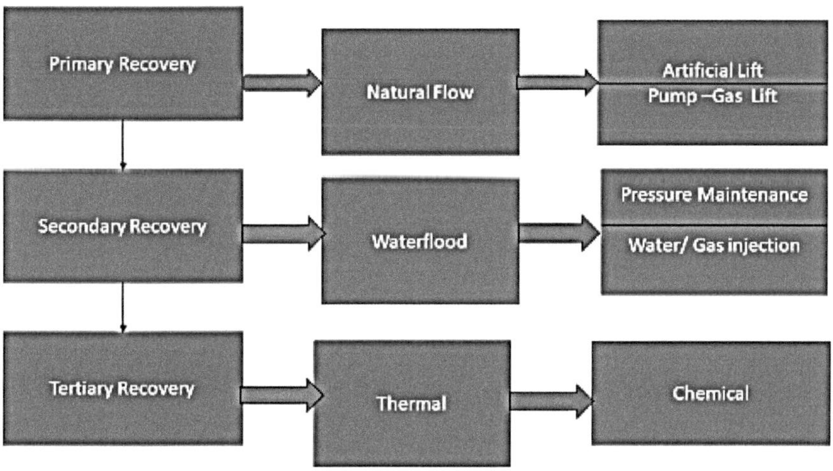

FIGURE 17.1 Different types of enhanced oil recovery techniques.

Utilization of CEOR procedures incorporates the way toward injecting various kinds of materials into a reservoir to improve general range productivity (Belhaj et al., 2020). CEOR strategies increment oil recovery by expanding the adequacy of water infused into the reservoir to dislodge the oil. Subject to the sort of CEOR measure, synthetics infused with the water slug adjust the liquid or potentially liquid stone interaction in the reservoir (Gbadamosi et al., 2019) and reduce the interfacial tension (IFT) between the imbibing liquid and oil or an augmentation in the viscosity of injected subterranean insect fluid for improving mobility and conformance control. In addition, the infused synthetic compounds cause a change in the wettability and increase the permeability of the reservoir (Sun et al., 2017).

CEOR techniques are polymer flooding, surfactant flooding, and alkaline flooding (Samanta et al., 2012). Polymers influence oil recovery by increasing the viscosity of the injected fluid (Gbadamosi et al., 2019). Surfactant and alkali lose their proficiency during their stream in permeable media because of adsorption marvels (Gbadamosi et al., 2019). Hence, various methods of chemical flood infusion were formulated, considered, and applied for EOR measures. A short while ago, the utilization of foam enhanced by surfactants and polymers for improved strength and mobility as a control that improves oil recovery was considered (Zaitoun et al., 2012; Bera et al., 2013). In this section, distinctive EOR techniques applied to shale are additionally evaluated.

17.2 CHEMICAL EOR IN CONVENTIONAL RESERVOIR

17.2.1 POLYMER FLOODING

When waterflooding (Fig. 17.2) of an oil reservoir demonstrates a lack because of viscous fingering wonders finishing in early water achievement, polymer flooding might be presented. The cycle of polymer flooding includes the infusion of high molecular weight water-solvent polymers alongside the water slug to build the viscosity of the injectant (Mandal et al., 2013). The gradual thickness of the injectant improves the portability and conformance control of the infused slug and destroys viscous fingering wonders. Thus, early water advancement typically experienced in the waterflooding measure is smothered and a steady oil recovery factor is achieved (Wei, 2014). Polymer flooding has long been effectively updated in numerous oil fields, either on a pilot scale or on a commercial scale. Polymer flooding additionally improves oil recovery through a combination of mobility control, permeability reduction, and the viscoelastic properties of polymers.

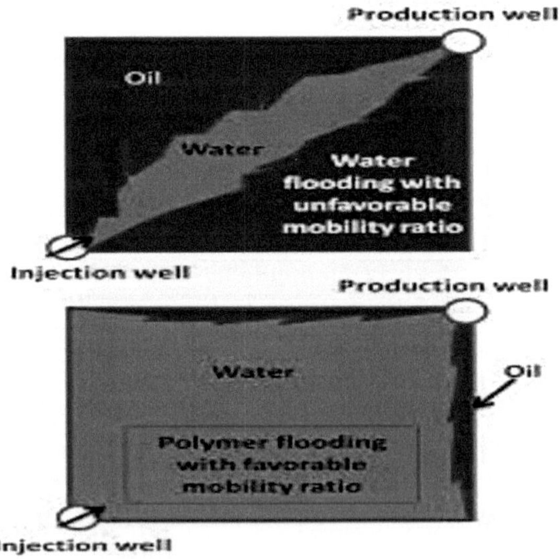

FIGURE 17.2 Fingering is effectively promoted by the unfavorable mobility ratio (above) and the good oil recovery, which is facilitated by the use of polymer flooding (below).

Source: Adapted from Abidin et al., 2012. Copyright 2012. Elsevier

17.2.2 SURFACTANT FLOODING

Surfactant flooding is a demonstrated EOR method (Fig. 17.3) utilized for assembling leftover oil caught in the reservoir (Abbas et al., 2018). The point of surfactant infusion into a reservoir for improving oil recovery factor is to adjust the liquid-liquid collaboration by diminishing IFT between the oil and salt water and the liquid-rock properties by means of wettability alteration of the permeable medium (Gbadamosi et al., 2019). A surfactant, also called a surface-dynamic specialist molecule, is amphiphilic in nature. This implies that surfactants are described by two practical gatherings inside their natural shell. These are the hydrophilic gathering, which is typically water-solvent, and the hydrophobic or nonpolar gathering, which is regularly oil-dissolvable (Gbadamosi et al., 2019). The lipophilic hydrophobic gathering is typically a long-chain hydrocarbon, fluorocarbon, siloxane chain, or short polymer chain, which might possibly be fanned. Then again, the lipophilic hydrophilic gathering is framed by moieties, with their arrangement relying upon the inalienable natural compound (Negin et al., 2017). What's more, surfactant flooding improves pore-scale dislodging proficiency through the system of IFT decrease, or wettability alteration (Gbadamosi et al., 2019).

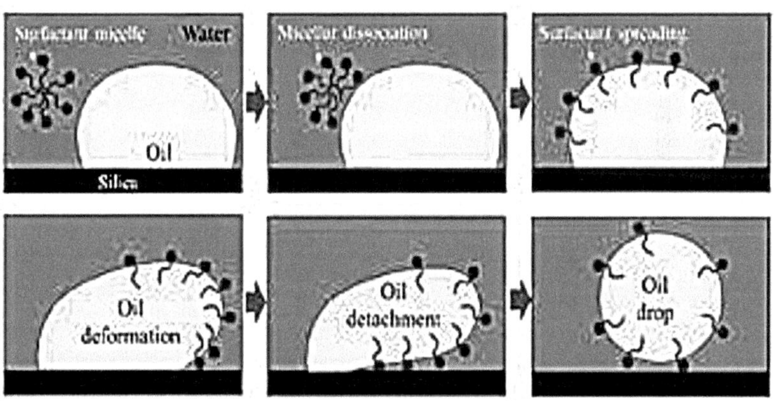

FIGURE 17.3 Schematic representation of the oil removal process when flooding the surfactant solution.

Source: Adapted from Tang et al., 2019. Copyright 2019. American Chemical Society

17.2.3 ALKALI FLOODING

Alkali flooding is an EOR method that uses an alkali base (an essential compound, an ionic salt of an alkaline parent metal, or a basic earth metal) to improve the oil recovery factor. The strategy is indistinguishable from other EOR techniques, as the synthetic substances that control the oil recovery are generated in situ during the EOR cycle through a saponification reaction (Gong et al., 2016), which is characterized as the proven reaction between a natural caustic and a sparingly soluble base on the frame cleaner. The natural corrosive is acquired from the acidic segment of unrefined petroleum. The created cleanser acts as an in situ surfactant to bring down IFT and emulsify the unrefined petroleum, improving oil recovery (Pei et al., 2013).

17.2.4 FOAM FLOODING

The restrictions of gas injection strategies, which incorporate viscous fingering and gravity, abrogate, foam, and a scattering of gas in fluid, with the end goal that the fluid stage is ceaseless and some piece of the gas stage is made irregular by a slender fluid film called lamellae, arose as a promising answer for improving oil recovery proficiency (Yekeen et al., 2017). Foam controls oil recovery by showing two good components that are helpful to the EOR cycle. Initially, foams increment the clear viscosity of infused subterranean insects in this manner, advancing a more positive versatility proportion. Finally, the air pockets of the foam extend into permeable media, subsequently displaying specific liquid redirection from thief zones to bring down penetrability districts or unwept zones in the reservoir (Rafati et al., 2016). This makes the foam flooding measure more productive than water-flooding, gas infusion, and water-substituting gas infusion measures. Up to this point, foams have been named a gas infusion technique (Gbadamosi et al., 2019). Instances of customary foam flooding strategies incorporate CO_2 foams, nitrogen foams, and air foams. Hitherto, synthetic substances, for example, surfactants, proteins, and polymers have been utilized to incorporate stable foams with a longer half-life (Samin et al., 2017).

17.3 CHEMICAL EOR IN UNCONVENTIONAL RESERVOIR

The conventional reservoirs are exhausting altogether on an overall premise. To fulfill the need for hydrocarbon energy, it is fundamental to create unconventional assets (Ling et al., 2014). In this way, EOR techniques will assume an essential function to recover the hydrocarbons. With EOR procedures, for example, gas infusion, chemical injection, microbial infusion, and thermal recovery, the recovery factor can be expanded by 10–30% (Ling et al., 2014). In this way, diverse CEOR strategies applied to shale supplies were investigated. Additionally, low-permeability reservoirs contain a critical and developing part of the world's oil reserves, yet their exploitation is frequently connected with helpless recovery even after waterflood (Delamaide et al., 2014).

The EOR procedures in unconventional reservoirs require a lot of testing. Fundamentally enhanced oil recovery is a cycle to infuse liquid or synthetic substances into the supply to mobilize the remaining oil after the creation declination of essential or common recovery (Speight, 2014). To examine CEOR strategies in the unconventional reservoir increment, oil recovery by expanding the adequacy of the water infused into the reservoir to uproot the oil and to build oil recovery through wettability adjustment, versatility control, IFT decrease, age of foam or emulsion by surfactant, making of cleanser and increment pH by alkaline synthetic responding with oil, and redirecting or potentially hindering by polymer gel to improve clear effectiveness (Speight, 2014). As of late, analysts have been planning to take a shot at the appropriateness and comprehension of EOR in unconventional oil reservoirs. The superclose grid and high conductivity of the normal cracks may be the two most significant components that hinder the accomplishment of traditional EOR strategies (Speight, 2014).

Be that as it may, traditional CEOR strategies have their constraints. Polymers, whose fundamental recovery component is to expand the consistency of, infuse ants, and thus mobility, endure viscosity misfortune within the sight of reservoir salt waters and raised temperature conditions (Gbadamosi et al., 2019). Surfactants and soluble bases lose their proficiency during their stream in permeable media because of adsorption marvels (Gbadamosi et al., 2019).

17.3.1 SURFACTANT FLOODING

Surfactant infusion is used to lower the IFT and improve the wettability state, which is fundamental to recovering avoided and leftover oil in unconventional reservoirs (Pautz et al., 1992). The oil-wet nature of the Bakken arrangement makes the change from oil-wet to water-wet conceivable by injecting surfactant. Morsy et al. (2014) contemplated a progression of surfactant-added substances joined into incitement liquid for enhanced oil recovery. Some unique item tests included similarity with arrangement saline solution, emulsion inclination, recovery factors from unconstrained imbibition with raw petroleum and development brackish water, and similarity with proposed breaking liquids. This chapter shows that a surfactant innovation intended to enhance oil recovery from cracked shale developments is fit for these regular broken shale reservoirs (Ling et al., 2014). Under most conditions, surfactant-added substances in the break liquids will infiltrate into the exceptionally oil-immersed lattice or common crack area and quicken the extraction of the oil set up by fast imbibition. At that point, the surfactant "huff and puff" circle and the caught oil in the stone network can promptly move into the propped crack framework to be delivered (Price, 2000). The surfactant-added substances likewise have the upside of leaving the framework or characteristic break face water-wet to encourage oil development during creation, on the off chance that they are designed appropriately (Ling et al., 2014).

(A) MOBILITY RATIO CONTROL THROUGH EOR:

In this technique, surfactant infusion or in situ surfactant creation through an infusing arrangement or raw petroleum makes oil move. Since the polymer is costly, a tad of it, around 4–5%, is infused and afterward dislodged by water (Mandal et al., 2013). Mobility control in the chemical cycle is completed in three phases: (1) preventing the fingering of the chemical material with oil; (2) mobility control between the chemical arrangement and the limited chemical slug; (3) preventing the fingering of the water infusion front with the polymer-containing substance (Mandal et al., 2013).

(B) INTERFACIAL TENSION REDUCTION DURING EOR:

In the secondary and tertiary recovery strategies, the fundamental point of the cycle is to diminish the IFT between the water and oil in the permeable medium, and in the waterflooding measure, it is unimaginable for water to dislodge the entirety of the oil in the pore scale, as oil is caught by capillary forces (Negin et al., 2017). This capillary force is estimated by a dimensionless capillary number (*Nc*) characterized in condition as:

$$Nc = (\mu * \nu)/(\sigma * \cos\theta) \tag{17.1}$$

where μ is the displacing fluid viscosity, v is the displacing Darcy velocity, θ is the contact angle, and σ is the IFT between the dislodging liquid (water) and the displaced liquid (oil) (Gbadamosi et al., 2018). From condition (1), oil recovery can be improved in three different ways: (I) builds the displacing fluid viscosity (μ); (II) expands the infusion liquid velocity (ν); (III) decreases the IFT (σ). Expanding the pace of infusion liquid can cause the infusion strain to be more noteworthy than the break weight of the supply, subsequently cracking the reservoir rock (Wei et al., 2013). By and by, just the IFT decrease technique can be utilized to build *Nc* by 100 times (Wei et al., 2013).

An adsorbed film forms because of the interaction between the oil and the alkyl tail of the surfactant, along these lines, bringing down the IFT at the oil/water interface (Olajire, 2014). The decrease in IFT at the oil/water interface debilitates the capillary forces that keep down the caught oil, causing oil beads to handily stream out of the stone pores and structure an oil bank downstream (Olajire, 2014).

(C) WETTABILITY ALTERATION:

The slop of a strong surface for a specific sort of fluid within the sight of other immiscible fluids is called wettability (Mohammed and Babadagli, 2015; Cheraghian, 2015), and in a reservoir rock framework, the wettability of the stone surface decides and controls the position, appropriation, and stream of fluids inside the separate supply (Kovscek et al., 1993). This petrophysical property is of specific significance since it influences oil recovery boundaries, for example, capillary pressure, relative permeability, surface tension, and IFT (Christensen and Tanino, 2017).

Wettability alteration can be estimated by the utilization of various techniques, for example, surface imaging tests, zeta expected estimations, unconstrained imbibition, and contact point estimations. Most examinations to gauge the wettability change are finished using a contact point characterized by where the oil-water interface meets the stone surface (Gbadamosi et al., 2019). A surface with a contact angle (θ) > 90° is viewed as oil-wet, while (θ) < 90° is viewed as water-wet. Changing the wettability of a surface from oil-sodden to water-moist lessens the adhesive force of the capillarity and builds the oil porosity of the reservoir. From this, it very well may be inferred that the oil recovery in the water-wet reservoir can be accomplished more effectively than in the oil-wet reservoir. The system by which surfactant changes the wettability of traditional stone spores is alluded to as a cleaning instrument in which the surfactant desorbs the oil-wet layer (Gbadamosi et al., 2019). Desorption of the oil-wet layer by a surfactant changes the wettability of the surface and places it in a water-wet state (Gbadamosi et al., 2019).

17.3.2 ALKALINE FLOODING

Alkaline flooding enhances oil recovery through emulsification and entrainment. It is normally joined with different synthetics during the injection to accomplish a superior, improved oil recovery impact (Ling et al., 2014). Alkaline flooding has the benefit of minimal effort compared to other chemical flooding. The impediments are consumption and scaling issues brought about by inorganic alkalis. A harsh flooding EOR pilot test at Court Bakken heavy oil was directed in 2004 (Xie et al., 2008). This single-well test indicated that scathing infusion was viable for lessening lingering oil immersion and could accomplish 9% gradual oil recovery in the chosen pilot territory (Xie et al., 2008).

17.3.3 POLYMER FLOODING

Polymer-upgraded waterflooding improves oil recovery by adjusting mobility proportion, hindering and redirecting injecting from moving through high-permeability channels to improve volumetric range effectiveness, decrease directing, and advancement (Ling et al., 2014). Manrique et al. (2007) demonstrated compound infusion in Blue Buttes field. In any case, we can't discover distributed data on this infusion. Hinrichsen and Mottle (1989) inspected the infusion of scale and consumption inhibitors to restrain scaling in the well in Blue Buttes field. The pace of disappointments diminished from a pace of every 10 years to around every 2 years.

17.3.4 CO_2-EOR FLOODING

The impact of CO_2 infusion on Bakken (shale) development has not been checked by field application since the previously mentioned pilot tests focused on different arrangements (Ling et al., 2014). Numerous investigations have assessed the CO_2 injection potential in the Bakken arrangement. Hawthorne et al. (2013) proposed calculated strides for the CO_2 dislodging oil in Bakken, which include:

1. CO_2 streams into and through the cracks,
2. Unbroken stone lattice is exposed to CO_2 at crack surfaces,
3. CO_2 saturates the stone driven by pressure, conveying some hydrocarbons internally; in any case, the oil is additionally growing and expelling some oil out of the pores,
4. Oil moves to the mass of CO_2 in the cracks through growing and decreasing viscosity, and

5. The CO_2 pressure gradient gets more modest; oil creation is gradually determined by focus gradient dispersion from pores into the mass of CO_2 in the cracks.

Numerous investigations (Hoffman, 2012; Wang et al., 2010; Kurtoglu et al., 2013) have distinguished enormous oil fields as having great reservoir qualities for CO_2-EOR. Iwere et al. (2012) displayed and recreated gas and water injections in normally cracked Bakken and Three Forks arrangements. Their recreation results outline that normal exhaustion, constant CO_2 flooding, and waterflooding yield recovery elements of 6.4, 7.0, and 6.7% of the first oil sctup for the Middle Bakken formation (Ling et al., 2014).

17.3.5 NANOCHEMICAL EOR FLOODING

Nanotechnology is the control of issues with at least one measurement from 1 to 100 nm. It empowers new imaging and calculation strategies to more readily find, size, and portray the reservoir, screen the liquid stream, and distinguish liquid sorts (Vipulanandan et al., 2014). In combination with fluids, it tends to be an incredibly sensitive borehole sensor for pressure, temperature, and stress. Meanwhile, the nanometer-sized beads are bound to infiltrate into the small voids or breaks of the thick stone, and because of their huge explicit surface zone, the nanodroplets can spread totally on the stone surface and improve the contact proficiency between the nanoemulsion and the reservoir and the general application impact of the borehole liquid. Some research has shown an association between surfactants and nanoparticles (NPs) (Jin et al., 2018). Cationic surfactants, anionic surfactants, decidedly charged titanium dioxide NPs, and adversely charged silicon dioxides are utilized in the test (Chen et al., 2013). The outcomes show that if the surfactant and the NP have a similar charge, the electrostatic aversion drives more surfactants at the interface, and if the two show a synergistic impact, at that point, the surface tension is diminished. If the surfactant and NP have inverse charges because of electrostatic fascination, the NPs adsorb to the surfactant and structure agglomerate, decreasing the powerful grouping of the surfactant at the interface and rather expanding the surface tension (Chai et al., 2019). Different studies have explored the impact of cleaning addicts who are heightened by nanoemulsions and surfactants (Kim et al., 2011). It is accepted that the nanoemulsion, which has an amazingly low surface tension, high solubility, and a small molecule size, can successfully enter the pores and come completely into contact with the strong surface, which can lessen the surface tension and afterward change the surface wetting point. It fundamentally assimilates water from the core, diminishing capillary resistance and viably improving crack-cleaning effectiveness (Chai et al., 2019).

Recently, many examinations have considered that by adding NPs into synthetic arrangements, better execution of CEOR can be accomplished (Ju et al., 2006; Ismailov and Veliyev, 2011). Ogolo et al. (2012) played out some EOR tests utilizing various kinds of NPs with different sizes in diverse dispersing media, including refined water, saline solution, ethanol, and diesel. They revealed that alumina NPs scattered in refined water and saline solutions have a high propensity to expand oil recovery because of the decrease in oil viscosity (Ali et al., 2018). While the capacity of the two kinds of silica-based NPs scattered in ethanol to expand oil recovery was because of wettability alteration, nonetheless, some permeability issues were discovered when magnesium oxide and zinc oxide were scattered in refined water and saline solution, and this led to a helpless oil recovery (Ogolo et al., 2012). A review of the latest exploration progress in the NPs application for CEOR is given. Through this audit, the accompanying ends were reached (Ali et al., 2018):

1. Conventional CEOR techniques face a few difficulties and issues; some of them were very well understood by presenting nanoliquids,
2. The huge surface territory, stable scattering, and high contagiousness of NPs are the primary highlights of nanoliquids,
3. NPs have been scattered in different fluid stages (e.g., water, alcohols, surfactants, and polymers) and gave high EOR execution in various plans, nanoliquids, nanoemulsions, and nanocatalysts,
4. Different organic and inorganic sorts of NPs have been applied; silica has been featured as the most widely recognized and compelling NP, which gave an additional oil recovery up to 20% original oil in place (OOIP),
5. NPs affect a few EOR instruments, for example, separating pressure, wettability adjustment, IFT decrease, and versatility proportion,
6. A few boundaries, including NP size, NP type, NP fixation, temperature, salinity, and scattering media, have been distinguished to influence the nanoliquid execution.

17.4 RECENT TRENDS IN CHEMICAL EOR FOR UNCONVENTIONAL RESERVOIR

The new trend in chemical EOR includes the application of nanotechnology, the trial of new chemical substances such as ionic liquids, and the development of improved technology from previously known CEOR methods.

17.4.1 GEMINI SURFACTANT FLOODING

Gemini surfactant has a decent application prospect in the field of EOR because of the benefits of high interfacial action, low critical micelle concentration (CMC), and great wettability alteration (Ma et al., 2019). In this section, ethylenediamine, sodium 2-bromoethanesulfonate, and greasy liquor polyoxyethylene ether are utilized as crude materials to orchestrate another sort of anionic-nonionic gemini surfactant, whose hydrophilic gathering comprises two sections of nonionic and anionic gatherings through a mellow response of amination (Chung et al., 2018). The oil relocation, surface tension, wettability, and IFT of the incorporated anionic-nonionic gemini surfactants were assessed. Also, the anionic-nonionic gemini surfactant can improve oil uprooting productivity by 7.03% (Ma et al., 2019).

17.4.2 SURFACTANT/POLYMER FLOODING

The strategies depend on partial streams and the investigation of one-dimensional streams, and they are approved against two-dimensional reproductions of moving through heterogeneous permeable media (Farajzadeh et al., 2019). Our outcomes stress the critical part of the water/oil relative permeability bends for the plan of mobility control in polymer and surfactant/polymer flooding. The polymer consistency obtained by setting the stun front versatility proportion to one is the base viscosity to guarantee a steady removal front (Glasbergen et al., 2015). Also, we find that for surfactant-polymer (SP) arrangements with a superlow IFT decrease (Winsor type III), the necessary polymer viscosity is consistently more prominent than the oil viscosity (at low shear rates) (Liu et al., 2010). Notwithstanding, for Winsor type I arrangements, for oils with medium and huge viscosities, the nondirect state of the general permeability work prompts polymer viscosities that are not as high as those of the oil. For light oils, the consistency of the ASP arrangement should be essentially bigger than the oil viscosity (Farajzadeh et al., 2019).

17.4.3 SURFACTANT FOAMS FLOODING

Nowadays, surfactant-aided CO_2 flooding is being tried in Center East carbonate reservoirs (Al-Mutairi and Kokal, 2011). Attributable to its actual properties and its ability to set up various associations with oil over a wide sweep of pressures and temperatures, CO_2 is viewed as one of the most significant dislodging liquids in gas-based EOR innovation (Blunt et al., 1993; Mathiassen, 2003). Nonetheless, there are a few issues related to gas injection (Sagir et al., 2013). Among them, the best test with CO_2 gas infusion lies in its poor volumetric range effectiveness, inferable from its low viscosity and density. Lighter gas supersedes gravity, and an enormous segment of recoverable oil in the lower porousness locales can't be reached (Pal et al., 2017). This helpless compass abandons a broad measure of oil in the reservoir. In spite of the fact that the macroscopic range effectiveness of CO_2 is very high, its viscosity (~ 0.01 cP) is a lot lower than both water (~ 1.0 cP) and the majority of the unrefined oils (0.6–10 cP for traditional oils), which prompts numerous conformance and versatility concerns and insecurity on the removal front. Issues of poor volumetric compass productivity, gas diverting through high-porousness streaks, and gravity abrogation are continuous (Hanssen et al., 1994). One of the methodologies to address these difficulties is to use foam, which is a scattering of gas in a consistent fluid that brings down the mobility proportion. Boud and Holbrook (1958) unexpectedly exhibited that foam could be produced in an oil reservoir by successive infusions of watery surfactant arrangements and both miscible and immiscible gas drives to build its sweep proficiency. Notwithstanding, because of the absence of legitimate comprehension of the mobility control component by foam, the idea was not generally received (Li et al., 2010). By the way, as the comprehension of foam mobility control progressed, there have been many field trials of foam application from that point forward.

In some cases, customary business nonionic or anionic surfactants utilized in CO_2 foam-based recovery are unacceptable for application in the HTHS reserves (Pal et al., 2017). The cloud purposes of ethoxylated nonionic surfactants are reliably achieved under 100°C (Adkins et al., 2010), and the solvency of most nonionic surfactants diminishes in salt water as the salinity increases (Rosen and Kunjappu, 2012). There are reports of a few lab-scale tests and field preliminary studies utilizing anionic sulfate and sulfonate surfactants for high-salinity limestone reservoirs (Hirasaki et al., 2008; Levitt et al., 2006). Nonetheless, because of the electrostatic power of fascination, they frequently adsorb firmly on the decidedly charged limestone surfaces within the sight of broken down acidic CO_2 at high pressures (Lawson, 1978; Wang et al., 2015). Cationic surfactants, then again, show low adsorption on carbonate arrangements because of the electrostatic shock between the cationic head and the positive charge-bearing carbonate surface (Hirasaki et al., 2008; Lawson, 1978; Ahmadall et al., 1993). By the way, they are infrequently solvents in CO_2, despite the fact that there are reports of a couple of special cases (Smith et al., 2007).

To defeat these restrictions, ethoxylated nonionic to cationic switchable amine surfactants were planned and presented in a progression of sand-pack tests (Chen et al., 2012; Chen et al., 2014). Ethoxylated amines are switchable from being nonionic in salt water to cationic within the sight of an acidic fluid stage, for example, CO_2 (Elhag et al., 2014a). Responses between essential, secondary, or tertiary amines with a proper alkoxylation specialist produced these ethoxylated amines. Comparatively to the size of a hydrophobic chain of alkyl amines, the size of the hydrophilic gathering expanded with ethoxylation, which thus expanded the hydrophilicity (Chen et al., 2014). Due to the correct equilibrium in the quantity of carbons in their alkyl chains and the quantity of ethylene oxide (EO) bunches connected to the tertiary nitrogen in their top gatherings, ethoxylated alkyl amines of the structure $C_{12-14}N(EO)_x$ were found to fulfill a few basic prerequisites for viable CO_2-EOR. This surfactant was profoundly dissolvable in the CO_2 stage on the grounds that the nitrogen iota stayed unprotonated in this stage. While in a low-pH watery stage because of disintegrated CO_2, the emphatically charged protonated amine made the surfactants more hydrophilic and raised the cloud highlight to 120°C (Pal et al., 2017). Further, within the sight of CO_2, the adsorption of ethoxylated alkyl amines (broken down in brackish water) on limestone surfaces was essentially decreased because of the decidedly charged cationic head gathering. Hence, switchable ethoxylated amine surfactants can be considered as another age surfactant, which interestingly consolidates the high cloud purpose of ionic surfactants in water with high dissolvability in CO_2 for nonionic surfactants, settling foam developments at 120°C with negligible adsorption on limestone (Elhag et al., 2014b). Regardless, switchable surfactant tests are still in the essential stage, and much inside-out investigation should be accomplished for appropriate arrangement and acknowledgment of this CEOR procedure in real field applications (Pal et al., 2017).

17.4.4 ALKALINE-SURFACTANT-ALTERNATED-GAS/CO_2 FLOODING

The volumetric range efficiencies of CO_2 flooding for EOR are commonly low because of issues of viscous fingering and gravity abrogate. As alluded to as alkaline-surfactant-substituted-gas/CO_2 (ASSG) flooding, this strategy is basically the synergic mix of synthetic and immiscible CO_2 flooding

(Phukan et al., 2019). In this section, compound plans were distinguished through foam dependability tests dependent on their foaming capacity coefficients (Bageri et al., 2014). The chosen definitions were additionally tried for their abilities to decrease oil–water IFT to superlow esteem. With the best-performing details, the research facility scaled core flooding tests to assess their EOR potential (Guo et al., 2012). It was seen that a higher oil recovery of 14.26% OOIP was acquired by surfactant-rotated gas (SRG) flooding compared with 12.03% OOIP by immiscible CO_2 substituted with salt water (WSG) flooding (Phukan et al., 2019). The most noteworthy leftover oil recovery of 20% OOIP was obtained for ASSG flooding with the salt, surfactant, and dark alcohol in the chemical slug (Phukan et al., 2019).

17.4.5 APPLICATION OF NANOTECHNOLOGY

The use of nanotechnology has proffered answers for a portion of the issues attacking the oil and gas industry, going from oil investigation, penetrating and fulfillment, stream confirmation issues, hydraulic fracturing, and EOR (Gbadamosi et al., 2019). NPs have recently been discovered to have applications in CEOR because of their capacity to respond to amphiphilic and macromolecular segments of EOR synthetic compounds. They structure polymeric nanofluids, nanosurfactants, and nanobalanced out foam when responding with polymers, surfactants, and foams separately (Sun et al., 2017). The recently shaped suspension (or composite materials, all things considered) displays real and captivating properties, not the same as the individual segments of the material. In polymer molecules, the presence of NPs improves rheological conduct and forestalls debasement of the polymer particles. NP-settled foams are found to have a more extended half-life (Fakoya and Shah, 2017). Also, the lamellae bubbles are thicker, and consequently, thermodynamically steady. Nanosurfactants have superlow IFT because of the irreversible adsorption of NPs at the interface. Generally, nanosynthetic compounds have lower adsorption during the different tests directed to permeable media (Almahfood and Bai, 2018).

(A) HYDRAULIC FRACTURING:

Hydraulic fracturing is a typical and widely used method, especially in North America, because of innovative advances that have made it conceivable to explore petroleum gas and oil from unconventional reservoirs (thick sand, coalbeds, and shale arrangements) (Luca, 2013). High-volume water-powered cracking (with processes that are regularly a significant degree bigger than traditional breaking strategies) began in 1968. This has been enhanced by flat boring since late 1980 and the utilization of synthetics (known as smooth water cracking) since 1997 (Luca, 2013). Essentially, the hydraulic fracturing procedures generally utilize oil-based mud as opposed to water-based mud in the unconventional reservoir foam advances, which are hence more costly than water-based incitements and offer a choice to lessen the measure of water utilized in shale gas incitement. These are accessible through the business (Luca, 2013).

The arrangement of the cracking liquid is fundamental to not just breaking the stone but also to guarantee proficient transport and position of the proppant in the stone to keep it open and to explore the oil and gas. The fluid utilized is a vital part of water-driven cracking, not just concerning its specialized properties (rheology, development similarity, etc.), but additionally as to its natural effect. In reality, a portion of the major natural issues related to shale gas breaking today is because of the utilization of water: the significant levels of water that are utilized and lost underground, the need to handle reflux, the likely pollution of springs through spillage of synthetic substances utilized in the area, fracturing liquids, and so forth (Luca, 2013). As a rule, a cracking liquid can be viewed as the sum of three principal segments:

$$\text{Breaking Liquid} = \text{Base Liquid} + \text{Added substances} + \text{Proppant} \quad (17.2)$$

All the more as of late, the advancement of some unpredictable reservoirs (thick gas, shale gas, and coal crease methane) has provoked the business to reevaluate "waterless" break procedures as suitable choices to water-based cracking liquids. In these reservoirs, the communications between rock development and the burst liquids can influence hydrocarbon creation (Ribeiro and Sharma, 2013).

17.5 SOME OF THE VITAL ISSUES

i. New innovations and treatment techniques have given the oil enterprises the ability to limit the contamination issues and immense uses properly, so efficiency is high.
ii. The petrophysical, geochemical, and topographical properties of the supply are intricately linked to building up the system of shale assets.

iii. Laboratory investigations for EOR in unconventional reservoirs have been restricted.
iv. Some potential issues might be the upkeep of a low enough pH to keep them protonated.
v. For unconventional reservoir improvement, level wells with multi-stage are preferred when compared with vertical wells.
vi. Water treatment would be more effective with information on some data and properties of the water created in shale reservoirs.
vii. Environmental effects might be the most testing issues when thinking about possibilities for unconventional gas advancement.

Collections of hydrocarbons, for example, oil and gas, in normal conventional and unconventional reservoirs all through the world, the vast majority of which were moved from clean, fine-grained, dim, or dark natural rich sedimentary source rocks, were alluded to as natural rich shales (Davarpanah, 2018). Over the past many years, natural rich shale developments have been considered the source rocks in oil reservoirs; in other words, hydrocarbons began and moved into sandstone and limestone of different reservoir characteristics on the grounds that unconventional reservoirs have lower permeability than different reservoirs and have lower monetary volumes of oil and gas delivered in oilfields (Davarpanah, 2018). To create business amounts from unconventional reservoirs, a blend of expanded oil and gas costs and improved innovation of even penetrating and multistage breaking is required (Rabbani et al., 2018; Rowan et al., 2015; Thacker et al., 2015).

17.6 PERMIAN BASIN CASE STUDY: DESIGN AND EXECUTION OF A CEOR HUFF-N-PUFF TRIAL IN THE MIDLAND BASIN

Huff-n-Puff (HnP) procedures using different chemicals and liquids are an emerging EOR strategy in unconventional reservoirs to improve the creation and life of the resources. HnP processes using wettability-adjusting science infuse a planned volume of the definition into the creation well, and afterward the well is closed in for around 14 days to help advance the imbibition of the chemistry. These processes can be either single-minded or spotted. After the drenching time frame, the well is put back into production, and execution is estimated according to the gauge creation preceding treatment. Because of the capacity of this chemical to tirelessly modify wettability, there is potential for a supported improvement underway and an expansion in estimated ultimate recovery (EUR) (Bidhendi et al., 2019).

The applicant well is a 5200 ft flat well with 31 phases of using pressurized water broke organizations situated in the Midland Basin, and it is finished in the Wolfcamp-B development (Bidhendi et al., 2019). The very much was on creation for just about 4 years preceding the treatment and is counterbalanced by six vertical and one flat well, which are all delivering from a similar zone and going after similar hydrocarbon assets. The oil rate pretreatment was entirely steady at 63.5 bbls/day with a water cut of 36% (Bidhendi et al., 2019).

In light of the lab information, it was prescribed to infuse the PROE I plan at a convergence of 1000 ppm. Similarity and emulsion propensity testing were run with the accessible new infusion water (TDS ~4000 ppm) and field rough, the two of which demonstrated no issues (Bidhendi et al., 2019). HnP plan volumes are based on the reservoir, fruition, and creation information of the well. From this information, a break volume and depleted lattice liquid volume are assessed, which are utilized as the reasons for the HnP volume (Bidhendi et al., 2019). It is basic that the HnP infusion volume is sufficiently huge to contact enough grids to help advance adequate wettability alteration while adjusting the monetary practicality of the treatment. Blends of mathematical reenactment and experimental models have been created to give the monetary treatment volume (Bidhendi et al., 2019). Subsequent to completing the infusion, the very much was closed in for 3 weeks before the all-around got back to creation. This CEOR HnP was as of late finished, and it is untimely to make inferences (Bidhendi et al., 2019).

17.7 CONCLUSION

This investigation adds to understanding the monetary ramifications of executing CEOR strategies in shale oil advancement as another option and their effect on handling the erupting of abandoned gas. The materiality of the proposed system is first exhibited through a contextual analysis in which a creation unit can have up to some well-cushions.

i. It is indicated that these tweaked surfactant-based definitions can possibly improve oil recovery through steady wettability alteration. It has been seen that, among different components, the saltiness and hardness of the arrangement of saline solution, oil attributes, mineralogy of the reservoir, and the power of common cracks assume a significant function in the science exhibition.

ii. This chemistry can be executed in various structures, for example, CEOR HnP, frac-ensure, fruitions, and ceaseless flooding.
iii. A CEOR HnP and preliminary consummations utilizing the PROE I chemical are explored. The plan and fruitful execution of the HnP preliminary appeared, and the creation results will be conveyed in a future distribution. The preliminary culminations demonstrated a 39% expansion in the 180 total recoveries and an expanded oil cut versus the province-type bend.

ACKNOWLEDGMENTS

As a research scholar of University of Petroleum and Energy Studies, firstly, I would like to acknowledge the great technical and financial support getting from the Department of Petroleum Engineering and Earth Science and Research and Developed Department of University of Petroleum and Energy Studies. I am grateful to my Guide Dr. Annapurna Baruah (Assistant Professor, Department of Petroleum Engineering and Earth Science) and my Mentor Dr. S. M Tauseef (Assistant Dean of School of Engineering). Also, I would like to specially thank my friend, Shubham Saraf (Pursuing Doctorate Degree from Pandit Deendayal Petroleum University, India).

KEYWORDS

- **unconventional reservoir**
- **CO_2 injection**
- **foam injection**
- **shale oil**
- **coal bed methane**

REFERENCES

Abbas, A. H.; Sulaiman, W. R. W.; Jaafar, M. Z.; Gbadamosi, A. O.; Ebrahimi, S. S.; Elrufai, A. Numerical Study for Continuous Surfactant Flooding Considering Adsorption in Heterogeneous Reservoir. *J. King. Saud. Univ. Eng. Sci.* **2018,** *5,* 1–9.

Abidin, A. Z.; Puspasari, T.; Nugroho, W. A. Polymers For Enhanced Oil Recovery Technology. *Procedia Chem.* **2012,** *4,* 11.

Adkins, S. S.; Chen, X.; Nguyen, Q. P.; et al. Effect of Branching on the Interfacial Properties of Nonionic Hydrocarbon Surfactants at the Air–Water and Carbon Dioxide–Water Interfaces. *J. Colloid Interface Sci.* **2010,** *346* (2), 455–463.

Ahmadall, T.; Gonzalez, M. V.; Harwell, J. H.; et al. Reducing Surfactant Adsorption in Carbonate Reservoirs. *SPE Reserv. Eng. J.* **1993,** *8* (2), 117–122.

Ali, J. A.; Kolo, K.; Manshad, A. K.; Mohammadi, A. H. Recent Advances in Application of Nanotechnology in Chemical Enhanced Oil Recovery: Effects of Nanoparticles on Wettability Alteration, Interfacial Tension Reduction, and Flooding. *Egypt. J. Pet.* **2018**. DOI: 10.1016/j.ejpe.2018.09.006

Almahfood, M.; Bai, B. The Synergistic Effects of Nanoparticle-Surfactant Nanofluids in EOR Applications. *J. Pet. Sci. Eng.* **2018,** *171,* 196–210.

Al-Mutairi SA, Kokal SL. *EOR Potential in the Middle East: Current and Future Trends*, SPE EUROPEC/EAGE Annual Conferenceand Exhibition, May 23–26, Vienna, Austria, 2011. DOI: 10.2118/143287-MS

Bageri, A. S.; Sultan, A. S.; Kandil, M. E. *Evaluation of Novel Surfactant for Nitrogen-Foam-Assisted EOR in High Salinity Carbonate Reservoirs*, SPE EOR Conference at Oil and Gas West Asia Muscat, Oman, Society of Petroleum Engineers. SPE-169717-MS, 2014; p 12.

Belhaj, A. F.; Elraies, K. A.; Mahmood, S. M.; et al. The Effect of Surfactant Concentration, Salinity, Temperature, and pH on Surfactant Adsorption for Chemical Enhanced Oil Recovery: A Review. *J. Petrol. Explor. Prod. Technol.* **2020,** *10,* 125–137.

Bera, A.; Kumar, T.; Ojha, K.; Mandal, A. Adsorption of Surfactants on Sand Surface in Enhanced Oil Recovery: Isotherms, Kinetics and Thermodynamic Studies. *Appl. Surf. Sci.* **2013,** *284,* 87–99.

Bidhendi, M. M.; Kazempour, M.; Ibanga, U.; Nguyen, D.; Arruda, J.; Lantz, M.; Mazon, C. In *A Set of Successful Chemical Enhanced Oil Recovery Trials in Permian Basin: Promising Field and Laboratory Results*, Proceedings of the 7th Unconventional Resources Technology Conference, 2019. DOI: 10.15530/urtec-2019-881

Blunt, M.; Fayers, F. J.; Orr, F. M. Carbon Dioxide in Enhanced Oil Recovery. *Energy Convers. Manag.* **1993,** *34* (9), 1197–1204.

Boud, D. C.; Holbrook, O. C. Gas Drive Oil Recovery Process. US Patents: 2866507. 1958.

Chai, Y.; Li, X.; Jing, D. Application of Surfactants in Hydraulic Fracturing for Enhanced Oil/Gas Recovery. *Oil Gas Res.* **2019,** *4,* 161.

Chen, J.; Cao, J.; Gao, W.; Qi, L.-W.; Li, P. Environmentally Friendly Ionic Liquid-in-Water Microemulsions for Extraction of Hydrophilic and Lipophilic Components from Flos Chrysanthemi. *Analyst* **2013,** *138* (20), 5933.

Chen, Y.; Elhag, A. S.; Poon, B. M.; et al. Ethoxylated Cationic Surfactantsfor CO_2 EOR in High Temperature, High Salinity Reservoirs. In *SPE Improved Oil Recovery Symposium*, April 14–18, Tulsa, Oklahoma, 2012.

Chen, Y.; Elhag, A. S.; Poon, B. M.; et al. Switchable Nonionic to Cationic Ethoxylated Amine Surfactants for CO_2 Enhanced Oil Recovery in High-Temperature, High-Salinity Carbonate Reservoirs. *SPE J.* **2014,** *19* (2):249–259.

Cheraghian, G. Effects of Nanoparticles on Wettability: A Review on Applications of Nanotechnology in the Enhanced Oil Recovery. *Int. J. Nano Dimens.* **2015,** *6,* 443–452.

Christensen, M.; Tanino, Y. Water-flood Oil Recovery from Mixed-Wet Limestone: Dependence upon the Contact Angle. *Energy Fuels* **2017,** *31* (2), 1529–1535.

Chung, J.; Boudouris, B. W.; Franses, E. I. Surface Tension Behavior of Aqueous Solutions of a Propoxylated Surfactant and Interfacial Tension Behavior Against a Crude Oil, Colloids Surfaces A Physicochem. *Eng. Asp.* **2018,** *537,* 163–172.

Davarpanah, A. The Integrated Feasibility Analysis of Water Reuses Management in the Petroleum Exploration Performances of Unconventional Shale Reservoirs. *Appl. Water Sci.* **2018,** *8* (2). DOI: 10.1007/s13201-018-0717-7

Delamaide, E.; Tabary, R.; Rousseau, D. *Chemical EOR in Low Permeability Reservoirs*, SPE EOR Conference at Oil and Gas West Asia, 2014. DOI: 10.2118/169673-ms

Elhag, A. S.; Chen, Y.; Chen, H.; et al. Switchable Amine Surfactants for Stable CO_2/Brine Foams in High Temperature, High Salinity Reservoirs. In *SPE Improved Oil Recovery Symposium*, April 12–16, Tulsa, Oklahoma, 2014b. DOI: 10.2128/169041-MS

Elhag, A. S.; Chen, Y.; Reddy, P. P.; et al. Switchable Diamine Surfactants for CO_2 Mobility Control in Enhanced Oil Recovery and Sequestration. *Energy Proc.* **2014a,** *63,* 7709–7716.

Enhanced Oil Recovery. www.doe.gov. U. S. Department of Energy, 2020.

Fakoya, M. F.; Shah, S. N. Emergence of Nanotechnology in the Oil and Gas Industry: Emphasis on the Application of Silica Nanoparticles. *Petroleum* **2017**. DOI: 10.1016/j.petlm.2017.03.001

Farajzadeh, R.; Wassing, B. L.; and Lake, L. W. Insights into Design of Mobility Control for Chemical Enhanced Oil Recovery. *Energy Rep.* **2019,** *5,* 570–578.

Gbadamosi, A. O.; Junin, R.; Manan, M. A.; Yekeen, N.; Agi, A.; Oseh, J. O. Recent Advances and Prospects in Polymeric Nanofluids Application for Enhanced Oil Recovery. *J. Indus. Eng. Chem.* **2018**. DOI: 10.1016/j.jiec.2018.05.020

Gbadamosi, A. O.; Junin, R.; Manan, M. A.; et al. An Overview of Chemical Enhanced Oil Recovery: Recent Advances and Prospects. *Int. Nano Lett.* **2019,** *9,* 171–202.

Gbadamosi, A. O.; Kiwalabye, J.; Junin, R.; Augustine, A. A Review of Gas Enhanced Oil Recovery Schemes Used in the North Sea. *J. Pet. Explor. Prod. Technol.* **2018,** *5,* 1–15.

Glasbergen, G.; Wever, D.; Keijzer, E.; Farajzadeh, R. *Injectivity Loss in Polymer Floods: Causes, Preventions and Mitigations*, SPE Kuwait Oil and Gas Show and Conference, 2015. DOI: 10.2118/175383-ms

Gong, H.; Li, Y.; Dong, M.; Ma, S.; Liu, W. Effect of Wettability Alteration on Enhanced Heavy Oil Recovery by Alkaline Flooding. *Colloids Surf. A Physicochem. Eng. Asp.* **2016,** *488,* 28–35.

Guo, H.; Zitha, P. L. J.; Faber, R. Buijse: A Novel Alkaline/Surfactant/Foam Enhanced Oil Recovery Process. *SPE J. SPE* **2012,** *10,* 145043.

Hanamertani, A. S.; Pilus, R. M.; Idris, A. K.; Irawan, S.; Tan, I. M. Ionic Liquids as a Potential Additive for Reducing Surfactant Adsorption onto Crushed Berea Sandstone. *J. Pet. Sci. Eng.* **2018,** *162,* 480–490.

Hanssen, J. E.; Holt, T.; Surguchev, L. M. Foam Processes: an Assessment of their Potential in North Sea Reservoirs Based on a Critical Evaluation of Current field Experience. In *SPE/DOE Improved Oil Recovery Symposium*, April 17–20, Tulsa, Oklahoma, 1994. DOI: 10.2118/27768-MS

Hawthorne, S. B.; Gorecki, C. D.; Sorensen, J. A.; Harju, J. A.; Steadman, E. N.; Melzer, S. Hydrocarbon Mobilization Mechanisms from Upper, Middle, and Lower Bakken Reservoir Rocks Exposed to CO. *Soc. Pet. Eng.* **2013**. DOI: 10.2118/167200-MS

Hinrichsen, C. J.; Brindle, K. H. *Scale and Corrosion Control in High Temperature Oil Production and Steam Flood Operations*, 18th Annual Convention Proceedings, Vol. 2, 1989; pp 277–300.

Hirasaki, G.; Miller, C. A.; Puerto, M. Recent Advances in Surfactant EOR. In *SPE Annual Technical Conference and Exhibition*, Denver, Colorado, USA, 21–24 September, 2008.

Hoffman, B. T. Comparison of Various Gases for Enhanced Recovery from Shale Oil Reservoirs. In *SPE 154329, the Eighteenth SPE Improved Oil Recovery Symposium held in Tulsa*, Oklahoma, USA, 14–18 April, 2012.

Ismailov, F. S.; Veliyev, E. F. Nanofluid for Enhanced Oil Recovery. *JPSE* **2011**, *78,* 431–437.

Iwere, F. O.; Heim, R. N.; Cherian, B. V. Numerical Simulation of Enhanced Oil Recovery in the Middle Bakken and Upper Three Forks Tight Oil Reservoirs of the Williston Basin. *Soc. Pet. Eng.* **2012**. DOI: 10.2118/154937-MS

Jin, J.; Li, X.; Geng, J.; Jing, D. Insights into the Complex Interaction Between Hydrophilic Nanoparticles and Ionic Surfactants at the Liquid/Air Interface. *Phys. Chem. Chem. Phys.* **2018,** *20* (22), 15223–15235.

Ju, B.; Fan, T.; Ma, M. Enhanced Oil Recovery by Flooding with Hydrophilic Nanoparticles, *China Particuol.* **2006,** *4,* 41–46.

Kim, J. T.; Kim, C. A.; Zhang, K.; Jang, C. H.; Choi, H. J. Effect of Polymer–Surfactant Interaction on its Turbulent Drag Reduction. *Colloids Surf. A Physicochem. Eng. Asp.* **2011,** *391* (1-3), 125–129.

Kovscek, A. R.; Wong, H.; Radke, C. J. A Pore-Level Scenario for the Development of Mixed Wettability in Oil Reservoirs. *AIChE J.* **1993,** *39* (6), 1072–1085.

Kurtoglu, B.; Sorensen, J. A.; Braunberger, J.; Smith, S.; Kazemi, H. Geologic Characterization of a Bakken Reservoir for Potential CO_2 EOR. *Soc. Pet. Eng.* **2013**. DOI: 10.1190/URTEC2013-186

Lawson, J. B. The Adsorption of Non-Ionic and Anionic Surfactants on Sandstone and Carbonate. In *SPE Symposium on Improved Methods of Oil Recovery*, April 16–17, Tulsa, Oklahoma, 1978.

Levitt, D.; Jackson, A.; Britton, L. N.; et al. Identification and Evaluation of High-Performance EOR Surfactants. In *SPE/DOE Symposium on Improved Oil Recovery*, April 22–26, Tulsa, Oklahoma, 2006.

Li, R. F.; Yan, W.; Liu, S.; et al. Foam Mobility Control for Surfactant Enhanced Oil Recovery. *SPE J.* **2010,** *15* (4), 92842.

Ling, K.; Shen, Z.; Han, G.; He, J. A *Review of Enhanced Oil Recovery Methods Applied in Williston Basin*, Proceedings of the 2nd Unconventional Resources Technology Conference, 2014. DOI: 10.15530/urtec-2014-1891560

Liu, S.; Miller, C. A.; Li, R. F.; Hirasaki, G. Alkaline/Surfactant/Polymer Processes: Wide Range of Conditions for Good Recovery. *SPE J.* **2010,** *15* (2), 282–293.

Luca, G. An Overview of Hydraulic Fracturing and other Formation Stimulation Technologies for Shale Gas Production, *JRC Technical Report.* JRC86065, EUR 26347 EN, ISBN 978-92-79-34729-0 (pdf), ISSN 1831-9424 (online), 2013. DOI: 10.2790/99937

Ma, T.; Feng, H.; Wu, H.; Li, Z.; Jiang, J.; Xu, D.; Kang, W. Property Evaluation of Synthesized Anionic-Nonionic Gemini Surfactants for Chemical Enhanced Oil Recovery. *Colloids Surf. A Physicochem. Eng. Asp.* **2019,** 123800. DOI: 10.1016/j.colsurfa.2019.123800

Mandal, A. Chemical Flood Enhanced Oil Recovery: A Review. *Int. J. Oil Gas. Coal. Technol.* **2015,** *9,* 241–264 .

Mandal, A.; Samanta, A.; Ojha, K. Mobility Control and Enhanced Oil Recovery Using Partially Hydrolysed Polyacrylamide (PHPA). *Int. J. Oil Gas Coal Technol.* **2013,** *6,* 245–258.

Manrique, E. J.; Muci, V. E.; Gurfinkel, M. E. EOR Field Experiences in Carbonate Reservoirs in the United States. *Soc. Pet. Eng.* **2007**. DOI: 10.2118/100063-PA

Mathiassen, O. M. CO_2 as Injection Gas for Enhanced Oil Recovery and Estimation of the Potential on the Norwegian Continental Shelf. Ph.D. Thesis, Department of Petroleum Engineering and Applied Geophysics, Norwegian University of Science and Technology, Trondheim, 2003.

Mohammed, M.; and Babadagli, T. Wettability Alteration: A Comprehensive Review of Materials/Methods and Testing the Selected Ones on Heavy-Oil Containing Oil-Wet Systems. *Adv. Colloid Interface Sci.* **2015,** *220,* 54–77.

Morsy, S.; Zhou, J.; Lant, K.; Cutler, J.; Sun, H.; Qu, Q.; Shuler, P. Optimizing Surfactant Additives for Enhanced Well Stimulation in Bakken Formation. *Soc. Pet. Eng.* **2014**. DOI: 10.2118/168180-MS

Negin, C.; Ali, S.; Xie, Q. Most Common Surfactants Employed in Chemical Enhanced Oil Recovery. *Petroleum* **2017,** *3,* 197–211.

Olajire, A. A. Review of ASP EOR (Alkaline Surfactant Polymer Enhanced Oil Recovery) Technology in the Petroleum Industry: Prospects and Challenges. *Energy* **2014,** *77,* 963–982.

Ogolo, N.; Olafuyi, O.; Onyekonwu, M. Enhanced Oil Recovery Using Nanoparticles. In *SPE Saudi Arabia Section Technical Symposium and Exhibition*, 2012.

Pal, S.; Mushtaq, M.; Banat, F.; Al Sumaiti, A. M. Review of Surfactant-Assisted Chemical Enhanced Oil Recovery for Carbonate Reservoirs: Challenges and Future Perspectives. *Pet. Sci.* **2017,** *15* (1), 77–102.

Pautz, J.; Sellers, C.; Nautiyal, C.; Allison, E. *Enhanced Oil Recovery Projects Database.* National Institute for Petroleum and Energy Research/US Department of Energy; Bartlesville, Oklahoma, 1992.

Pei, H.; Zhang, G.; Ge, J.; Jin, L.; Ma, C. Potential of Alkaline Flooding to Enhance Heavy Oil Recovery Through Water-in-Oil Emulsification. *Fuel* **2013,** *104*, 284–293.

Phukan, R.; Gogoi, S. B.; Tiwari, P. Enhanced Oil Recovery by Alkaline-Surfactant-Alternated-Gas/CO_2 Flooding. *J. Petrol. Explor. Prod. Technol.* **2019,** *9,* 247–260.

Phukan, R.; Gogoi, S. B.; Tiwari, P. Effects of CO_2-Foam Stability, Interfacial Tension and Surfactant Adsorption on Oil Recovery by Alkaline-Surfactant-Alternated-Gas/ CO_2 Flooding. *Colloids Surf. A Physicochem.Eng. Asp.* **2020,** 124799. DOI: 10.1016/j.colsurfa.2020.124799

Price, L. C. *Origins and Characteristics of the Basin-Centered Continuous Reservoir Unconventional Oil-Resource base of the Bakken Source System, Williston Basin.* Energy and Environmental Research Center, University of North Dakota: Grand Forks, ND, 2000.

Rabbani, E.; Davarpanah, A.; Memariani, M. An Experimental Study of Acidizing Operation Performances on the Wellbore Productivity Index Enhancement. *J. Pet. Explor. Prod. Technol.* **2018**. DOI:10.1007/s13202-018-0441-8

Rafati, R.; Haddad, A. S.; Hamidi, H. Experimental Study on Stability and Rheological Properties of Aqueous Foam in the Presence of Reservoir Natural Solid Particles. *Colloids Surf. A Physicochem. Eng.* **2016,** *509*, 19–31.

Ribeiro, L.; Sharma, M. Fluid Selection for Energized Fracture Treatments. *SPE Hydraul. Fract. Technol. Conf.* **2013**. DOI: 10.2118/163867-ms

Rosen, M. J.; Kunjappu, J. T. *Surfactants and Interfacial Phenomena*, 4th ed.; Wiley: Hoboken, 2012.

Rowan, E. L.; Engle, M. A.; Kraemer, T. F.; Schroeder, K. T.; Hammack, R. W.; Doughten, M. W.. Geochemical and Isotopic Evolution of Water Produced from Middle Devonian Marcellus Shale Gas Wells, Appalachian Basin Pennsylvania. *AAPG Bull.* **2015,** *99* (2), 181–206.

Sagir, M.; Tan, I. M.; Mushtaq, M.; et al. Synthesis of a New CO_2 Philic Surfactant for Enhanced Oil Recovery Applications. *J. Dispers. Sci. Technol.* **2013,** *35* (5), 647–654.

Samanta, A.; Bera, A.; Ojha, K.; Mandal, A. Comparative Studies on Enhanced Oil Recovery by Alkali–Surfactant and Polymer Flooding. *J. Pet. Explor. Prod. Technol.* **2012,** *2,* 67–74.

Samin, A. M.; Manan, M. A.; Idris, A. K.; Yekeen, N.; Said, M.; Alghol, A. Protein Foam Application for Enhanced Oil Recovery. *J. Dispers. Sci. Technol.* **2017,** *38,* 604–609.

Smith, P. G.; Dhanuka, V. V.; Hwang, H. S.; et al. Tertiary Amine Esters for Carbon Dioxide Based Emulsions. *Ind. Eng. Chem. Res.* **2007,** *46* (8), 2473–2480.

Speight, J. G. *Handbook of Petroleum Product Analysis*, 2nd ed.; John Wiley & Sons Inc.: Hoboken, NJ, 2014.

Speight, J. G. Properties of Organic Compounds. *Environ. Org. Chem. Eng.* **2017,** 203–261. DOI: 10.1016/b978-0-12-804492-6.00005-8

Sun, X.; Zhang, Y.; Chen, G.; Gai, Z. Application of Nanoparticles in Enhanced Oil Recovery: A Critical Review of Recent Progress. *Energies* **2017,** *10,* 345.

Tang, X.; Xiao, S.; Lei, Q.; Yuan, L.; Peng, B.; He, L.; Pei, Y. Molecular Dynamics Simulation of Surfactant Flooding Driven Oil-Detachment in Nano-Silica Channels. *J. Phys. Chem. B* **2019**. DOI: 10.1021/acs.jpcb.8b09777

Thacker, J. B.; et al. Chemical Analysis of Wastewater from Unconventional Drilling Operations. *Water* **2015,** *7* (4), 1568–1579

Thomas, S. Enhanced Oil Recovery—An Overview. *Oil. Gas. Sci. Technol. Rev. l'IFP* **2007,** *63,* 9–19.

Vipulanandan, C.; Mohammed, A.; Qu, Q. Characterizing the Hydraulic Fracturing Fluid Modified with Nano Silica Proppant, 2014.

Wang, J.; Han, M.; Fuseni, A. B.; et al. *Surfactant Adsorption in Surfactant/Polymer Flooding for Carbonate Reservoirs*, SPE Middle East Oil & Gas Show and Conference, March 8–11, Manama, Bahrain, 2015. DOI: 10.2118/172700-MS

Wang, X.; Luo, P.; Er, V.; Huang, S.-S. S. Assessment of CO_2 Flooding Potential for Bakken Formation, Saskatchewan. *Soc. Pet. Eng.* **2010**. DOI: 10.2118/137728-MS

Wei, B. Advances in Polymer Flooding. In *Viscoelastic and Viscoplastic Materials*; El-Amin, M. F., Ed.; IntechOpen, 2014.

Wei, B.; Romero-Zerón, L.; Rodrigue, D. Oil Displacement Mechanisms of Viscoelastic Polymers in Enhanced Oil Recovery (EOR): A Review. *J. Pet. Explor. Prod. Technol.* **2013,** *4* (2), 113–121.

Xie, J.; Chung, B.; Leung, L. Design and Implementation of a Caustic Flooding EOR Pilot at Court Bakken Heavy Oil Reservoir. In *SPE/PS/CHOA 117221/PS2008-323, the 2008 SPE International Thermal Operations and Heavy Oil Symposium held in Calgary*, Alberta, Canada, 20–23 October, 2008.

Yekeen, N.; Idris, A. K.; Manan, M.; Samin, A. M.; Risal, A. R.; Kun, T. X. Bulk and Bubble-Scale Experimental Studies of Influence of Nanoparticles on Foam Stability. *Chin. J. Chem. Eng.* **2017,** *25,* 347–357.

Zaitoun, A.; Makakou, P.; Blin, N.; Al-Maamari, R. S.; Al-Hashmi, A.-A. R.; Abdel-Goad, M. Shear Stability of EOR Polymers. SPE-141113-PA. *SPE J.* **2012,** *17,* 335–339.

CHAPTER 18

Chemical EOR in India

KRISHNA RAGHAV CHATURVEDI and TUSHAR SHARMA

Department of Petroleum Engineering & Geological Sciences, Rajiv Gandhi Institute of Petroleum Technology, Jais, Amethi, India

ABSTARCT

While India has several oilfields undergoing pressure depletion, the application of chemical enhanced oil recovery (C-EOR) methods has been limited to few due to unfavorable costs, unavailable technology, and the lack of motivation from policy makers. The Directorate General of Hydrocarbons investigates best practices, spurred by insufficient focus on EOR despite its potential. This work explores a few success stories include Oil and Natural Gas Corporation (ONGC)'s commercial EOR activities in the oilfields of Gujarat, including Gandhar and Viraj, low salinity water flooding in Bombay High, as well as gas injection studies some fields of the Indian North-East. The role of chemical EOR to boost oil recovery in the fields of Rajasthan by Cairn India (now Vedanta) has also been reported in this work. Since EOR's inception in Gujarat in the 1980s, Cairn India's success in the Rajasthan fields, notably the Mangala project, has positioned India as a global EOR leader. Similarly, ONGC's involvement in multiple EOR projects enhances India's appeal in this domain. India's crude oil reserves, primarily light and sweet, present economic opportunities. EOR techniques encompass alkaline surfactant flooding, immiscible hydrocarbon water alternating gas injection, in situ combustion, and simultaneous water and gas injection. Subsequently, in 2018, fiscal incentives were approved to stimulate investment and technology development for aging oil and gas reserves, anticipating substantial hydrocarbon releases over the next two decades. Fiscal incentives involve chess reduction on oil produced through EOR projects. ONGC spearheads

Advancements in Chemical Enhanced Oil Recovery.
Tushar Sharma, Krishna Raghav Chaturvedi, Tarek Ganat, and Imtiaz Ali (Eds.)

various EOR technologies in Gujarat, Assam, and Mumbai. Public sector undertakings, ONGC and IOCL, collaborate on an EOR carbon capture, utilization, and storage (CCUS) project. Oil India Limited explores CO_2-assisted EOR, and Cairn Oil & Gas invests in large-scale EOR programs. Challenges for chemical EOR in India include limited petroleum engineering education and expensive analytical equipment. Nevertheless, ONGC's collaboration with Baker Hughes for a significant EOR project signifies ongoing efforts. The Indian government's policies aim to reduce oil imports and meet climate goals through EOR. The future potential for chemical EOR in India aligns with the nation's need for increased oil output and decreasing fresh finds. The government's keen interest, incentives, and policies underscore the pivotal role of EOR in securing India's energy self-reliance, addressing geopolitical and energy security concerns. Finally, the challenges and opportunities in the implementation of chemical EOR have also been explored in this work.

18.1 INTRODUCTION

India has a sizable number of mature fields of various sizes, some of which have been productive for decades. On the other hand, EOR has almost certainly received insufficient attention (Alvarado and Manrique, 2010). The Directorate General of Hydrocarbons, India's upstream oil and gas regulator, has initiated an investigation into EOR best practices for Indian fields, which would include stakeholder engagement. EOR was launched in India in the early 1980s, with fields in Gujarat (Baskaran et al., 2014). The Santhal, Balol, and Lanwa factories in Gujarat have commercialized thermal recovery technology. Furthermore, Oil and Natural Gas Corporation (ONGC) is exploring commercial EOR activities in other Gujarat fields, including Gandhar and Viraj, as well as gas injection in Borholla, Assam (Pratap and Gauma, 2004). However, Cairn India's achievement in the Rajas fields using chemical EOR has propelled India to the forefront of the global EOR effort. The operator was fully aware of the vital function of EOR during the field's early phases of development. Mangala, the largest producer in Cairn India, began production in 2009, while the pilot project commenced operations in 2011. To evaluate incremental recovery, the operator executed a thorough EOR pilot for the Mangala field, which included a water flood, a polymer flood, and an alkaline surfactant polymer (ASP) flood. Chemical EOR was previously established and is now being applied in other Rajasthan sectors such as Bhagyam and Aiswariya, following the success of the Mangala EOR project (Kumar Singh et al., 2021). This has resulted in the plateau production of the field remaining greater than anticipated during the field's early growth phase. The operator has transferred the lessons they gained in Managla to other districts of Rajasthan (Kumar et al., 2016). It has developed to become one of the largest such endeavors in the world. According to the operator's current assessment, these three fields contain a combined recoverable oil EOR resource base of 300 million barrels. At the moment, all of India's EOR fields are on land and include sandstone/clastic reservoirs, but the majority of mature resources are offshore and contain carbonates. This complicates the installation of EOR in the Indian offshore. According to media sources, ONGC is now engaged in 28 EOR projects, and the Indian government is considering auctioning off certain EOR properties, which is encouraging. All of these characteristics contribute to India's attractiveness as an EOR destination (Bera et al., 2021).

18.2 CURRENT ECONOMIC OPPORTUNITIES IN INDIA

India's crude oil reserves are predominantly light and sweet, ranging in specific gravity from 38° API offshore to 32° API in a variety of onshore basins (Baskaran et al., 2014; Bera et al., 2021). In 2009, India produced around 880,000 barrels of oil per day (bbl/d) from more than 3600 operating oil wells. EOR is a word that refers to procedures that extract greater oil from a petroleum reservoir than can be recovered through primary and secondary operations. The word "primary recovery" is commonly used to refer to output that is fuelled by the energy present in the reservoir, such as gas under pressure or natural water drive. Secondary recovery is sometimes used interchangeably with tertiary recovery as it typically involves water injection or flooding. The technologies in this collection are grouped into four categories: heat recovery, gas miscible recovery, chemical flooding, and microbiological flooding. EOR techniques used in India include alkaline surfactant flooding (ASF) in Jhalora, Rajasthan, Kalol, Gujarat, and Ankleshwar (Tiwari et al., 2008), Gujarat, immiscible hydrocarbon water alternating gas injection (HC-WAG) in the Gandhar fields of southern Gujarat, in situ combustion/air injection, and simultaneous water and gas injection (SWAG) in the Mumbai High. Immiscible CO_2, in situ microbial stimulation, and paraffin control microorganisms are among the EOR pilots presently under development in western onshore India. It is intended to inject air into light oil in the western offshore, to inject immiscible flue gas in the western onshore, and to inject immiscible CO_2 in the eastern onshore as well as monitor the impact of in situ CO_2 on the performance of existing chemical additives (Mishra and Pandey, 2017). Due to India's limited domestic production options, the Indian government and other oil exploration and production (E&P) organizations foresee a future rise in geographical dependency and are hence interested in increased oil recovery techniques. Sanand oil field was found in 1962 and production began in 1969 in Gujarat's Ahmedabad-Mehsana tectonic block (Dass et al., 2008). In 1985, polymer flooding was employed in five separate spot injection processes, recovering 23% of proved oil in situ (as stated in DGH India's press release). ONGC has completed pilot studies on polymer flooding on the Kalol field in Gujarat's Ahmedabad asset and is now conducting miscible gas flooding and CO_2 injection pilot trials in Ankleshwar. Jhalora oil field was discovered in 1967 and production began in 1978 in Gujarat's Ahmedabad-Mehsana tectonic block (Pratap et al., 1997). Jhalora has a total production capacity of 3.41 million barrels of oil from 29 wells. In the Jhalora field, the use of ASP flooding in a K-IV well with sandstone rock leads to a 23% improvement in displacement efficiency. Additionally, ASP flooding is being deployed in the Viraj oil field in the Cambay basin, where it has increased by 18% in OOIP recovery. Additionally, polymer flooding is being tested in the Mangala field in the Barmer basin to successfully reduce the water cut (Shankar et al., 2018).

18.3 INDIA'S ENGAGEMENT

According to Oil Minister Dharmendra Pradhan, the Indian government approved fiscal incentives in September 2018 to spur investment and technology development for reclaiming ageing oil and gas reserves, which are expected to release an estimated Rs 50 lakh crore of hydrocarbons over the next two decades ("Government to auction unmonetised oil and gas fields of ONGC, OIL: Dharmendra Pradhan- The New Indian Express," n.d.). The government would deduct half of the Rs 4500 per tonne cess on oil produced through EOR and IOR projects in ONGC-nominated fields and Vedanta Ltd's Rajasthan block. Gas would receive a royalty cut of 75%. "As a consequence, new investment and technology will flood the industry, boosting output by Rs 50 lakh crore over the next decade," the minister said at a news conference here. The output and recovery rate of conventional oil and gas fields decline as they age. EOR and IOR procedures such as polymer injection must be used to speed recovery. "Technological advancements have the potential to dramatically increase hydrocarbon reserve recovery in mature/ageing areas." Over the next two decades, a 5% increase in the rate of recovery of original in-place volume in oil production is estimated to result in the production of 120 million tonnes of additional oil. "In the case of gas, we predict a 3% improvement in recovery rate on initial in-place volume, resulting in an additional 52 billion cubic meters of gas output over the next two decades," he said. The policy, which contains a sunset clause, will be in force for 10 years from the date of its announcement. On the other hand, fiscal incentives for such projects will be available for 120 months from the commencement of production. If increased recovery actions are undertaken, prizes will be made promptly upon accomplishment of the established threshold. The policy's various stages have been assigned deadlines. Pradhan stated that the Union Cabinet approved a policy framework to promote and incentivize enhanced recovery (ER)/improved recovery (IR)/ unconventional hydrocarbon (UHC) production methods/technologies to increase the recovery factor of existing hydrocarbon reserves to augment domestic oil and gas production. The policy will apply to all contractual regimes and geographical regions mentioned in the policy. "We anticipate that this policy approach will boost new investment, economic activity, and job creation." "The strategy is expected to accelerate the adoption of new, innovative, and cutting-edge technologies, as well as the establishment of technical collaborations to increase existing field output." He alluded to this. The CEO of Vedanta Ltd's Oil & Gas business made a statement on the decision. The judgement, according to Sudhir Mathur, will contribute to the development of India's oil and gas industry. "This plan will attract significant investment and a wave of best-in-class technologies geared at boosting India's hydrocarbon recovery," he continued. While enhanced oil and gas recovery (EOR) and enhanced gas recovery (EGR) are classified as UHC production methods, shale oil and gas production, tight oil and gas production, oil shale production, gas hydrate production, and heavy oil production are also classified as UHC production methods. UHC E&P, as well as ER and IR, are all capital-intensive, technically complex, and physically demanding. According to Pradhan, they require supported infrastructure, logistical help, economic incentives, and an enabling environment. The policy's strategic objective is to foster collaboration between universities and research institutions, as well as between industry and academia, in order to create a supportive ecosystem. "As well as to promote and encourage E&P contractors to employ ER/IR/ UHC methods/approaches," he explained.

18.3.1 EFFORTS OF THE INDIAN OIL AND NATURAL GAS CORPORATION

In Gujarat, Assam, and Mumbai, ONGC is presently implementing a range of EOR technology. The necessity of deploying EOR to improve output comes as ONGC crude oil production fell 7% to 8.20 million tonnes equivalent from April to June 2019, boosting India's reliance on oil imports to 85.2%, up from 83.8 % the previous year. In 2018/2019, ONGC's total crude oil output fell to a 16-year low of 21.042 million tonnes, down from a record high of 26.485 million tonnes in 2004/05 (ONGC, 2017).

18.3.2 OTHER PUBLIC SECTOR UNDERTAKINGS (PSUS) IN INDIA

In India, ONGC and IOCL will begin work on an enhanced oil recovery (EOR) carbon capture, utilization, and storage (CCUS) project. This CCUS initiative's goal is to reduce carbon emissions while increasing oil recovery from domestic resources. The ONGC and the Indian Oil Corporation Limited (IOCL) have teamed to build a CCUS project in India with the objective of reducing carbon emissions while enhancing oil recovery. In this context, the two Indian state-owned conglomerates have chosen to pursue EOR utilizing CO2 collected at IOCL's Koyali Refinery in western India. The MOU aims to boost domestic hydrocarbon output while assisting the government in meeting its COP21 carbon emission reduction targets. According to ONGC, CCUS is an extremely effective kind of EOR that is gaining popularity as a critical component of achieving carbon neutrality. The MOU, according to ONGC, would lay the groundwork for the two companies to collaborate on mutually beneficial CO2-EOR projects as part of their CCUS strategy. The parties have committed to a common goal: resolving the country's energy security and climate change concerns. As part of the CCUS project, the partners want to construct a CO2 collection facility at the Koyali Refinery. The plant will employ carbon capture technology. Additionally, the collaboration will focus on developing a commercial strategy for expanding domestic oil output through the use of CO2-EOR at Gujarat's Gandhar oil field. Additionally, the partners will endeavor to incorporate the CCUS program into a national emission reduction strategy aimed at supporting India in achieving its low-carbon development goals.

According to ONGC, the technical feasibility of CO2-EOR for the Gandhar oil field has been proven, meaning an additional 10% oil recovery. "The project will add new dimensions to CCUS's national strategy and revitalize ONGC's mature oil assets," the company said in a statement. The learning curve associated with this program will provide the groundwork for future CCUS implementation in India. The success of CCUS in India would not only improve local oil production, but will also assist the government meet its Paris climate agreement-mandated Nationally Determined Contributions (NDCs) of reducing the emission intensity of GDP by 33–35 % by 2030."

Additionally, Oil India Limited, a state-owned public sector corporation with its headquarters in Assam, is conducting research in its field on the use of carbonated water and ASP floods ("IndianOil, OIL tie up for CO2-assisted enhanced oil recovery | IBEF," n.d.). Oil India Limited has expanded its reliance on water injection in a number of oil fields, including NHK, MRN, SLM, JRN, and Chabua, to maintain pressure. Flooding using a mixture of water and polymer is a well-established method of EOR. About 11 water injection stations (WIS) are now active, pumping a total of 9800 kilolitres of fresh water daily through around 40 water injection wells to maintain pressure in various reservoirs. Assam fields currently have 10 WIS. Indian Oil and Oil India Ltd (OIL), like ONGC, have opted to participate in an EOR project in Assam that would utilize CO2. The two companies will work on the creation of a -CCUS facility to help mitigate climate change. According to Indian Oil, injecting CO2 into oil reservoirs for EOR may be an exciting option for enhancing CCUS activity in support of climate change-related carbon storage goals while simultaneously increasing oil well productivity. The business proposes to transfer CO2 collected from the flue-gas stacks of the Digboi Refinery's Hydrogen Generation Unit and gas turbine power plant to OIL's Nahorkatiya and Dikom oil fields in Assam, roughly 50–60 kilometers from Digboi ("IndianOil, OIL tie up for CO2-assisted enhanced oil recovery - The Hindu BusinessLine," n.d.).

18.3.3 PRIVATE SECTOR INITIATIVES IN INDIA

Cairn Oil & Gas is India's largest E&P company for oil and gas. It is wholly owned by Vedanta Limited. Cairn supplied 25% of India's domestic crude oil production in fiscal year 2018–2019 and aims to supply 50% of the country's oil and gas by 2025. Cairn has built the world's longest continuously heated and insulated pipeline, completed the world's largest polymer flood project for EOR, and is now building the world's largest ASP project. Cairn has a world-class resource base, with a present stake in 58 blocks in India, comprising 41 blocks won under the OALP Round I auction, 5 blocks awarded under the OALP Rounds II and III auctions, and 2 blocks awarded under the Discovered Small Fields (DSF) Round-II auction. The states of

Andhra Pradesh, Assam, Tamil Nadu, Tripura, Rajasthan, Maharashtra, and Gujarat are home to these blocks. Cairn's portfolio has grown to 65,000 square kilometers, ushering in the next phase of the country's hydrocarbon growth. Cairn India NSE 0.81% is prepared to start on the world's largest EOR program, investing over $760 million in Rajasthan's Thar desert, where the business made its maiden oil discovery. As part of the strategy, Cairn India CEO Mayank Ashar revealed that the company will drill 100 exploration and appraisal wells and 350 development wells in 2016. The EOR investment will benefit the Mangala and Bhagyam oil fields in the Rajasthan block, helping to reverse natural decline and maintain the high production ratio necessary to fulfil Indian demands (the company's Mangala oil field now provides around 23% of India's domestic crude needs) (Kumar et al., 2016). The Bhagyam field will be able to generate an additional 40–50 million barrels of untapped oil with an appropriate EOR deployment. The project is expected to cost USD 260 million in total, with incremental production costs ranging between USD 5 and USD 6 per barrel (Shankar et al., 2018).

Cairn Oil & Gas, a subsidiary of Vedanta Limited, has granted Baker Hughes, a GE business, a development contract in Rajasthan to drill around 300 new wells and conduct a chemical EOR program. The project is BHGE's largest integrated project in India, increasing the company's foothold in the country and supporting the government in meeting its aim of reducing dependency on imports. BHGE will deliver a complete range of oil field services and equipment to Cairn Oil & Gas and Vedanta Limited over the next 24 months to aid them in unlocking huge untapped reserves in the Mangala, Bhagyam, and Aishwariya (MBA) fields. This includes both technology and processes for promptly and cost-effectively constructing new wells, as well as solutions for increasing the output of existing wells. The project will be phased. BHGE is developing a solution for ASP flooding to maximize recovery from existing wells, a first in India and one of the most cutting-edge kinds of chemical EOR currently accessible. BHGE's efforts encompass both chemistry development and the installation of chemical recovery units. "By accelerating operations in the MBA fields, we hope to boost output and meet the country's energy requirements," Sudhir Mathur, CEO of Vedanta Limited's Cairn Oil & Gas business, explained, "our partner was awarded the contract based on their ability to develop cost-effective solutions that increase output." "BHGE has been a long-standing partner of Cairn, and this project will strengthen our relationship even further." "This is a critical link that highlights both our technological superiority and the value of our integrated portfolio." "Our cutting-edge technological solutions, which range from reservoir modeling and well design to chemical injection and facility construction, seek to give our clients with unrivalled levels of production." As stated by Ashish Bhandari, President and CEO of Baker Hughes in India and South Asia, a GE Company, "this is a crucial strategic project that will help India achieve its goal of decreasing oil and gas imports while increasing domestic energy production to meet growing demand." The 3-year endeavor is scheduled to commence in the latter half of 2018. The activity is expected to aid in the growth and support of regional supply chain partners, as well as the development of regional worker competencies (Kumar Singh et al., 2021).

18.4 CHALLENGES FOR CHEMICAL EOR IN INDIA

There are several challenges that have inhibited the growth of chemical EOR in India. They have been provided as Figure 18.1.

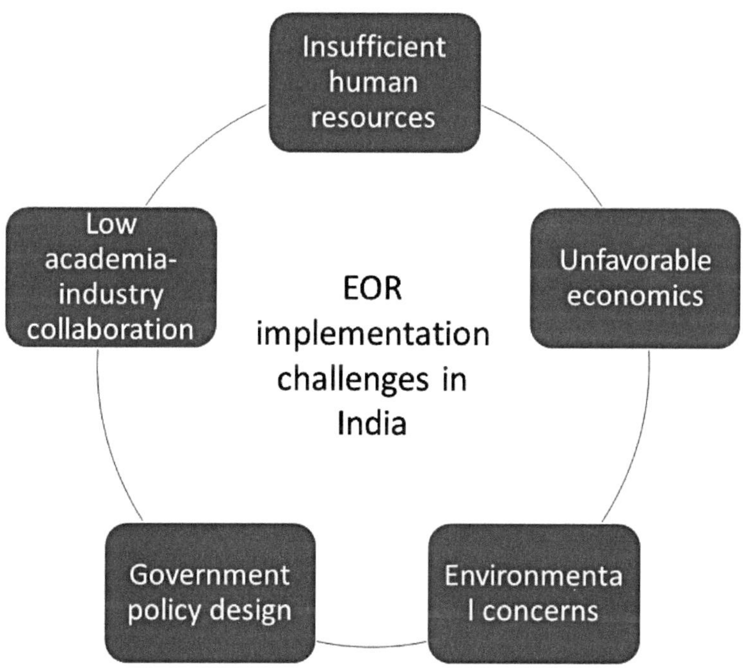

FIGURE 18.1 Challenges preventing long-term viable implementation of chemical EOR methods in India.

Petroleum engineering is a unique subject that is not taught alongside advanced research at all higher educational institutions in India, including the

Indian Institute of Science (IISc), the India Institute of Technology (IIT), the Indian Institute of Science, Education, and Research (IISER), the Council of Science and Industrial Research Institutes (CSIT), the National Institute of Technology (NIT), and central and state universities. Petroleum engineering courses are available at just a few government and commercial universities, and research facilities and opportunities are few. Petroleum engineering research and development are continued by employees with nonpetroleum engineering backgrounds. Although prior knowledge in physics, chemistry, mathematics, biology, and computer engineering is critical for successfully implementing a petroleum engineering project in the EOR domain, a subject-matter expert is always preferable for ultimate decision-making. Unfortunately, few petroleum engineers have maintained a domain expertise in EOR within India (Bera et al., 2021).

Furthermore, several analytical equipments are required for the analysis of a successful EOR implementation method, unfortunately, such equipments are highly expensive (Table 18.1). This has limited ongoing research and constrained field-specific performance evaluation, constraining their applicability in wider oil field applications (Chaturvedi et al., 2021; Hu et al., 2016).

TABLE 18.1 Most Essential Equipment for Oil Field Chemical Characterization.

Equipment	Purpose
Core flooding apparatus	Oil recovery investigations, water shutoff, and gas storage studies (Al-Abri et al., 2019; Chaturvedi and Sharma, 2020)
IFT tensiometer	Determination of interfacial tension (Bagalkot et al., 2018; Rezk and Foroozesh, 2019)
Contact angle assembly	Wettability alteration studies (Kumar et al., 2021)
Dynamic foam analyzer	Stability and structure of foams (Farhadi et al., 2016)
Ion chromatography	Efficacy of low-salinity water flooding (Al-Saedi et al., 2019)
Rheometer	Rheology and viscoelasticity of chemical solutions (Zhang et al., 2019)

18.5 INDIA'S FUTURE POTENTIAL FOR CHEMICAL EOR

Given the pressing need to increase output from Indian oil fields and the dwindling number of fresh finds, the Indian government is severely concerned about energy security, which is crucial for preserving the region's geopolitical balance ("Govt working on policy to incentivise enhanced oil recovery, says Pradhan - The Hindu BusinessLine," n.d.; ONGC, 2017). While India has a multiplicity of foreign suppliers, escalating tensions with China and the Indian government's goal for self-sufficiency have encouraged Indian hydrocarbon companies to examine chemical EOR technology as a way to boost local output and reduce India's dependency on foreign oil (Dasgupta, 2018). To that end, the Indian government has given many incentives to promote chemical EOR and is encouraging universities/research institutes to include EOR into their curriculum in order to speed information dissemination. This is encouraging for chemical EOR's future, but it will require researchers to create novel EOR procedures for use in Indian oil fields. The advantages of promoting OER in India are manifold (Figure 18.2).

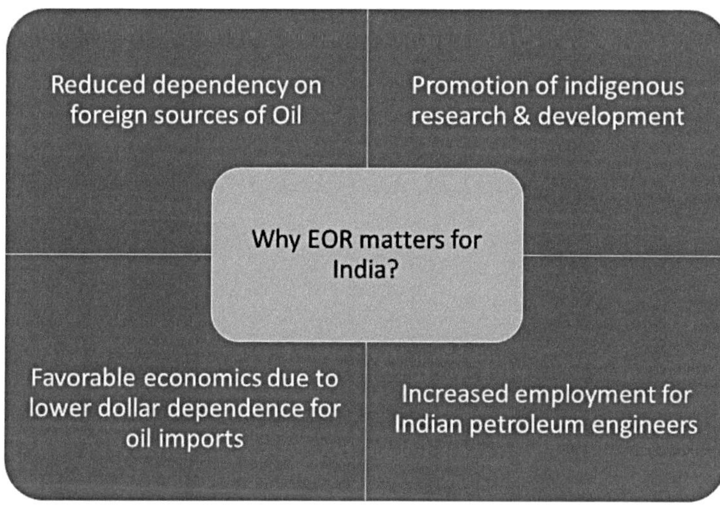

FIGURE 18.2 The role of EOR in making India, energy self-reliant.

Hopefully, several Indian GOVERNMENT policy decisions have tended to promote EOR implementation in the recent years (Bera et al., 2021). The time, for sure, they are changing.

KEYWORDS

- **chemical EOR**
- **Cairn Energy**
- **polymer flooding**
- **ONGC**
- **human resources**

REFERENCES

Al-Abri, H.; Pourafshary, P.; Mosavat, N.; Al Hadhrami, H. A Study of the Performance of the LSWA CO 2 EOR Technique on Improvement of Oil Recovery in Sandstones. *Petroleum* **2019,** *5*, 58–66. https://doi.org/10.1016/j.petlm.2018.07.003

Al-Saedi, H. N.; Brady, P. V.; Flori, R. E.; Heidari, P. Insights into the Role of Clays in Low Salinity Water Flooding in Sand Columns. *J. Pet. Sci. Eng.* **2019,** *174*, 291–305. https://doi.org/10.1016/j.petrol.2018.11.031

Alvarado, V.; Manrique, E. Enhanced Oil Recovery: An Update Review. *Energies* **2010,** *3*, 1529–1575. https://doi.org/10.3390/EN3091529

Bagalkot, N.; Hamouda, A. A.; Isdahl, O. M. Dynamic Interfacial Tension Measurement Method Using Axisymmetric Drop Shape Analysis. *MethodsX* **2018,** 5, 676–683. https://doi.org/10.1016/j.mex.2018.06.012

Baskaran, V. K.; Dani, K. C.; Kumar, K. P.; Urkude, A. M.. Implementation of Enhanced Oil Recovery Techniques in India: New Challenges and Technologies. In *Society of Petroleum Engineers - SPE EOR Conference at Oil and Gas West Asia 2014: Driving Integrated and Innovative EOR*. Society of Petroleum Engineers, 2014; pp 227–242. https://doi.org/10.2118/169680-ms

Bera, A.; Vij, R. K.; Shah, S. Impact of Newly Implemented Enhanced Oil and Gas Recovery Screening Policy on Current Oil Production and Future Energy Supply in India. *J. Pet. Sci. Eng.* **2021,** *207*, 109196. https://doi.org/10.1016/J.PETROL.2021.109196

Chaturvedi, K. R.; Narukulla, R.; Amani, M.; Sharma, T.. Experimental Investigations to Evaluate Surfactant Role on Absorption Capacity of Nanofluid for CO2 Utilization in Sustainable Crude Mobilization. *Energy* **2021,** *225*, 120321. https://doi.org/10.1016/j.energy.2021.120321

Chaturvedi, K. R.; Sharma, T. Carbonated Polymeric Nanofluids for Enhanced Oil Recovery from Sandstone Reservoir. *J. Pet. Sci. Eng.* **2020,** *194*, 107499. https://doi.org/10.1016/j.petrol.2020.107499

Dasgupta, A. India's Strategy in the Indian Ocean Region: A Critical Aspect of India's Energy Security. *Jadavpur J. Int. Relat.* **2018,** *22*, 39–57. https://doi.org/10.1177/0973598418757817

Dass, C.; Jain, M. A.; Dhawan, A. K.; Misra, T. R. Monitoring of Polymer Flooding Project at Sanand Field of India. In *Society of Petroleum Engineers - SPE Indian Oil and Gas Technical Conference and Exhibition 2008: "The Changing Landscape Emerging Opportunities in the Indian E and P Industry*." OnePetro, 2008; pp 322–330. https://doi.org/10.2118/113552-MS

Farhadi, H.; Riahi, S.; Ayatollahi, S.; Ahmadi, H.. Experimental Study of Nanoparticle-Surfactant-Stabilized CO2foam: Stability and Mobility Control. *Chem. Eng. Res. Des.* **2016,** *111*, 449–460. https://doi.org/10.1016/j.cherd.2016.05.024

Government to Auction Unmonetised Oil and Gas Fields of ONGC, OIL: Dharmendra Pradhan- The New Indian Express, n.d.

Govt Working on Policy to Incentivise Enhanced Oil Recovery, says Pradhan - The Hindu Business Line [WWW Document], n.d. URL https://www.thehindubusinessline.com/economy/policy/govt-working-on-policy-to-incentivise-enhanced-oil-recovery-says-pradhan/article9585786.ece (accessed Jan 14, 2022).

Hu, Z.; Azmi, S. M.; Raza, G.; Glover, P. W. J.; Wen, D. Nanoparticle-Assisted Water-Flooding in Berea Sandstones. *Energy Fuels* **2016,** *30*, 2791–2804. https://doi.org/10.1021/acs.energyfuels.6b00051

IndianOil, OIL Tie Up for CO2-Assisted Enhanced Oil Recovery - The Hindu BusinessLine [WWW Document], n.d. URL https://www.thehindubusinessline.com/companies/indianoil-oil-tie-up-for-t-co2-assisted-enhanced-oil-recovery/article30535341.ece (accessed Jan 14, 2022).

Indian Oil, OIL Tie Up for CO2-Assisted Enhanced Oil Recovery | IBEF [WWW Document], n.d. URL https://www.ibef.org/news/indianoil-oil-tie-up-for-co2assisted-enhanced-oil-recovery (accessed Jan 14, 2022).

Kumar, P.; Raj, R.; Koduru, N.; Kumar, S.; Pandey, A. Field Implementation of Mangala Polymer Flood: Initial Challenges, Mitigation and Management. In *Society of Petroleum Engineers - SPE EOR Conference at Oil and Gas West Asia, OGWA 2016*. Society of Petroleum Engineers, 2016. https://doi.org/10.2118/179820-MS

Kumar, R. S.; Al-Arbi Ganat, T.; Sharma, T. Performance Evaluation of Silica Nanofluid for Sand Column Transport with Simultaneous Wettability Alteration: An Approach to Environmental Issue. *J. Clean. Prod.* **2021,** 303, 127047. https://doi.org/10.1016/j.jclepro.2021.127047

Kumar Singh, A.; Raju Vegesna, P.; Prasad, D.; Chandrashekar Kachodi, S.; Lohiya, S.; Srivastava, D.; Raj, R.; Koduru, N.; Mishra, S.; Barua, D.; Pandey, A.; Oil, C. Successful Implementation of Polymer Flood in Aishwariya Field, Rajasthan, India - Concept to Full Field. In *ADIPEC*. OnePetro, 2021. https://doi.org/10.2118/207943-MS

Mishra, S., Pandey, A. Modelling the Impact of Reservoir CO2 on Polymer Flood Performance in Aishwariya Field. In *ADIPEC*. SPE, 2017. https://doi.org/10.2118/186880-MS

ONGC, 2017. EOR in Indian Context.

Pratap, M.; Gauma, M. S. Field Implementation of Alkaline-Surfactant-Polymer (ASP) Flooding : A Maiden Effort in India. In *SPE Asia Pacific Oil and Gas Conference and Exhibition, APOGCE*. OnePetro, 2004, pp 97–101. https://doi.org/10.2118/88455-MS

Pratap, M.; Roy, R. P.; Gupta, R. K.; Singh, D. Field Implementation of Polymer EOR Technique - A Successful Experiment in India, in: SPE Annual Technical Conference and Exhibition. OnePetro, 1997. https://doi.org/10.2118/38872-MS

Rezk, M. G.; Foroozesh, J. Effect of CO2 Mass Transfer on Rate of Oil Properties Changes: Application to CO2-EOR Projects. *J. Pet. Sci. Eng.* **2019,** *180*, 298–309. https://doi.org/10.1016/j.petrol.2019.05.053

Shankar, V.; Jain, S.; Mishra, A.; Seth, S.; Rautela, C. M.; Jha, B. C. Polymer Injection Pilots in Bhagyam Oil Field and Full Field Polymer Development Planning, in: Society of Petroleum Engineers - SPE EOR Conference at Oil and Gas West Asia 2018. OnePetro, 2018; pp 26–28. https://doi.org/10.2118/190472-MS

Tiwari, D.; Marathe, R. V.; Patel, N. K.; Ramachandran, K. P.; Maurya, C. R.; Tewari, P. K. Performance Of Polymer Flood In Sanand Field, India - A Case Study, in: SPE Asia Pacific Oil and Gas Conference and Exhibition 2008 - "Gas Now: Delivering on Expectations." OnePetro, 2008; pp 372–380. https://doi.org/10.2118/114878-MS

Zhang, Y.; Gao, M.; You, Q.; Fan, H.; Li, W.; Liu, Y.; Fang, J.; Zhao, G.; Jin, Z.; Dai, C.. Smart Mobility Control Agent for Enhanced Oil Recovery During CO 2 Flooding in Ultra-Low Permeability Reservoirs. *Fuel* **2019,** 442–450. https://doi.org/10.1016/j.fuel.2018.12.069

CHAPTER 19

Conclusion

KRISHNA RAGHAV CHATURVEDI and TUSHAR SHARMA

Department of Petroleum Engineering & Geological Sciences, Rajiv Gandhi Institute of Petroleum Technology, Jais, Amethi, India

ABSTARCT

In the modern age, Enhanced Oil Recovery (EOR) practices will continue to hold an importance to improve the quantum of subsurface resource mobilization. These EOR techniques includes gas EORs, Thermal EORs and Chemical EORs. This chapter explores the dynamics of the global EOR industry, emphasizing the crucial role of producing oil from existing reserves efficiently. The decline in oil production rates since 2012 has led to the necessity of exploring deeper waters and mature oil fields. However, the U.S. has been a leader in EOR, producing billions of barrels in 2009, and the Middle East continues to hold vast untapped oil reserves. The industry's portfolio includes mature oil fields and unconventional assets, with Chevron, Husky Energy, Lukoil, BP, and Petronas leading EOR activities. Despite the higher cost of tertiary recovery methods, increasing global demand for petroleum products and legislative incentives drive the growth of EOR. However, factors such as government policies, environmental constraints, and the competition from natural gas impact the future trajectory of the industry. The role of the United States, exemplified by legislative incentives like the 1980 Crude Oil Windfall Profit Tax, showcases how policy decisions can stimulate EOR facility growth. The chapter also explores the reasons behind the recent decline in growth of EOR, citing factors like limited modern discoveries and the lengthy timeline for EOR project development. The decreasing cost of EOR since 2014 is contrasted with the rapid decline in costs for other ventures like shale and offshore drilling. However, the EOR's

Advancements in Chemical Enhanced Oil Recovery.
Tushar Sharma, Krishna Raghav Chaturvedi, Tarek Ganat, and Imtiaz Ali (Eds.)

growth could be revitalized through collaborative efforts, increased financial incentives, and technological advancements. Furthermore, from a technical perspective, the role of exergy analysis, which is a powerful approach for analyzing materials and energy streams, as well as the inherent limitations of energy conversion processes (second law efficiency) becomes important. Given favorable exergy analysis results, from an economic, environmental, and energy standpoint, chemical EOR methods have a wider scale applicability. The findings emphasize the potential of polymer injection in addressing energy transition challenges and reducing the CO2 footprint of oil production. While acknowledging the challenges, it is predicted that an increase in EOR output by the mid-2020s, with EOR accounting for around 4% of global production by 2040. The comprehensive exploration of EOR dynamics and the integration of exergy analysis make this chapter a valuable resource for understanding the industry's present and future.

19.1 THE WAY AHEAD

Even, in the modern era where discourse is dominated by environmental and ecological sensibilities, there is an urgent need to boost oil recovery (Jing and Davarpanah, 2021; Muggeridge et al., 2014; Raffa et al., 2016). The world's EOR industry is driven by the amount of oil that can be extracted from proven reserves and the rate at which it can be extracted (Alvarado and Manrique, 2010). Thermal EOR, gas EOR, and chemical EOR all work together to figure out how much oil can be found (Thomas, 2008). When the amount of oil extracted from the Earth's crust reached its peak in 2012, the growth rate was 36.6% per year. In 2013, the R/P ratios of petroleum resources in Asia Pacific, Europe, and Eurasia were years and years old. This shows how important it is for future oil exploration to look for oil in deeper waters and older reservoirs. The U.S. oil production in enhanced oil recovery (EOR) was the largest in the world in 2009. In 2009, the United States made billion barrels of EOR oil. Even in 2022, the Middle East will have billions of tonnes of oil that can be drilled. The portfolio of mature oil reservoirs and unconventional petroleum assets, as well as the costs of developing EOR projects, are the main parts of the industry. The International Energy Agency says that Chevron, Husky Energy, Lukoil, BP, and Petronas made up a lot of the world's most EOR activity in 2013. There is a good chance that the EOR business will do well because non-OECD countries are expected to use more than a million barrels of crude oil per day by 2018 (Doman, 2016). As a result, the oil and gas industry will increasingly turn to EOR technology to help them make more oil. As a result, the price of crude oil is having a big impact on the international EOR business. Making tertiary recovery technologies like EOR is more expensive than using traditional primary and secondary oil recovery methods. The EOR industry will benefit from more refineries around the world because the demand for petroleum products like gasoline and diesel is going up all over the world at the same time as the world's population grows and people use more energy. When looking at the future of the EOR industry around the world, there are a lot of things to think about, like government policy, environmental constraints, the introduction of carbon dioxide EOR technology, and the competition from natural gas as a fuel source, among other things (Maggio and Cacciola, 2012; Osman et al., 2021).

It is important to note that compared to other methods of oil recovery, EOR tends to have a lower long-term cost than other types of energy generation, but it may require a large investment and a long payback time (Alvarado and Manrique, 2010; Thomas, 2008). It has always depended on getting help or making a smart decision before EOR could be set up. Today, governments or national oil companies are subsidizing or prioritizing more than 80% of the world's EOR production in an effort to get the most out of their investments in natural resources at home. An important thing to note here would be the role of the United States which shows how legislative incentives can encourage the growth of EOR facilities in a country like its own (Azzolina et al., 2016; Núñez-López and Moskal, 2019). The 1980 Crude Oil Windfall Profit Tax, which was put in place in response to the possibility of less oil being produced in the United States in the 1980s, helped the EOR industry by lowering its tax burden. It was recently changed in the United States to include a 12-year tax credit of \$35/tCO_2 for CO_2 that was kept during EOR activities. This credit is now in Section 45Q of the US Tax Code. A similar occurrence was also found to take place in other countries of the world where nations have prioritized domestic production instead of importing oil from foreign sources.

But why hasn't EOR increased as much during the last few years? We for certain, cannot say why, but some possibilities include the following. When available subsurface resources that are modern discoveries are few, EOR is more likely to be used. These issues are becoming fewer and further apart these days. As a result, the upstream company is looking for projects that will create money in the near future. EOR projects require a long time to create, test, and execute, and they only add to the output of a field nearing the end of its life. EOR has developed into a specialized area within the oil and service business, with the essential skills, technology, and expertise being underutilized.

Also since 2014, the cost of EOR has reduced significantly. Other ventures, like as shale and offshore drilling, have seen a more rapid decline in cost. EOR technology cannot compete with other investments at the moment (McCoy and Rubin, 2009).

However, for this to change, a number of events would have to occur rapidly. For instance, governments and enterprises might collaborate to screen fields and determine if EOR can be used in areas with abundant resources. EOR may be used in countries where it has never been used before, and increased financial incentives for EOR may be offered, as well as technical developments that lower the amount of chemicals necessary for injection. While the rise of shale gas in the United States, along with contributions from Brazil and Canada, has meant that EOR has limited room to expand at the moment, in the near future, economics and lack of alternative funding can stifle the growth of shale resources, boosting adoption of EOR in the developing world, especially India. By the mid-2020s, EOR, on the other hand, will begin to generate higher volumes. Additional regions and nations that have developed into established producing areas are expected to support EOR projects at this stage. EOR output increased from 2.7 million barrels of oil per day to more than 4.5 million barrels per day over this time period. By 2040, it is predicted to account for around 4% of global production (Alvarado and Manrique, 2010).

Here, the role of exergy analysis becomes very essential and a comprehensive study in which the energy efficiency and CO_2 emissions of polymer and surfactant-EOR systems, respectively, were quantified using a technique based on the exergy-return-on-exergy-investment concept has been published after detailed investigation (Farajzadeh et al., 2021). Exergy is the beneficial work that a system does when exposed to a certain thermodynamic situation (Farajzadeh, 2019). Exergy expenses connected with water-based oil recovery are mostly due to the circulation of water required for the extraction process to occur. When water consumption is reduced by more than 90%, the injection and lift pumps consume more than 70% of the total energy invested, resulting in an enhanced CO_2 footprint for the generated oil (Farajzadeh et al., 2019b). By lowering the volume of created water and the associated exergy investment, injection of polymer with or without surfactant has been shown to dramatically lower the CO_2 footprint of mature water flood projects. When comparing a barrel of polymer injection oil to a barrel of baseline water flood oil in the field samples analyzed in this chapter, the polymer injection barrel emits 2–5 times less CO_2. Due to the high exergy required to synthesize synthetic polymers and surfactants, the unit exergy investment required for oil production may in some cases exceed the unit exergy investment required for water flooding (Farajzadeh et al., 2021, 2019a). The injection of polymers into high water cut reservoirs is proposed as a method of aiding in the resolution of two important energy transition challenges: (1) satisfying global energy demand through greater oil recovery and (2) decreasing the CO_2 footprint of oil production (more and cleaner energy).

Exergy analysis is a powerful approach for analyzing materials and energy streams, as well as the inherent limitations of energy conversion processes (second law efficiency). While this is true, when calculating the "real" embodied exergy of process equipment, such as a pump, it is frequently necessary to consider both the embodied and manufacturing exergy of the raw material (steel or another material used in the pump, for example), as well as the machining of that raw material. This strategy frequently results in an overestimation of the exergy value of equipment, that is, the price per unit exergy of equipment is abnormally high, out of step with the price per unit exergy of fossil fuels. This may not result in

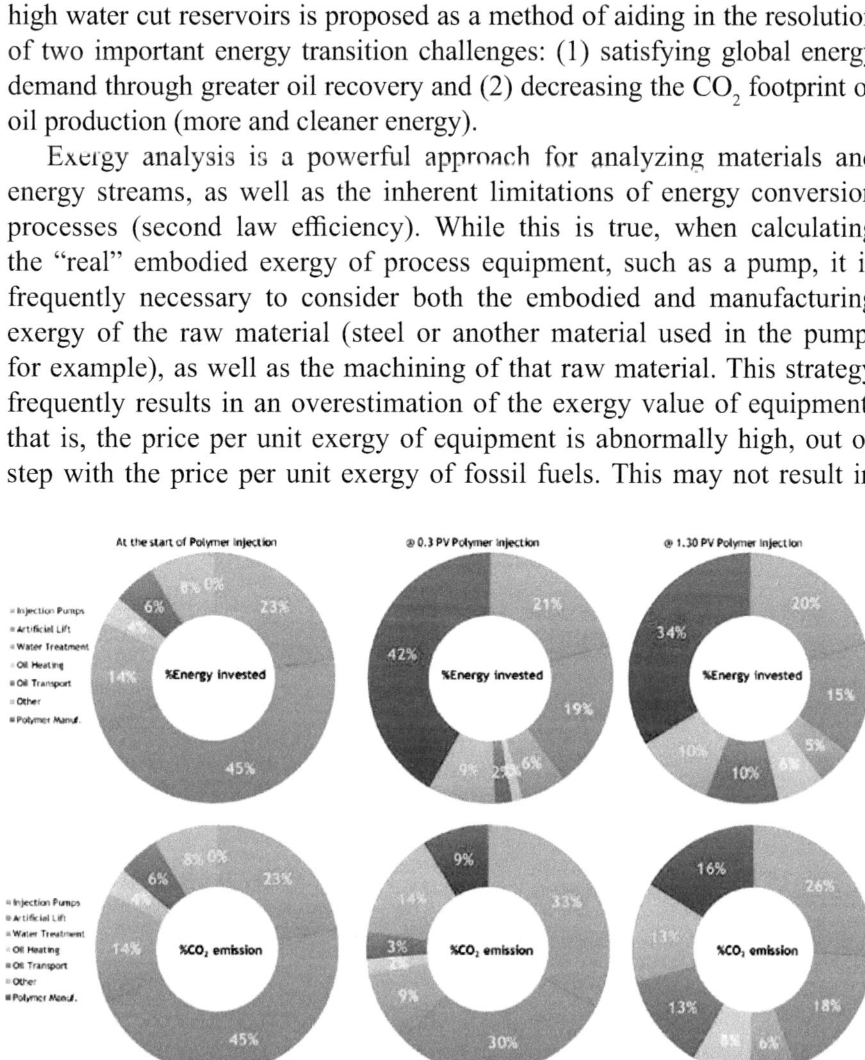

FIGURE 19.1 The observations of exergy fractions invested and CO_2 emission at different time intervals for water and polymer injection cases.

Source: Reprinted with permission from (Farajzadeh et al., 2021). http://creativecommons.org/licenses/by/4.0/.

significant errors in operations with high exergy streams (such as traditional oil production), but it can create considerable problems when recovery factors become sensitive to the exergy of the equipment (Eftekhari et al., 2012). For example, a significant number of wells must be dug to gather unconventional oil and gas resources from tight formations. Drilling is a costly process that consumes little direct energy due to its high altitude. While the direct exergy consumption of drilling may be altered in response to the price of fossil fuels per unit exergy, the indirect exergy consumption of drilling must be extensively examined and analyzed before it can be used in any way (Hassan et al., 2019).

According to the concept of exergy return on exergy investment (ERoEI), the goal of that research was to develop an evaluation approach for evaluating the life cycle effect of polymer and surfactant improved oil recovery (IOR) (Farajzadeh et al., 2021). The following are some of the most important findings from the investigation:

- With the use of the exergy concept, it is possible to evaluate the overall efficiency of chemical EOR operations throughout the course of their whole operational life cycle.
- In particular, when it comes to chemical EOR, the exergy recovery factor decreases when the time horizon is extended. In other words, the amount of process exergy required to create the exergy increases in direct proportion to the passage of time.
- The manufacture of hydrolyzed polyacrylamide (HPAM) polymers is expected to require 123.6 MJ/kg of practical exergy, according to current estimates. Natural gas emits 3.25 kilogram CO_2/kg of natural gas, crude oil emits 4.72 kilogram CO_2/kg of crude oil, and coal emits 6.35 kilogram CO_2/kg of coal, respectively, whereas crude oil emits 4.72 kilogram CO_2/kg of coal.
- It is usually found that the injection of polymer results in greater overall energy efficiency than flooding with water when comparing the energy efficiency of an oil recovery system to flooding with water. Polymer flooding reduces energy investment while increasing oil production (exergy gain) in the project's time-averaged energy investment to produce one barrel of oil via polymer flooding, which is less than that of extended water flooding, due to the handling of polymer flooding.
- Specifically, when the energy expenditure for specific chemical EOR projects is compared to the energy expenditure for water flooding, the CO_2 intensity of the oil produced from these projects

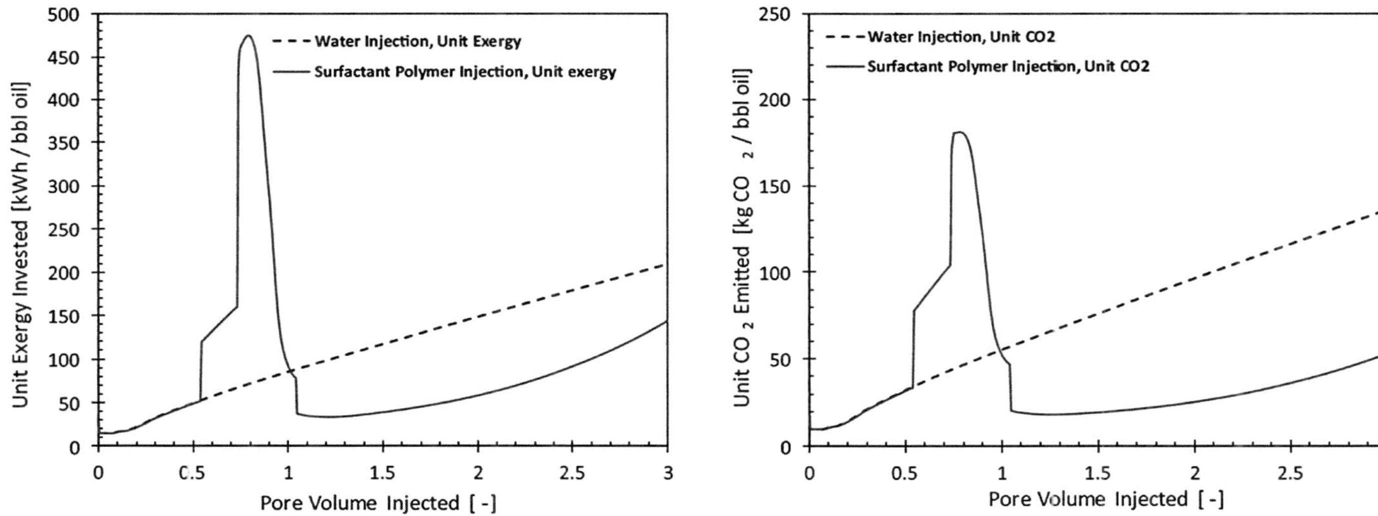

FIGURE 19.2 Plot of unit energy consumed (left) and instantaneous unit CO_2 emitted (right) for water injection (dashed line) and surfactant-polymer injection (solid line) cases.

Source: Reprinted with permission from (Farajzadeh et al., 2021). http://creativecommons.org/licenses/by/4.0/.

is significantly lower than the CO_2 intensity of the oil produced by water injection during periods of widespread water restrictions

- Polymer injection into reservoirs with a high water cut rate has the potential to address two important energy transition challenges: (1) satisfying global energy demand through higher oil recovery; and (2) decreasing the CO_2 intensity of oil generated during the extraction process. It is possible that polymer injection into reservoirs with a high water cut rate will alleviate both of these issues (more and cleaner oil).

In addition to the incremental gain, the amounts of compounds injected into the formulation are important factors in determining the energy efficiency attained as a result of surfactant EOR. During instances of acute water scarcity it must also be noted, on the other hand, oil produced by surfactant-enhanced oil recovery (SEOR) is frequently more ecologically benign than oil produced through conventional water injection techniques. Thus, from an economic, environmental, and energy standpoint, chemical EOR methods have a wider scale applicability

KEYWORDS

- **chemical EOR**
- **exergy analysis**
- **polymer flooding**
- **water injection**
- **energy consumption**

REFERENCES

Alvarado, V.; Manrique, E.; Enhanced Oil Recovery: An Update Review. *Energies* **2010,** *3*, 1529–1575. https://doi.org/10.3390/EN3091529

Azzolina, N. A.; Peck, W. D.; Hamling, J. A.; Gorecki, C. D.; Ayash, S. C.; Doll, T. E.; Nakles, D. V.; Melzer, L. S. How Green is my Oil? A Detailed Look at Greenhouse Gas Accounting for CO2-Enhanced Oil Recovery (CO2-EOR) Sites. *Int. J. Greenh. Gas Control* **2016,** *51*, 369–379. https://doi.org/10.1016/j.ijggc.2016.06.008

Doman, E. L. International Energy Outlook 2016, U.S. Energy Information Administration, 2016.

Eftekhari, A. A.; Van Der Kooi, H.; Bruining, H. Exergy Analysis of Underground Coal Gasification with Simultaneous Storage of Carbon Dioxide. *Energy* **2012,** *45*, 729–745. https://doi.org/10.1016/j.energy.2012.07.019

Farajzadeh, R. Sustainable Production of Hydrocarbon Fields Guided by Full-Cycle Exergy Analysis. *J. Pet. Sci. Eng.* **2019,** *181*, 106204. https://doi.org/10.1016/j.petrol.2019.106204

Farajzadeh, R.; Kahrobaei, S.; Eftekhari, A. A.; Mjeni, R. A.; Boersma, D.; Bruining, J. Chemical Enhanced Oil Recovery and the Dilemma of More and Cleaner Energy. *Sci. Rep.* **2021,** *11*, 1–14. https://doi.org/10.1038/s41598-020-80369-z

Farajzadeh, R.; Wassing, B. L.; Lake, L. W. Insights Into Design of Mobility Control for Chemical Enhanced Oil Recovery. *Energy Rep.* **2019a,** *5*, 570–578. https://doi.org/10.1016/j.egyr.2019.05.001

Farajzadeh, R.; Zaal, C.; van den Hoek, P.; Bruining, J. Life-Cycle Assessment of Water Injection Into Hydrocarbon Reservoirs Using Exergy Concept. *J. Clean. Prod.* **2019b,** *235*, 812–821. https://doi.org/10.1016/j.jclepro.2019.07.034

Hassan, A. M.; Ayoub, M.; Eissa, M.; Musa, T.; Bruining, H.; Farajzadeh, R. Exergy Return on Exergy Investment Analysis of Natural-Polymer (Guar-Arabic Gum) Enhanced Oil Recovery Process. *Energy* **2019,** *181*, 162–172. https://doi.org/10.1016/j.energy.2019.05.137

Jing, H.; Davarpanah, A. Implications of Chemical-Thermal Enhanced Oil Recovery Methods in Shale Reservoirs. *Cogent. Eng.* **2021,** *8*. https://doi.org/10.1080/23311916.2021.1920564

Maggio, G.; Cacciola, G. When will Oil, Natural Gas, and Coal Peak? *Fuel* **2012,** *98*, 111–123. https://doi.org/10.1016/j.fuel.2012.03.021

McCoy, S. T.; Rubin, E. S. The Effect of High Oil Prices on EOR Project Economics. In *Energy Procedia*; Elsevier, 2009; pp 4143–4150. https://doi.org/10.1016/j.egypro.2009.02.223

Muggeridge, A.; Cockin, A.; Webb, K.; Frampton, H.; Collins, I.; Moulds, T.; Salino, P. Recovery Rates, Enhanced Oil Recovery and Technological Limits. *Philos. Trans. R. Soc. A Math. Phys. Eng. Sci.* **2014,** *372*. https://doi.org/10.1098/rsta.2012.0320

Núñez-López, V.; Moskal, E. Potential of CO2-EOR for Near-Term Decarbonization. *Front. Clim.* **2019,** *1*, 5. https://doi.org/10.3389/fclim.2019.00005

Osman, A. I.; Hefny, M.; Abdel Maksoud, M. I. A.; Elgarahy, A. M.; Rooney, D. W. Recent Advances in Carbon Capture Storage and Utilisation Technologies: A Review. *Environ. Chem. Lett.* **2021,** *19*, 797–849. https://doi.org/10.1007/s10311-020-01133-3

Raffa, P.; Broekhuis, A. A.; Picchioni, F. Polymeric Surfactants for Enhanced Oil Recovery: A Review. *J. Pet. Sci. Eng.* **2016,** *145*, 723–733. https://doi.org/10.1016/j.petrol.2016.07.007

Thomas, S. Enhanced Oil Recovery - An Overview. *Oil Gas Sci. Technol.* **2008,** *63*, 9–19. https://doi.org/10.2516/OGST:2007060

Index

D

E

F